The New Generation of Computable General Equilibrium Models

AF812758

Federico Perali · Pasquale Lucio Scandizzo
Editors

The New Generation of Computable General Equilibrium Models

Modeling the Economy

Springer

Editors
Federico Perali
Department of Economics
University of Verona
Verona
Italy

Pasquale Lucio Scandizzo
Economics Foundation—CEIS
University of Rome Tor Vergata
Rome
Italy

ISBN 978-3-030-09633-5 ISBN 978-3-319-58533-8 (eBook)
https://doi.org/10.1007/978-3-319-58533-8

© Springer International Publishing AG, part of Springer Nature 2018
Softcover re-print of the Hardcover 1st edition 2018
This work is subject to copyright. All rights are reserved by the Publisher, whether the whole or part of the material is concerned, specifically the rights of translation, reprinting, reuse of illustrations, recitation, broadcasting, reproduction on microfilms or in any other physical way, and transmission or information storage and retrieval, electronic adaptation, computer software, or by similar or dissimilar methodology now known or hereafter developed.
The use of general descriptive names, registered names, trademarks, service marks, etc. in this publication does not imply, even in the absence of a specific statement, that such names are exempt from the relevant protective laws and regulations and therefore free for general use.
The publisher, the authors and the editors are safe to assume that the advice and information in this book are believed to be true and accurate at the date of publication. Neither the publisher nor the authors or the editors give a warranty, express or implied, with respect to the material contained herein or for any errors or omissions that may have been made. The publisher remains neutral with regard to jurisdictional claims in published maps and institutional affiliations.

Printed on acid-free paper

This Springer imprint is published by the registered company Springer International Publishing AG part of Springer Nature
The registered company address is: Gewerbestrasse 11, 6330 Cham, Switzerland

Preface

This book grew out of an initial collaboration between a team from the University of Rome "Tor Vergata" and its spin-off Openeconomics srl (http://www.openeconomics.eu), and one from the University of Verona and its spinoff Economics Living Lab (http://econlivlab.eu). This prompted the creation of a working group and then a workshop within the Association of Italian Development Economists (SITESIDEAS: http://www.sitesideas.org/) in January 2017. The workshop brought about a number of interesting papers, but more importantly, uncovered the interest cultivated by a growing group of SITES associates, who continued to collaborate and correspond after the workshop. Some members of the group met again at the SITES Summer School in Prato in June 2017. The papers originally presented were in part modified and some other papers were added as a consequence of further contacts and collaborations.

The book aims to present state-of-the-art theory and practical applications of CGEs and social accounting matrices (SAM) focusing on recent advances and techniques, but also reaching back to basic assumptions and theoretical tenets for a class of models that are becoming ever more diffused as the bread and butter of policy analysis. The focus of the models presented is on estimation and policy impact analysis, within a pragmatic vision of the underlying economic theory that echoes the fact that the practical reasons of model successes reside in their capacity to provide consistent and credible counterfactuals to the effects produced by the changes induced by the policies, programs and projects to be assessed.

The book is divided into 3 parts and 12 chapters. Part I, consisting of only one chapter (Chapter "General Equilibrium Modelling: The Integration of Policy and Project Analysis"), presents an introduction to CGE modeling, focusing on the integration of policy and project assessment, as the frontier toward which CGEs have been evolving for the past 20 years. The chapter discusses the basic theoretical models that lay behind the CGEs and their SAM cores, with special emphasis on the fundamental differences that emerge on their interpretation under alternative economic theories, and assumptions on the cause–effect relations hypothesized. The chapter also reasons and comments on some of the latest trends of CGE-SAM models, and their ever-increasing extensions and applications to macro and micro areas of economic policy, with a view to integrate the different layers of an economy in a comprehensive structural representation.

Part II of the book presents a sample of Methodology and Estimation Issues, concerning both special problems of dynamic representation of the economic system (Chapter "Demand-Driven Structural Change in Applied General Equilibrium Models") and estimation and modeling problems mainly related to micro-macro integration (Chapters "Micro-Macro Simulation of Corporate Tax Reforms" –"Analysis of Local Economic Impacts Using a Village Social Accounting Matrix: The Case of Oaxaca"). This part focuses on solutions to incorporate special structural features and changes in both SAMs and CGEs, attempting to overcome the straightjacket of the economy snapshots given by the national accounting systems. Chapter "Demand-Driven Structural Change in Applied General Equilibrium Models" presents new results from CGE estimates and simulations on demand-driven structural changes, embedded in the model as an effect of changing tastes over time. Chapter "Micro-Macro Simulation of Corporate Tax Reforms" discusses the potentialities of integrating microsimulation models and CGE-SAMs and presents a microsimulation analysis of a recent corporate tax reform in Italy. The micro model simulates corporate tax liabilities according to the prevailing fiscal rules and is updated and used on a regular basis by the Italian Central Institute of Statistics for revenue forecasting and policy analysis. Chapter "Estimating an Energy-Social Accounting Matrix for Italy" describes the estimate of an energy model for Italy that integrates some of the information of a comprehensive technology optimization model for energy (TIMES (The "TIMES", Integrated MARKAL-EFOM System) with a detailed SAM. Chapter "Analysis of Local Economic Impacts Using a Village Social Accounting Matrix: The Case of Oaxaca" presents the results of a research project aimed at applying the SAM technique to small inhabited areas within a hierarchically ordered set of national, regional, and local accounts. The application described uses national and regional statistics as well as survey methods of estimation to be able to use a "local" SAM for the village of Oaxaca in Mexico. This social accounting approach to local economic development applies to the local economies of the disparate economic realities of other continents and, thanks to its ease of operation and interpretation, can be used both as an impact evaluation tool for large projects and as a policy evaluation platform for local and national politicians.

Part III of the book, Static and Dynamic CGEs and Policy Applications, addresses the theory and the application of state-of-the-art CGEs to policy problems. These range from recent CGE applications to policy choices and investment planning in Kenya, Italy, Mauritius (Chapter "A CGE Model for Productivity and Investment in Kenya"–"A CGE Model for Mauritius Ocean Economy"), to micro-macro analysis, policy reforms, Euro devaluation, and regional dynamics (Chapter "A Micro-Macro Simulation Model Applied to the French Economy: The Case of a Euro's Real Depreciation"–"A Regional Dynamic General Equilibrium Model with Historical Calibration: A Counterfactual Exercise"). While the case studies reported cover a wide spectrum of methods, models, and policy questions, they have in common a focus on specific policy questions, rather than an attempt to build a model with special structural or time-varying characteristics. Thus, while the simulations presented are shaped by the questions asked, the models developed present general features of their own that transcend the specific policies examined and can be interpreted in a broader framework. This is the case, for example, of the Kenya and Mauritius case studies (respectively, Chapters "A CGE Model for Productivity and Investment in Kenya" and "A CGE Model for Mauritius Ocean Economy"), where the analysis of possible development strategies unveils models that can address more general questions about the role of productivity growth and investment. On the other hand, the study assessing the impact of climate change uses a fine spatial resolution for the European Mediterranean countries to measure the differential impacts of both climate and physical process models such as those describing the allocation of land use, crop growth, and flood risk at the local level. Such an integrated approach coupled with an appropriate treatment of spatial heterogeneity produces information that is highly relevant to both planners and the business community. The proper treatment of heterogeneity is fundamental not only to understand differences in both behavioral responses and policy impacts across regions, but also across aggregate family types. This is clearly shown in the study devoted to the ex ante socio-economic evaluation of the impact of the CAP reform on Italian agriculture and the whole economy using a micro-funded general equilibrium model that differentiates the impact at the household level and for each interest group involved in the policy process. The political economy analysis of the consequences of the reform incorporates the political positions of farmers and agro-food industries, consumers, and unions and estimates the impact of each scenario on each stakeholder. The policy analysis permits both an understanding of the possible social conflicts arising from the implementation of the reform and a unique ranking of the policy alternatives. If the interest is in estimating the impact of a policy change at the disaggregate household level, then an integrated micro-macro simulation model needs to be implemented to evaluate the distributional effects as it has been implemented in the study devoted to the estimation of the impact on the French economy of a real depreciation of the Euro. The research finds that a 10% real depreciation of the Euro stimulates the aggregate demand by increasing exports and reducing imports, which increases real GDP by 0.7% and reduces the unemployment rate in the economy by 2 percentage points. At the individual level, the study reveals that the macroeconomic shock reduces poverty and, to a lesser extent, income inequality. The regional dynamic general equilibrium model introduces a novel historical calibration technique based on two regional SAMs for the Italian region Valle D'Aosta for the years 1963 and 2002 that ensures that the modeled tendencies perfectly reproduce the actual observed growth patterns. The dynamic general equilibrium model provides an original and powerful tool for historical counterfactual analysis not available using standard dynamic general equilibrium models. The model is used to compare the growth path followed by the region during the period of interest with a counterfactual scenario intended to evaluate how the region would have performed in the case of a contraction of the transfers from the national government to the regional government and the families.

The works collected in this book represent a joint effort to take macro CGE models closer to a realistic description of the response behavior of families and enterprises to project and policy changes. In real-world situations, market imperfections and failures often require the interventions of "visible hands" to reach feasible and stable equilibria involving nonlinear (shadow) price schemes, where prices can vary across agents, permitting the efficient management of externalities, transaction costs and non-convex technologies or budgets. The ability to make these theoretical challenges tractable is one of the most fascinating items of the future research agenda of both theoretical and applied general equilibrium analysts.

This book should be useful as reading and teaching material in graduate courses in economics, especially those focusing on development theory and practice. If nothing else, it should convince the reader that computable general equilibrium modeling is a dynamic subject, need not be confined to specialists, does not have to produce black boxes, and can be very helpful in addressing many interesting questions of political and economic relevance. While theoretical and empirical controversies on the foundations of general equilibrium are still sharp and partly unresolved, the essays presented show that designing, estimating, calibrating, and using CGE models may help economists to raise policy-relevant questions and shape them in a meaningful way and to suggest effective and implementable policy solutions.

Verona, Italy Federico Perali
Rome, Italy Pasquale Lucio Scandizzo

Contents

Part I
Introduction

General Equilibrium Modelling: The Integration of Policy and Project Analysis

Federico Perali and Pasquale Lucio Scandizzo

Abstract This chapter presents an overview of frontier topics of general equilibrium that are especially important to effectively integrate the policy and project dimensions of the equilibrium analysis. Project evaluation as a new frontier for modelling implies a general view of the traditional benefit-cost calculations that researchers can now afford implementing thanks to the recent computational developments that can host more realistic assumptions about model closures. A differential representation of general equilibrium permits also to unveil the opportunity cost structure associated with alternative resource uses of both policy and project evaluations. This extension enriches the policy content of both the micro and macro level of the equilibrium analysis. It takes advantage of the fact that in a modern policy and project analysis the micro-macro link exactly aggregates from the individual to the family, community, which is often the level of feasibility and impact analysis of large projects, and society level using micro and macro behavioural models that are closely integrated.

Keywords General equilibrium modelling · Policy analysis
Project evaluation · Micro-Macro simulations

JEL Codes C68 · R13

1 Introduction

General equilibrium (GE) modelling as a methodology to analyse broad policy issues, has been around for many years, at least since the pioneering efforts of Wassily Leontief, Hollis Chenery and Leif Johansen in the 60s. The revival which we are

F. Perali
Department of Economics, University of Verona, Verona, Italy

P. L. Scandizzo (✉)
Centre for Economic and International Studies (CEIS), University of Rome Tor Vergata, Rome, Italy
e-mail: scandizzo@uniroma2.it

© Springer International Publishing AG, part of Springer Nature 2018

F. Perali and P. L. Scandizzo (eds.), *The New Generation of Computable General Equilibrium Models*, https://doi.org/10.1007/978-3-319-58533-8_1

witnessing today, however, is based on several new facts and advancements of both theory and practice. First, the extensive experimentation with computable general equilibrium (CGE) models in the past 50 years has been instrumental in generating a greater degree of understanding of both the potential and the limitations of both GE ideas and CGE models. Second, the advancement of computational techniques and the power of modern computers have made possible to construct more transparent models, more easily penetrable by numerical techniques and, as a consequence, much less "black boxes" that their earlier progenitors. Third, the combination of CGEs and Social Accounting Matrices (SAMs) has become a standard that allows to treat the GE model as an extension, however complex, of national accounting. Fourth, the greater availability of microdata has widened the horizon of the SAM-CGE possible coverage, extending their reach to seemingly elusive phenomena, such as income distribution and employment, trade and migration flows, factor markets and their spatial mobility, the environment and climate change. For example, labor and workforce accounts measuring labor force in terms of hours, occupations, full versus part time, type of household, gender, skills, wages within the frame of Industry-Occupation matrices are increasingly available for a high level of sectoral detail and for the smallest administrative territorial units. The evaluation of the impact of policy programs or environmental shocks on well-beings is now enriched by satellite accounts, statistically consistent with national accounts, that collect and order information about human, social, cultural and political dimensions of economic and social life. Common examples are satellite accounts for the environment, or tourism/migration/commuting, unpaid household work or related to different forms of capital besides the traditional financial and physical capital such as human capital, natural capital in the form of amenity indices, social capital, cultural and political capital. This information adds value to a modern analysis of an economy not simply because it allows representing an "augmented" reality where the effective productivity, for example, of a unit of physical capital accounts for the fact that it is invested within a community that is also endowed with a high or low level of human and social capital. New techniques, based on sophisticated statistical and mathematical algorithms, have become available to estimate and calibrate model parameters, by incorporating and integrating information from macro and micro data, using time series, surveys as well other model estimates. This advanced computing capacity makes it easier to handle highly detailed information sets and large-scale models thus opening new prospects for inferential and causal analysis within a general equilibrium context.

As we learn more about their potential and hidden messages, CGEs have become the tool of selection for economists and policy makers to perform evaluative simulations within a context of coherent and transparent hypotheses on the technology, the behavior of the economic agents and the status and the evolution of the external environment and the representative exogenous variables. They have become the only point of encounter of macroeconomic policies with project evaluation, where they promise to perform a critical function to connect two frameworks that typically don't mingle and often risk contradicting each other.

In this study, we look at the basic design of modern CGEs and some of their more interesting variants, with a special focus on the emerging connection between the policy and the project level. We present and discuss several different attempts to operationalize the CGE context to analyse the connection between policies and projects and, in some cases, the corresponding macro-micro nexus. For this, we develop model structures that correspond to a common framework and aim to both clarify and simplify the intricacies of the CGE procedures.

2 Project Evaluation as a New Frontier for Modelling

The individual assessment of investment projects, as developed by economic theory, is based on the consideration of quantitative traits related to financial and economic "profitability" of the project. These are measured by the difference between the so-called "benefits" and "costs" of the project with and without the project under consideration. The costs are generally concentrated in the investment or construction phase of the project, while the benefits are almost exclusively part of the subsequent operational phase. Since the individual assessment is based on the characteristics of the project, it regards as "given" the external conditions of the overall economic system, which are synthetically represented by so-called shadow prices used. For this reason, the costs and benefits that depend on the interaction between the project and its economic environment are typically neglected in whole or in part in the cost-benefit analysis, particularly regarding the effects of the stimulation of economic activity prevailing during the construction phase. CB Analysis also neglects—as each project is evaluated independently of the other—any interdependencies with other projects, thus creating the risk of making a mistake that will tend to be greater, the greater will be the size and degree of complexity of the group of projects selected for funding. Finally, none at the so-called "external effects", i.e. the provision of public goods and environmental impact of the project are considered. These effects are particularly relevant in the case of public projects, which themselves, ultimately, are a vehicle for improving the physical and economic environment of the country.

As we said, the quantitative measurement of the costs and benefits of investment projects is based on the dichotomy: construction—operational phase. This dichotomy is part of an approach that does not consider the multiplicative effects of investment on factor employment. In fact, the benefits of traditional investment analysis arise especially during operations through increasing production, driven in turn by an increase of fixed assets. The costs are concentrated in the construction phase, because it is at this stage that fixed assets are built by committing productive resources in the hope of future benefits. The very concept of productive investment is therefore defined by the dichotomy between anticipation of costs and of realization of benefits according to a time profile that constitutes one of the fundamental determinants of the profitability of the project.

The ability to calculate the values of equilibrium prices, quantities, household incomes and other variables of interest in complex multi-sectoral models is on the other hand a recent achievement of applied economics. It is based on the specification of mathematical structures which reflect the rigorous definitions of economic equilibrium, developed by Kenneth Arrow, Gerard Debreu, Michio Morishima and others, and other simplifications and approximations necessary to allow the calculation of the equilibrium values.

In a series of important research attempts, in large part conducted at the World Bank, several generations of computable general equilibrium models (CGE) since the late 70's were developed and gradually became important and useful tools for policy analysis. In these models, social accounting matrices (SAM) became the core of the representation of general equilibrium as a circular flow of production, consumption and incomes, with prices in all markets as the equilibrating variables. Solving algorithms started with fixed point (Scarf and Hansen 1973) and mathematical programming procedures (Norton and Scandizzo 1981; Walbroeck and Ginsburg 1986) and gradually developed into nonlinear equation systems and local or global search solution methods (Devarajan et al. 1997). At present, while the macro-econometric models prevailing in the 1970s have all but disappeared from the economic practice, CGEs are increasingly used around the world, both in their static and dynamic versions, as tools to analyze economic policy options.

This chapter presents a family of general equilibrium models, which extend the results already obtained in several earlier and recent contributions by Scandizzo (1980, 1995, 2014, 2016; Scandizzo and Ferrarese 2015), with the main objective to evaluate the effects of alternative investment programs. These contributions consider both directly demand and supply systems and in connection with the different "upstream" and "downstream" links that characterize the production structure and the multiplicative effects that occur through the movement of prices and consumption. The resulting assessment methodology can account for changes that projects bring on economic activity level, technological changes that they incorporate and environmental impacts which they generate.

3 Some General Equilibrium Concepts

The main concepts of the category that goes by the name of general economic equilibrium can be found only with considerable effort in the economic literature. This is because the notion of equilibrium depends on the historical context in which it is used, and the model of the economy to which it refers.

Classical economists such as Adam Smith, David Ricardo, J.S. Mill, and Karl Marx believed that the value was determined by the cost of production and the absence of profits. Both conditions can be regarded as characteristics of an equilibrium condition: the equality of prices to the cost of production, in fact, ensures that individual producers do not wish to change their plans, while the absence of profits implies the absence of competitive pressure from new companies trying to enter in the markets. This equilibrium can be described as "general" if it extends to all markets.

Although at first sight satisfactory, especially for its simplicity, this classical view of equilibrium reveals two weaknesses. First, equality between prices and unit costs of production, if acceptable as a condition of balance for goods and services produced, says nothing about the value of primary production factors and especially labor. The state of equilibrium is not then "general" because it does not extend to factor markets. Secondly, because consumers are not involved either in the equality of prices and costs, or in the absence of profits, they also appear to be excluded from the equilibrium described that turns out, therefore, to be wholly partial.

A general equilibrium model in the modern sense of the word must have some essential requirements, both in terms of the equilibrium condition and that of "being general". Equilibrium should, in fact, result from supply and demand equality, but it must also assume that consumers and producers are, individually, where they want to be, that is, on their individual curves of supply and demand. This means in practice that every solution must depend parametrically on taste, technology and the initial distribution of goods. Secondly, the equilibrium should be "general". This implies that no price (except the numeraire) can be considered a purely exogenous variable. If this were the case, in fact, the corresponding market could not be in balance except by chance and the description of the model would be incomplete. In addition, equilibrium between demand and supply should cover not only the goods produced, but also the primary factors of production such as land, capital, labor and other resources that characterize the initial endowment.

Both in the classical description, and in the newer ones, general equilibrium is finally typically characterized as a set of conditions of real balance of a closed economy. This implies that the demand functions are homogeneous of degree zero in prices (no money illusion) and that therefore it is possible to apply an appropriate normalization rule, such as the choice of a simple or composite commodity (a numeraire) whose price is conventionally equal to unity.

Given these characteristics, the concept of equilibrium is also associated with efficiency and to the question of its existence, which was taken up by Arrow and Debreu (1954), as well as McKenzie's (1959) in a series of celebrated contributions. Their work, although focused on a rather narrow sub-problem, essentially proved that under certain conditions, given a set of demand and supply equations of individual agents, aggregate demand and supply could be equated by a set of non negative prices. This was a non trivial result, that was contingent on a series of rather restrictive assumptions, but was obtained through a mathematical powerful and unifying instrument (the fixed point theorem) that was in itself shining for originality and simplicity. The result had two drawbacks, however. First, it did not cover nor it proved to be a feasible base for finding circumstances under which the equilibrium was unique. Second, as proved in a series of important and somewhat astounding later contributions by Sonnenschein (1972, 1973), Debreu (1974) himself, and Mantel (1974), the base of the existence proof was an aggregate excess demand function, which, although resulting from the aggregation of individual demand and supply, was not bound by the limitations deriving from the postulates of rationality. In what has been called "the everything goes" conclusion, in fact, it was proved that such a function, even though the result of individual rational behavior, is not characterized by any special mathematical property. Thus, the existence of general equilibrium seemed to be quite independent of its "micro-foundations", as a consequence of an essential weakness of the microeconomic "rationality" assumptions, which were proved to be not sufficiently discriminating to impose anything resembling rationality on aggregate behavior.

A further point arises from the consideration of the causal chains contained, or implied by the process of reaching the equilibrium, or re-establishing it after a perturbation (the comparative static problem). From the point of view of the underlying causal chain, a general equilibrium model, from the original Walrasian formulations to the latest computable forms, does not in itself indicate any direction of causality, as relations between its variables are fully simultaneous. If full employment is considered to be the crucial element of discrimination between the classical and Keynesian approach, it is clear that this condition does not characterize necessarily a solution that meets the conditions of general equilibrium. If it is true, indeed, that such a solution cannot contain involuntary unemployment, given that the supply of labor and other resources depend entirely on household preferences and prices, it is also true that the level of employment in the solution found is not necessarily the maximum possible, given the fact that there may be multiple equilibria. Even when uniqueness of equilibrium is guaranteed by ad hoc conditions, a higher employment level could be achieved by changing the attitudes of consumers (and among them we can mention the expectations) technology or deployment of resources. If the change in autonomous expenditure that sets in motion the Keynesian causal chain is interpreted, as it seems legitimate to do, as an exogenous change in preferences, technologies or distribution, the general equilibrium model is therefore fully compatible with income stabilizing fiscal policies.

More generally, the level of employment of a solution of a specific model depends both on the characteristics of the solution (if it is not unique), and the characteristics of the model. The latter consist of the structure (number of equations, functional forms, variables included and excluded etc.) as well as parameters, that is, variables whose value depends on the model but is set exogenously. A causal chain of Keynesian type, then, is the sequence of changes caused by an exogenous variation of one or more parameters. For example, if the level of domestic demand depends on the percentage of wealth held by the richest 5% of the population and that percentage changes after the imposition of a 1% tax, the consequent change in demand will result in a new parametric balance that may result in less than full employment. Stabilizing fiscal policy will then consist in determining the value of another parameter: an exogenous variable in the model, but subject to political control, such as government spending, to reconstruct a situation which is as close as possible to that which preceded the distributive variation.

4 The "Closures"

The incompleteness of the classical general equilibrium model has been overcome in modern models since the days of Walras, through the introduction of both labor and leisure in the household behavioral function and the inclusion of all factors, including labor, in the original allocation of resources. However, several problems remain for a realistic representation of the economic system. First, the model does not consider the formation of savings and investment. Secondly, it only considers real quantities and prices and therefore does not include the supply and demand of money and other financial resources. Finally, it describes a closed economy and thus ignores the possibilities for international trade as well as domestic and foreign currency and relations between domestic and international prices.

These three areas are all theoretical "holes" of general economic equilibrium, in the sense that their "closing" forces us to deal with the problem of reconciling the micro with the macro-economy, making choices that may be justified by personal beliefs and ideological reasons, as well as by empirical evidence. In some sense, this is equivalent to choose between Keynesian and monetarist theories in one of their many meanings.

Consider the simple case of the introduction of money. If we accept the classical scheme, we can also introduce money through a quantitative equation. This equation says that the monetary value of income is proportional to the amount of money exogenously supplied. Substituting this equation into a normalizing equation that assigns to an arbitrary good a unit price, we get a system with two characteristics. (a) There is a commodity called money, the price of which is fixed to unity, and whose application is due only to the fact that it is necessary to carry out transactions. (b) The general price level (understood as the arithmetic average of the weighted prices with quantity quotas) is proportional to the amount of money supplied. The system resulting from the introduction of money through the quantitative equation is then "dichotomous" in the sense that its real part continues to determine the general price level.

Once money is introduced, however, the issue of savings and investment arises: a funding activity becomes possible based on the availability of some traders to surrender their temporary surpluses of money to other operators that are characterized by temporary deficits. In the neoclassical story, aggregate surplus represents excess savings, which are matched to excess investment to achieve equilibrium. The bank money, which is only a particular type of numeraire, promises in addition to pay investors who save. The "price" of these promises is greater the smaller the interest owed by debtors to creditors. The interest rate thus becomes the variable balancing savings and investment. The introduction of the savings-investment balance at the aggregate level allows to give more substance to the activities of the banking sector, which is not limited to distribute a "currency, but acts as an intermediary between families and businesses.

Finally, we consider the introduction of the external sector. As for the other two "closures", the inclusion of international trade in the aggregate is not difficult, because it is, in fact, the simple addition of a macro-enterprise: the "external sector", that transforms exports into imports (and vice versa) with a given technology (transactions at world prices).

Considering for simplicity only imports that are finished products, household budgets will be divided between spending on the domestic market and in foreign markets, while the balance sheets of firms will in turn be fed either by domestic sales, and/or by those in the foreign market. Equilibrium conditions will not be changed, except for the addition of the condition of balance in value between imports and exports (balance-of-payments constraint).

The extension of this model through Keynesian assumptions does not pose any special problem. The liquidity preference can be incorporated either at the aggregate level or directly in the demand functions that describe the behaviour of households. The dependence of the demand for money on the interest rate, however, forces us to introduce at the same time, both an investment demand schedule (through an appropriate function of corporate behaviour) as a function of the interest rate, and a saving function. The latter, barring multi temporal complications, can be introduced directly into the household budget constraints by assuming, in the Keynesian tradition, that saving is a function of income, but not of the interest rate. The extension to international trade can now be performed in a not dissimilar way from the one already described above for the neoclassical model.

The model obtained differs from previous one in that it reflects the basic differences between the neoclassical and Keynesian macro-economic structure, since the interest rate has to perform the task to balance money demand and supply in the Keynesian model. These differences, however, are not such as to affect the simultaneity characterizing both models, which are both the combination of hypotheses of aggregate type (e.g. quantitative equation, investment function) on a disaggregated, Walrasian type structure. Much more important are the differences relating to the causal chains that can be associated to the exogenous variables or parameter changes. It is these changes that have profound consequences on the use of two models for the valuation of investments.

5 The New Frontier of the CGE Models

5.1 General Equilibrium as a Model Foundation

What is "general equilibrium"? One is tempted to reply that general equilibrium describes a condition where all markets are in equilibrium, both in the sense that all markets are cleared (demand equals supply) and all agents fulfil their plans. However, while this definition certainly appears simple and direct, it is neither complete, nor satisfactory. It is intrinsically incomplete, since general equilibrium, unlike partial equilibrium, in addition to material balances and subjective fulfilment, requires that the distribution of wealth is consistent with resource allocation. It is not satisfactory, because market clearance depends on a flow condition, i.e. it can be satisfied only for one particular interval of time. If we take the year as the reference time frame, for example, there may be several markets that require more or less than a year to be cleared. Inventories and other capital goods bridge the gap, both as flows and accumulating stocks, between the production and consumption timelines and play a special role in both static and dynamic CGEs.

In order to address the time matching and the stock-flow problem, general equilibrium modelling must address four different circles of causation: (i) between demand and supply of goods and services on one hand, and prices and incomes on the other; (ii) between the formation of incomes from demand and supply of factors of production and their prices, (iii) between the initial resource endowment and the redistribution caused by productive choices and institutional transfers, (iv) between investment and savings and the rates of return to all forms of capital. The precise way in which these four circles interact is still not clear, especially for what concerns the link between flow and stock variables, although Stone and Brown (1962) formalized the main flow balance relations in a form essentially consistent with the Keynesian model in the so called Social Accounting Matrix (SAM). As Taylor (2010) persuasively argues, computable general equilibrium models (CGE), mainly developed because of research efforts at the World Bank in the '70s, are a spinoff of the application of Input-Output matrices and SAMs, more than any attempt to compute Walrasian equilibria. Even in their advanced, present day form, they tend to reflect a basic indeterminacy of capital accounting, deriving both from lack of consensus on capital theories and on best practices of accounting. They also evoke an intrinsic dualism between a core set of social accounts and a complementary, highly variable set of behavioural and technical equations.

Figure 1 summarizes the fundamental variables of general equilibrium, as conceived in most CGE models. The figure also shows the causal links, according to the two extreme versions of the classical and Keynesian theory. Following the arrows connected by solid lines, and starting from the top, we can follow the classical chain, in which the productive capacity determines the level of employment (which is the one that maximizes the profits of the entrepreneur). This in turn determines the level of production, and prices of these factors determine in turn the level of prices of goods, income and consequently the level of consumption. In the Keynesian version, on the other hand, it is the level of consumption that determines monetary income and then, through employment, the level of production, by establishing a series of effects on product and factor prices with feedbacks on incomes and consumption. The Keynesian causal sequence differs from the neoclassical one, in both the origin and the direction of change, but eventually recovers parts of the neoclassical relationships through income and price feedbacks on consumption.

Although the representation of equilibrium just described contains all the essential ingredients of a "general" equilibrium, it is not sufficiently analytic, because it is limited to considering only the "final" variables. This form of the model can be called "reduced", because it is constituted by a set of relationships among key variables, (i.e. variables that cannot be suppressed without depriving the model from its "generality") and cannot be reduced to a form with fewer variables.

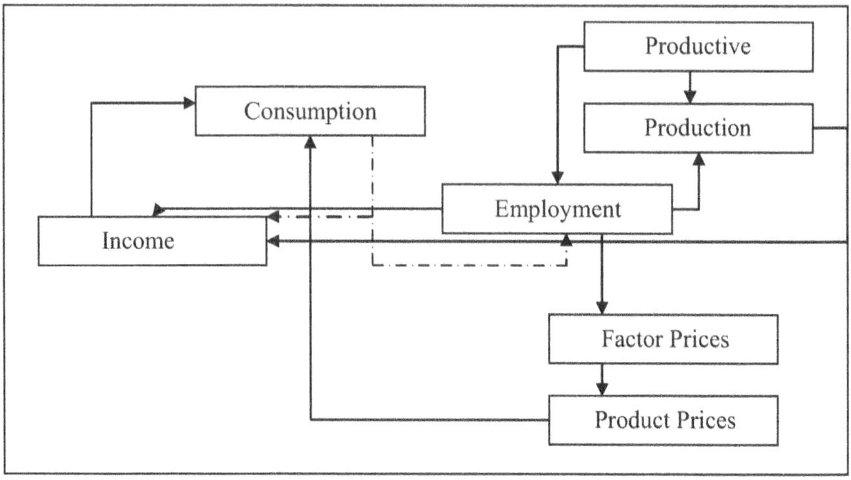

Fig. 1 The basic economic model

In order to formulate a structural model, one has to explicitly introduce some relations and variables that carry the assumptions on the causal chain moving the model itself from disequilibrium to equilibrium. The easiest way to introduce these structural elements is shown in Fig. 2 where supply and demand for factors and products are included as four additional structural variables. A variable "changes in the stock of capital" includes new capital accumulated through investment, including inventories, and allows supply and demand flows to differ from production flows.

The supply-demand equilibrium is achieved through four causal relations according to which, in particular: (a) factor supply and demand are determined by households, "given" price and income levels, (b) the supply of goods and the demand factors is determined by the firms, given price levels. Assume further that (c) the price level of goods and services is determined by the firms to cover the costs of production (or to maximize profits).

These relationships, together with those already present in the reduced form, complete the model whose equilibrium depends on a series of simultaneous relations. In these relations, the "causal links" (the level of the prices "cause" the level of demand, of supply, etc.) only describe subjective relationships of the type: each consumer determines the quantities requested of the goods assuming that the prices and its income are given. There are, however, no objective causal links, except for the productive capacity-to-production link, which is objective because if it is true that prices determine the level of supply and demand, it is also true that in equilibrium prices are themselves to be determined.

A comparative static exercise consists in disrupting the balance described in Figs. 1 and 2 by introducing an exogenous shock in one of the variables subject to simultaneous determination or "given" assumptions in the equilibrium situation. Depending on the variables perturbed, a sequence of different reactions will be generated, that

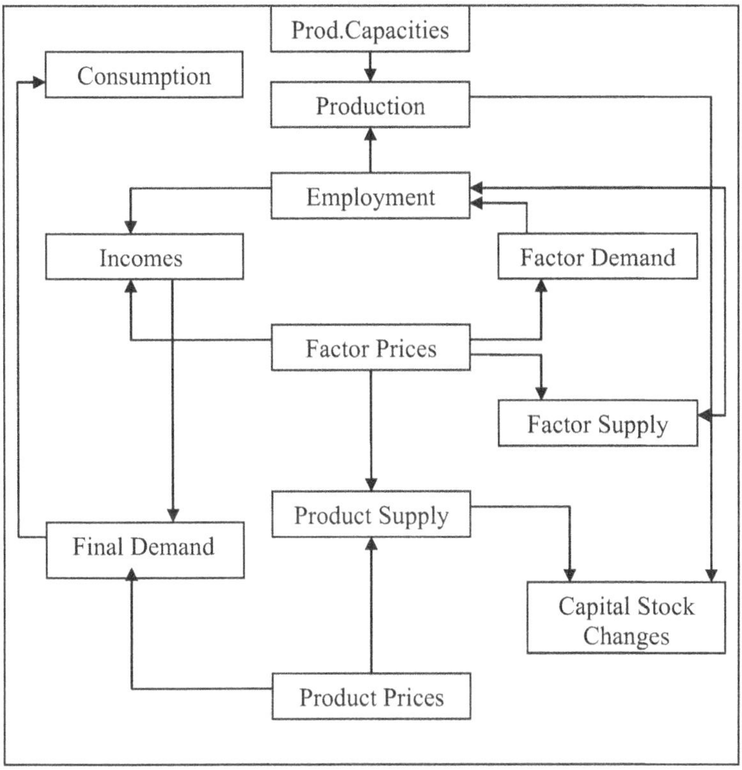

Fig. 2 The CGE models and their causal chains

can represent a different theory of achieving the general economic equilibrium. For example, an increase in production capacity due to technical progress will tend, through the reaction of firms that maximize profits, to increase demand for factors and therefore employment, supply of goods and production. This will put in motion a causal chain of classic type, which moves from production to consumption. Conversely, an increase in demand for goods, due to an exogenous variation in consumer preferences (or producers as regards investment goods) will tend to result in a causal chain of Keynesian type in which the increase in global demand ultimately causes an increase in production.

In both cases, however, to the initial cause-effect sequence, which moves in a different direction depending on the original exogenous impulse, follows an adjustment phase based on the reciprocal interaction between the variables. Thus, for example, given an increase in production capacity, the first reaction of producers will be to modify their production plans, increasing demand for factors and employment. However, the supply of factors constrains the possibilities of expanding employment and, through the increase in income of the factors themselves and therefore of families, to expand the demand for goods. After the first impact on the variables directly linked to it, the exogenous shock and the causal chain connected to it will then be "absorbed" by the mechanism of general economic equilibrium that will restore the simultaneousness of the interactions between variables and, ultimately, the equilibrium.

5.2 From Partial to General Equilibrium

The transition from partial equilibrium to general equilibrium is a delicate moment in the construction of the model, both because it identifies some crucial points for its solutions, and because it represents an important point and subject to frequent misunderstandings of the very notion of economic equilibrium. First of all, it is clear that can defined as "partial" any equilibrium that fails to represent one or more markets of the economy in question. The reciprocal of this statement is not true, however, that is to say that an equilibrium that involves all the markets is not necessarily "general". In addition to containing equations for all markets of real goods and services, in conditions of greater or lesser aggregation, in fact, a general equilibrium must also contain all the crucial variables of an economic system and, in particular, the quantities produced and consumed of the goods, the employment of factors, the prices of goods, services, factors and the incomes.

Let us consider for example the condition of Marshallian equilibrium between supply and demand:

$$Q(P) = \sum_{i=1}^{k} D(P, Y_i) \tag{1}$$

where Q(P) is a vector of quantities offered for the individual markets of the economy as a function of the n vector prices (P) and D(P, Yi) is the vector of quantities demanded by the i-th family as a function of the n prices and its income.

The equilibrium in (1) is partial for two reasons. First, it does not account for all markets of the economy. Even if it includes all the markets of the goods, it excludes the markets of the factors. Second, it does not account for the formation of incomes Y, which represent parameters and not variables of the equilibrium described.

We now add a market for the services of the factors, in the form:

$$Z_d(P_f) = Z_s(P_f) = Z \tag{2}$$

where $Z_s(P_f)$ and $Z_d(P_f)$ are, respectively, two quantity vectors demanded and supplied of the services of the m factors as functions of the corresponding m prices of the P_f vector.

The equilibrium described by (1) and (2) can now encompass all the markets of the economy, since it contains equations for both the goods and services markets, but it is not general because it is not yet able to give account of the determination of income.

If we add the equation:

$$Y = \Omega P'_f Z \tag{3}$$

where Ω is a k $\times$ m array of factor endowments, and Y is a m $\times$ 1 vector of household incomes, the equilibrium described has finally become "general". It corresponds, in fact, to a situation where all prices and incomes are simultaneously determined as a function of: (a) the preferences of the families (summarized in the product demand and the supply of factors), (b) technology (synthesized in the functions of supply of products and demand of the factors) and, (c) the endowment of the factors (synthesized by the matrix Ω).

The generality of the equilibrium therefore requires the model to be able to simultaneously determine all the quantities subject to transaction on the market, the corresponding prices and all incomes in a compatible way.

Note that the model (1)–(3) is not the only possible model of general economic equilibrium, even though it can be said that it corresponds to the "nucleus" of the Walrasian model and also of many of the newer versions. A model that reflects an alternative theory is instead based on the equality prices-production costs and the absence of profits.

For this model, we can also start with the equation:

$$P = C(P, P_f) \tag{4}$$

where C is an n $\times$ 1 vector of unit costs as a function of prices of goods and factor services. Since (4) is a theory of equilibrium formation in the goods market, to obtain the generality it is necessary to add a balance equation for the market of the factors services and an equation of income formation.

A common condition to general equilibrium models is given by the so-called "Walras Law", which ensures that the purchasing power emerging from the sum of the incomes generated according to the mechanism of the model coincides with the value of the goods consumed. In the system (1)–(3) this follows from the fact that the individual demand functions respect the budgetary constraint:

$$P'C_i(P, Y_i) \leq Y_i \quad \mathrm{i} = 1, 2, \ldots, \mathrm{k} \tag{5}$$

i.e. the expenditure of the i-th agent of consumption cannot exceed its income. Summing up yields:

$$P' \sum_{i=1}^{k} C_i(P, Y_i) - \sum_{i=1}^{k} Y_i \leq 0 \tag{6}$$

A sufficient condition for (6) to hold as an equality, given (5), is therefore that for each consumer the budgetary constraints are respected as a strict equality.

From the point of view of the firms, Walras law also requires that the cost of production is equal to the value of total production and that there are no profits, beyond what the market ensures as remuneration to the productive factors in the ownership of the entrepreneur (capital, know how, etc.). If there is more than one firm, however, this implies that the condition of absence of extra-profits is also valid for each one of them, because of the constraint that production costs cannot exceed revenues.

It should be noted that if household demand functions and firms' supply functions are derived from patterns of behavior that ensure that budgetary constraints are respected as strict equalities, Walras law is automatically verified. Conversely, if the patterns of behavior only ensure that the budgetary constraints are respected, possibly as inequalities, Walras law must be imposed as a constraint and, consequently, it ensures that the budgetary constraints are, ultimately, stringent (i.e. respected as equalities) for each operator.

It is also worth noticing that casting the model in terms of demand and supply functions allows to see more clearly the differences between partial and general equilibrium analysis. The former, in fact, proceeds from the Marshallian framework entirely based on behavioral functions, in a framework broadly consistent with revealed preferences, with response parameters such as demand and supply elasticities. General equilibrium, on the other hand, requires going beyond the purely behavioral functions by specifying their connections with income formation. For households and firms, this can be done by specifying the budget constraints, while underlying utility and/or objective functions can be invoked only to change the elasticities in response to large changes in prices or incomes, and thus can be neglected if these changes are sufficiently small.

5.3 A Simple Generalization of a CGE Structure: The Differential Model

A general formulation of a computable general equilibrium model can be developed by using a differential mathematical structure, as Leif Johansen, and after that several other scholars proposed in 1960. Unlike these earlier proposals, however, our formulation is not conceived as a linear approximation, but as a general structure underlying any model, starting from a reference point, which can be considered a general equilibrium. In the following, the subscript t refers to a particular time at which the parameters have been estimated and the model is run. We start with a commodity balance equation:

$$dX_t = A_t dX_t + dA_t X_t + dY_t + dC_t - dM_t \tag{7}$$

where:

A_t an n × n social accounting matrix

dX_t $n \times 1$ vector of variations of productive activities;
dA_t n × n parameter changes for the SAM
dY_t $n \times 1$ vector of variations of investment levels;
dC_t $n \times 1$ vector of variations of quantities consumed;
dM_t $n \times 1$ vector of variations of (net) quantities imported.

The production factors are governed by the following equations:

$$dZ_t^d = F_t dX_t + dF_t X_t + G_{dt} dP_{ft} + dG_{dt} P_{ft} \tag{8}$$

$$dZ_t^s = G_{st} dP_{ft} + dG_{st} P_{ft} + G_t dY_{t-1} + dG_t Y_{t-1} \tag{9}$$

$$dZ_t^d = dZ_t^s = dZ_t. \tag{10}$$

The first of these equations represents factor demand on the part of the firms. In it:

dZ_t^d $m \times 1$ vector of variations of demand levels for the m factors on the part of the firms;
dZ_t^s $m \times 1$ vector of variations of supply levels for the m factors;
F_t $m \times n$ matrix of output shares for the m factors;
G_{dt} $m \times m$ matrix of factor demand price coefficients;
G_{st} $m \times m$ matrix of factor supply price elasticities;
G_t $m \times n$ matrix of factor supply variations as a function of past investment.

Expression (8) states that factor demand depends, given the technology, the linear parameter levels and the parameter changes, on the production levels of the various sectors as well as on factor prices.

Equation (9), on the other hand, represents factor supply with:

dZ_t^s $m \times 1$ vector of variations of supply for the m factors on the part of their owners (households, firms etc.);
dP_{ft} $m \times 1$ vector of factor price variations;
dY_{t-1} $n \times 1$ vector of investment levels in the previous period.

Factor supply, according to Eq. (3) is thus determined, given the preferences of the agents involved, the linear parameters and the parameter variations, by factor prices and by the variation in productive capacity determined by past investment.

Equation (10), that represents the condition to maintain equilibrium between factor demand and factor supply, implies, together with (8) and (9):

$$dP_{ft} = G_{\Pi t}^{-1} \left(G_t dY_{t-1} - F_t dX_t - dF_t X_t + dG_{st} P_{ft} - dG_{dt} P_{ft} + dG_t Y_{t-1} \right), \quad G_{\Pi t} = G_{st} - G_{dt}. \tag{11}$$

Because $G_{\Pi t}$ and G_t are both positive matrices, the market theory contained in Eqs. (8)–(10) implies that increases of production and investment determine, respectively a positive tension on past prices and a negative tension on future prices. These forces, however, can be attenuated or even reversed by parameter changes. These changes, in turn, can be of two types: (i) they can correspond to changes required in the linear parameters to maintain a close approximation to a corresponding non-linear solution, (ii) they can be due to exogenous causes, such as for example the introduction of a new technology, the changes in the underlying preferences of the households etc. This means that if the new equilibrium solution is sufficiently close to the original equilibrium, parameter changes (i) can be neglected, and if there are no outside shocks to account for, changes (ii) can be assumed to be zero.

For example, assume that the labor demand function derives from profit maximization under a Cobb-Douglas technology. In this case, labor demand of the i-th sector is a unit elasticity function and the corresponding parameter of the G_{dt} matrix is $\overset{\circ}{g}_{idt} = \frac{\overset{\circ}{L}_{it}}{\overset{\circ}{P}_{Lit}}$ or the ratio between labor employment and wage in each sector at the initial equilibrium point. The parameter of the dG_{dt} matrix is[1]:

(11) $dg_{idt} = (g^1_{idt} - \overset{\circ}{g}_{idt})(\frac{P^1_{Lit} - \overset{\circ}{P}_{Lit}}{P^1_{Lit}})$, where the superscript 1 indicated the new equilibrium point.

From now on, we proceed on the hypothesis that parameter changes can be neglected, and/or, equivalently, that the model parameters are continually updated based on equations such as (11). Under these conditions, factor income variations are governed by the equation:

$$dV_t = P'_{ft} dZ_t + Z'_t dP_{ft}. \tag{12}$$

Or, upon substitution of (7) and (10):

$$dV_t = \left(\left\{P'_{ft}\right\} + \left\{Z'_t\right\} G^{-1}_{\Pi t}\right) F_t dX_t - \left\{Z'_t\right\} G^{-1}_{\Pi t} G_t dY_{t-1} \tag{13}$$

where the curls indicate the matrix obtained by diagonalizing the corresponding vector. Product prices vary with factor prices:

$$dP_t = \Phi(I - A')^{-1} F' dP_{ft} \tag{14}$$

where Φ is a diagonal n x n matrix of flexibility parameters w_i (i = 1, 2, ..., n). In particular,we assume $w_i = 0$ for internationally tradable goods and $w_i = 1$ for the other goods. Dropping for simplicity the t subscript from the parameter matrices, and substituting (12) into (14), we obtain:

$$dP_t = \Phi(I - A')^{-1} F' G^{-1}_{\Pi} (F dX_t - G_t dY_{t-1}) \tag{15}$$

[1]Since price elasticity equals 1, we have, equivalently: $dg_{idt} = (g^1_{idt} - \overset{\circ}{g}_{idt})(\frac{P^1_{Lit} - \overset{\circ}{P}_{Lit}}{P^1_{Lit}}) = (g^1_{idt} - \overset{\circ}{g}_{idt})(\frac{L^1_{it} - \overset{\circ}{L}_{it}}{L^1_{it}})$.

Consumption levels are function of prices and incomes:

$$dC_t - dM_t = dC_t^* = \Lambda dP_t + \Gamma dW_t \iota_k \tag{16}$$

where ι_k is a k,1 sum vector, $\Lambda = \sum_h^k \Lambda_h$, Λ_h (h=1,2, …, k) being an n, n matrix of linearized demand price elasticities, Γ is a n, k diagonal matrix of linearized Engel elasticities $dW = \Omega dV$ is a k,1 vector of households and other institutions' incomes, and Ω is a 1, k vector of income distribution shares, with generic element ω_{hj} (h=1, …, k; j=1,2, …, m). Note that expression (16) refers to aggregate consumption for each commodity. With many households and other institutions represented, we can write:

$$dC_t^* = dc_t^* \iota_k, \tag{17}$$

where dc_t^* is an n, k matrix having as a generic element $\left\{dc_{tih}^*\right\}$, i.e. consumption of the i-th commodity by the h-th institution (h = 1, 2, …, k). Each institution has thus its own system of demand functions, according to the expression:

$$dc_{th}^* = \Lambda_h dP + \Gamma_h dW_h \tag{18}$$

Using (14) and (16):

$$\begin{aligned} dC_t - dM_t = dC_t^* = & \left[\mathrm{T}\Omega\left(\left\{P'_{ft}\right\} + \left\{Z'_t\right\} G_\Pi^{-1}\right) + \Lambda\,\Phi\left(I - A'\right)^{-1} F' G_\Pi^{-1}\right] F dX_t + \\ & - \left[\mathrm{T}\Omega\left\{Z_t\right\} + \Delta\,\Phi\left(I - A'\right)^{-1} F'\right] G_\Pi^{-1} G_t dY_{t-1}. \end{aligned} \tag{19}$$

Since Λ is a negative definite matrix, in accordance with consumer theory, both the effect of increases in production and of previous investment may be either positive or negative, according to whether the income or the price effect prevails. If production increases, in fact, this tends to increase factor prices with positive effects on their incomes, but also with the consequence of increasing the prices of goods and services in the economy. Final demand is thus pushed upward by the income effect and downward by the price effect. Vice versa, the increase in productive capacity determined by previous investment, by shifting outward factors' supply, tends to reduce their prices. This causes a reduction of factor income, but a parallel reduction of prices of intermediate and final goods. The sign of the net result of these shifts will ultimately depend on the relative magnitude of the behavioral and technological parameters involved.

Substituting (19) into (7) and solving for dX_t, we obtain an equation that synthesizes the effect of investment (or any other exogenous shock) on total demand:

$$dX_t = (I - A - \theta_t)^{-1} (dY_t + \psi dY_{t-1}) \tag{20}$$

where[2] $\theta_t = \left[\Gamma\,\Omega\left(\left\{P'_{ft}\right\} + \{Z'_t\}\,G_{\Pi}^{-1}\right) + \Lambda\,\Phi\left(I - A'\right)^{-1} F' G_{\Pi}^{-1}\right] F$

and $\psi_t = -\left[\Gamma\,\Omega\left\{Z'_t\right\} + \Lambda\,\Phi\left(I - A'\right)^{-1} F'\right] G_{\Pi}^{-1} G_t$

We come now to the question of closure. In an open economy, the following equation will hold:

$$P'_t\left[(I - A)\,dX_t - dC_t - dY_t\right] + \left[X'_t\left(I - A'\right) - C'_t - Y'_t\right] dP_t = e P_t^{*\prime}\left(dE_t - dM_t\right) + P_t^{*\prime}\left(E_t - M_t\right) de \tag{21}$$

In (21) e denotes the exchange rate and we have assumed no variation in international prices P_t^*.

From the commodity balance conditions in (7), the first term respectively on the left and the right-hand side must be equal so that they will cancel. As a consequence, we can write:

$$\left[X'_t\left(I - A'\right) - C'_t - Y'_t\right] dP_t = P_t^{*\prime}\left(E_t - M_t\right) de \tag{22}$$

Assume first that there is no change in the exchange rate. Then Eq. (21) will hold by a simple aggregation of the commodity balance equations and can be written as:

$$\left[dS_t - P'_t dY_t\right] = e P_t^{*\prime}\left(dE_t - dM_t\right) \tag{23}$$

where $dS_t = P'_t\left[(I - A)\,dX_t - dC_t\right]$ denotes the aggregate variation of domestic savings. This equation states the familiar equality condition between the domestic savings gap and the foreign exchange gap. If we assume that dY_t is exogenous, the first and simplest closure implies the further assumption that the equality between the left and the right-hand side in (23) is maintained by a balancing variation of savings, for an exogenous variation of the current account.

Alternatively, we can include an equation describing the formation of savings and let the current account balance the equality in (23). This can be done by assuming that the current account may absorb the shock of a positive or negative variation of the domestic savings gap, without the need to adjust the exchange rate or use the variation of the exchange rate to absorb the shock in all or in part.

---

[2]Note that, in order for (12) is derived from the equation: $dX_t = (A_t + \theta_t)\,dX_t + (dY_t + \psi dY_{t-1})$, where $A_t + \theta_t$ constitutes an "augmented" input-output coefficient matrix, which takes into account not only the transactions in the base year (through the initial i-o matrix A_t, but also of the endogenous income and price effects (through the matrix θ_t). This augmented i-o must contains all the information to construct a social accounting matrix (SAM) that corresponds to a general equilibrium and must also be balanced, both in terms of transactions and coefficients (summing to 1 for each column), since the budget constraint will continue to hold for each sector when the endogenous incomes and prices are taken into account.

A second, necessary condition to ensure the "generality" of the economic equilibrium described is the so called Walras Law, according to which the value added from production must equal factor income. Starting from this budget constraint for the base year:

$$P_t'(I - A)X_t = P_{ft}'Z_t, \tag{24}$$

the variations analysed by running the model must respect the equation:

$$P_t'(I - A)dX_t + X_t'(I - A)'dP_t = P_{ft}'dZ_t + Z_t'dP_{ft}. \tag{25}$$

Given Eqs. (24) and (25), Walras Law is automatically respected in the case of a closed economy. If we introduce a class of internationally tradable goods, however, the situation becomes more complex, since their prices do not vary in response to domestic demand and or supply variations, thus providing no mechanism for equilibrium except through exchange rate variations.

Indicating with e_t the exchange rate that converts international prices into domestic currency, and assuming no import or export taxes/subsidies, we can write:

$$dP_{1t} = P_1^* de_t + e_t dP_1^* \tag{26}$$

$$dP_{mt} = P_{mt}^* de_t + e_t dP_{mt}^* \tag{27}$$

$$dP_{1t} = A_{11}'dP_{1t} + A_{12}'dP_{2t} + F_1'dP_{ft} + F_{1m}'dP_{mt} \tag{28}$$

$$dP_{2t} = A_{2t}'dP_{1t} + A_{22}'dP_{2t} + F_2'dP_{ft} + F_{2m}'dP_{mt} \tag{29}$$

where:

- dP_{1t} is an $n \times 1$ vector of variations of domestic prices of internationally traded goods (IT);
- P_1^* is an $n \times 1$ vector of international prices expressed in foreign currency (e.g. dollars) for (IT);
- P_{mt}^* is a variation of the international price of the aggregate intermediate inputs imported, expressed in foreign currency;
- de_t is the variation of the exchange rate (in domestic currency per unit of foreign currency);
- dP_{2t} is an $n - n_1$, 1 vector of internationally non-tradable goods (NT);
- dP_{mt} is the variation of the domestic price of the aggregate intermediate imports;
- $A_{ij,}' F_i'$ are appropriate sub-partitions, respectively of matrices A' and F'.

With this new specification, we consider the fact that in this model (and in general in the SAM representation of the economy), imports of intermediate goods are treated as a factor of production. Unlike the domestic factors, however, they have an exogenously set price (since they are an IT), that can only be affected by a variation of the exchange rate, and earn an income that accrues to the rest of the world. We thus introduce two extra-equations to determine the domestic prices of internationally tradable goods (IT), which makes demand prices, i.e. prices faced by consumers, potentially different from supply prices, i.e. prices that reflect production costs according to Eq. (27). Solving Eqs. (27) and (28) obtains explicit expressions for factor and NT prices. In this case, factor prices will reflect their productivity at world prices. If these prices are above the factor prices that would clear the factor markets (from Eq. (10)), there will be unemployment, while an excess demand for factors will occur in the opposite case. More specifically, solving (26)–(29), and ignoring for simplicity the intermediate imports, we obtain:

$$dP_{ft} = \Xi(I - A'_{11})dP_{1t} + A'_{12}\Xi dP_{2t} \tag{30}$$

$$dP_{2t} = [(I - A'_{22} - F'_2\Xi A'_{12})^{-1}A_{21} + F'_2\Xi(I - A'_{11})]dP_{1t} \tag{31}$$

where Ξ is $F_1'^{-1}$ if the number of factors is the same as the number of IT goods, a generalized inverse of F'_1 if the number of factors is less than the number of IT goods, and the inverse of a submatrix of F'_1 with the size equal to the number of goods otherwise.

Substituting (31) into (30):

$$dP_{ft} = [\Xi(I - A'_{11}) + A'_{12}\Xi(I - A'_{22} - F'_2\Xi A'_{12})^{-1}A_{21} + F'_2\Xi(I - A'_{11})]dP_{1t}. \tag{32}$$

This expression shows how factor prices reflect the opportunity costs of producing IT goods and how these costs vary with a domestic price variation reflecting international prices. Expressions (30) and (31) also suggest that if factor markets are perfectly integrated, and world prices and/or the exchange rate do not change, there will be no changes in domestic prices. In practice, however, factor markets will generally display a certain degree of segmentation and imperfect mobility. Factor prices will thus only reflect marginal productivities at world prices for the factors whose markets are more closely related to IT production.

To satisfy Walras law, by achieving equilibrium in the balance of trade, the variation de_t of the exchange rate must satisfy the equation:

$$de_t = -\frac{\left[dX'_{1t}\left(I - A'_{11}\right) + A'_{21}dX'_{2t}\right]P_1^*e_t - \left(Z'_{1t}dP_{ft} + P'_{ft}dZ_{1t}\right)}{\left[X'_{1t}\left(I - A'_{11}\right) + A'_{21}X'_{2t}\right]P_1^*}. \tag{33}$$

This equation, obtained by applying Walras Law to the system (25)–(28), expresses the amount of devaluation (or revaluation) necessary to re-equilibrate the economic system in response to a variation of the equilibrium values of the quantities dX_{1t}, dX_{2t} e $dZ_{1t} = F'_1dX$ and of factor prices dP_t^f. According to Eq. (33) such a devaluation should be equal to the variation of the net benefit from the increase in production of IT goods at international prices, in domestic currency, per unit of value of the production of the same goods in foreign currency.

The CGE model presented above can be represented by the following social accounting matrix (SAM) in differential form.

	Products= dX	Factor employment= dZ	Factor Incomes= dV	Institutions' income= dY	Factor prices= dP_f	ITs Domestic Prices dP_1	NTs Domestic Prices=dP_2	Exchange rate= de	Previous Shocks= dY_{t-1}
dX	A	0	0	Γ	0	0	Λ		0
dZ	F	0	0	0	0	0	0		0
dV	0	$\{P_f\}$	0	0	$\{Z\}$	0	0		0
dY	0	0	Ω	T	0	0	0		0
dP_{f1}	0	0	0	0	0	Ξ	0	0	0
dP_{f2}	G_{Π}^{-1} $(G_s - G_d)$	0	0	0	0	0	0	0	$G_{\Pi}^{-1}G_t$
dP_1	0	0	0	0	0	0	0	P_1^*	0
dP_2	0	0	0	0	F_2'	U_{21}	A_{22}'	0	0

Indicating this modified matrix with the symbol B, we can write:

$$d\xi_t = Bd\xi_t + HdY_{t-1}. \tag{34}$$

where $d\xi_t$ is a column vector of products, factors, income and prices unknown $(dX_t, dZ_t, dV_t, dY_t, dP_{ft}, dP_t)$ and dY_{t-1} is a vector of effects of previous investment or other exogenous changes.

Notice that B is a particular SAM matrix that includes both the primal and the dual variables. This matrix is not singular, since it is conditional to the small change represented by the exogenous shock. This implies, for example, that the product columns list all costs to produce the corresponding commodity from both intermediate goods and factors (as in the original SAM), but also that factor employment columns include the increases in factor costs due to price changes.

In general, therefore, one or more columns and rows of B have to be considered exogenous and can be the object of an exogenous shock. For example, indicating with dY_t a vector of exogenous shocks to the capital formation- investment account, and with B_Y the matrix B without the row and the column corresponding to the capital formation account Y, and assuming $HdY_{t-1} = 0$, we can write:

$$d\xi_t = (I - B_Y)^{-1}dY_t. \tag{35}$$

Expression (35) indicates the CGE solution in response to an exogenous investment shock, consisting of an increase in the demand of goods and services from producing sectors that are directly able to contribute to capital building. Because of the general equilibrium structure of the matrix, the effects of these exogenous shocks will include the impact on the increased costs of endogenously priced factors and intermediate inputs, thereby taking into account opportunity costs from alternative resource use.

This extension of CGE fundamentals thus adds the analysis of opportunity costs to the traditional CGE-based policy analysis. This extension enriches the policy content of both the micro and macro level of the equilibrium analysis. It takes advantage of the fact that in a modern policy and project analysis the micro-macro link exactly aggregates from the individual to the family, community, which is often the level of feasibility and impact analysis of large projects, and society level using micro and macro behavioural models that are closely integrated. The next sections illustrate this assertion.

5.4 An Exactly Aggregable Micro-Macro Link

In the traditional simulation literature linking the micro and macro level of policy analysis the micro level of analysis refers to the household while the macro level refers to society as a whole. This representation neglects several layers of aggregation of high policy relevance. Each household is a collection of individuals with their own preferences and levels of well-being that are employed in both marketable and non-marketable household production activities. The household enterprise is per se a miniature economy that can be studied within an equilibrium framework. The recent advances in the collective theory of the household (Chiappori 1992; Chiappori and Ekeland 2011; Chiappori and Lewbel 2015) makes it possible to identify preferences of each member of the household and distributive effects within the household so that, for example, adults and children can be regarded as social classes of the family micro-society. It is then natural to describe "input-output" transactions within the household using a social accounting framework (Matteazzi et al. 2017).

Until the recent past, the micro-macro link was missing also at the community or village level (Taylor and Adelman 2006; Taylor 2012; Taylor and Filipski 2014). This limitation was mainly due to lack of complete statistical data at low administrative levels. Non-survey methods working top-down from macro input-output tables produce approximation errors that increase the larger is the zooming at the micro level. Trade information is especially exposed to this imprecision. One of the main problems in the assembly of local economy tables consistent with the system of regional input-output tables is obtaining inter-community commodity flows, along with their respective zones of influence. Moreover, it is especially difficult to account for the possibility of existing simultaneous import and export of the same product (cross hauling).

When the level of disaggregation or the local economy of interest does not coincide with administrative units, as for example in the case of a natural park, an industrial district or a large project, it is preferable to supplement the published statistics available at the local level with representative business surveys about the input-output structure of local industries and the associated trade flows with the surrounding zone of influence and the rest of the world (Taylor 2012). To save survey costs, household consumption information may be inferred from sufficiently representative national income, consumption or living standard surveys.

Figure 3 describes the micro-macro link between the general equilibrium model at the macro level of the economy and the general equilibrium model at the micro level

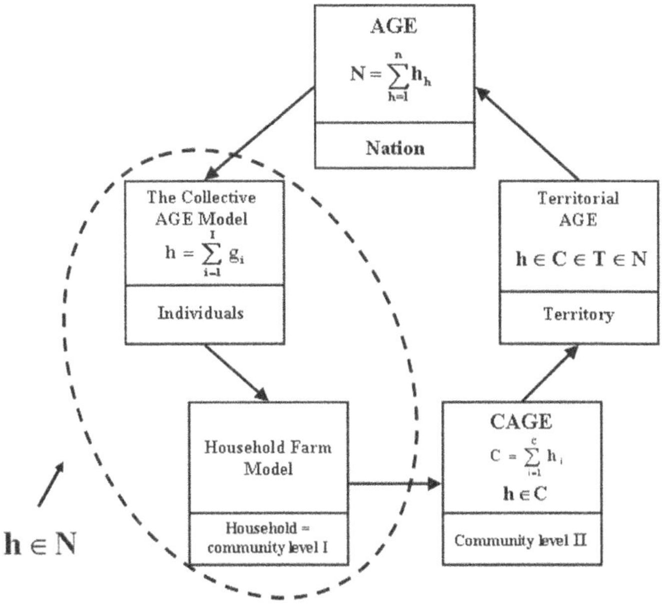

Fig. 3 A "general" micro-macro link

of the household economy that accounts for individual preferences and well-beings. The dashed set diagram emphasizes the fact that the primitive macro-micro link is the one aggregating all household individuals into the family seen as a mini-society. The *h* household farm, or enterprise, level can be interpreted as a first community level aggregating each *g* member of the household using the collective theory of the household (Chavas et al. 2017). Then, households at the micro level aggregate up to the macro-level of the whole economy. As shown in the right panel of the graph, households can aggregate also at the intermediate level of a community, such as a village, or of a territory such as a natural park, an industrial district or a region.

The micro dimension of the "general" representation of the micro-macro link described in Fig. 3 is specular to the general equilibrium macro dimension. In the traditional microsimulation literature, the approach is partial in the sense that the focus is limited either to labor supply or consumption, health, housing, education, marketable production, home production or other issues that are analyzed in separate modules.

To implement this approach, data must be general too. The multi-topic approach implemented by the World Bank Living Standard Measurement Study (LSMS) integrates information about the household, marketable and non-marketable household production (when the time use module is included) and the service and business community is an appropriate example. In less developed economies, where agriculture still contributes significantly to domestic production, it is relevant to record who does what in the family, both in agricultural and household related activities to construct a reliable input-output matrix of agriculture (or other marketable family business) and of home activities. This aspect is recommended in the Wye Group Handbook (2007) that also stresses the relevance to record individually disaggregated data and the consumption of goods assignable to specific members of the household, if the interest is to implement the micro-macro link starting from the individual to the family, community and society level.

Using surveys with a design that integrates consumption, production, and time use information as in the LSMS (Grosh and Glewwe 2000) or the ISMEA case (Finizia et al. 2004) and reporting individual specific information, it is possible to implement collective household enterprise models within an equilibrium framework (Caiumi and Perali 1997; Matteazzi et al. 2017) representing the base micro level depicted in Fig. 3. These studies extend the traditional farm household model (Singh et al. 1986; De Janvry et al. 1991) to encompass recent advances in collective theory. The model represents production and consumption-leisure choices along with the rule governing intra-household resource allocation to analyse the income and wage responses of each family member and recover their level of well-being. The household is treated as an equilibrium model whose accounts are based on a collective household accounting matrix, with the social dimension being the wife/husband classes. The micro data permit the joint estimation of behavioral parameters characterizing consumption, farm production and household production choices of farm-households. These estimates are used to construct the micro farm-household model. The household enterprise, be it a farm or a firm, is by analogy the micro-level mirror image of the macro-economy.

At the household level, production and consumption decisions are non-separable. The collective approach permits deducing the welfare levels of individual household members thus making it possible to account for gender and inter-generational differences in the evaluation of policy impacts and individual responses to policy changes in the labor or capital markets.

Because of non-separability, farm production and household consumption are estimated jointly. The econometric methodology consists first in estimating household production and deriving the price of the aggregate non-marketable domestic product, and, secondly, in estimating the production and consumption side of the household economy conditional on the estimated instrumented domestic price.

The specification of the micro econometric model takes the following behavioral aspects into account: non-separability of the farm, household and home activities, time allocation and associated labor supply between on-farm, off-farm and on-home production, corner solutions related to the choice of input use such as capital and labor and optimal portfolio of production activities, rule governing the intra-household distribution of resources permitting the recovering of individual preferences and welfare levels of the husband, the wife and, possibly, the children, estimation of shadow wages considering family labor as a quasi-fixed input.

The design of this estimation strategy is clearly general in the sense that it estimates consumption, leisure and both marketable non-marketable production jointly so that the econometric model can be transferred as such within the structure of a general equilibrium model without the need of a traditional calibration (Matteazzi et al. 2017). Interestingly, the modelling of zero consumption, labor or other factor choices at the micro-level can also be easily mirrored at the macro-level adopting Löfgren and Robinson's (1997, 1999) mixed complementarity approach.

When integrated surveys are not available, as in the case of many developed societies that prefer to maintain the higher level of detail of separate consumption, income, wealth, labor and time use surveys, statistical matching techniques should be applied (Wolff et al. 2012; Dalla Chiara et al. 2016). Such integrated data bases are suited for the analysis of standard of living and for the estimation of complete collective demand systems describing consumption, domestic production and leisure/labor choices (Caiumi and Perali 2015). If the available micro data-base is not integrated, then the micro level of analysis is by force partial.

Clearly, the production side of the economy is not represented and should be econometrically summarized using micro-data from business surveys possibly incorporating a design that records input-output and trade transactions or macro-data recording input use per sector as in the KLEM-style data bases (Jorgenson 2007; Jorgenson and Samuel 2014). Both data sources would be consistent with the aggregate account data of the corresponding input-output data.

Farm or non-agricultural enterprise household models can be aggregated into a local general-equilibrium framework representing a village, larger communities or counties linked together spatially through trade (Taylor and Adelman 2006; Taylor 2012; Taylor and Filipski 2014) as illustrated in Fig. 3. The aggregation process may continue bottom-up and stop at the desired level of policy analysis.

In this section, we stressed the importance of a "general" and integrated data design with an input-output structure and sufficient individual level information so that a single source of information may feed both the micro and macro behavioural model. Micro data would then exactly aggregate to the macro level (Jorgenson et al. 1980; Savard 2003; Magnani and Mercenier 2009) so that it would be possible to consistently zoom in from the macro to the micro level or to implement a micro-macro zooming out statistically consistent across levels of aggregation. It would also be possible to estimate exactly aggregable micro-econometric models of consumption, labour supply and production and the associated set of parameters along with their standard errors that would permit a more flexible representation of economic behaviour as compared to the standard CES forms used in traditional applied general equilibrium analyses. In addition, there would be no need for a traditional calibration procedure, aiming at deriving the set of parameters from a given SAM or borrowing them from other external econometric studies, because the set of parameters of interest would be produced "in-house" along with their confidence sets (Jorgenson et al. 2013; Taylor 2012; Taylor and Filipski 2014; Matteazzi et al. 2017).

Another virtue of integrating micro-econometric modelling with applied general equilibrium is the possibility to obtain confidence intervals for the equilibrium outcomes of changes in economic policies. Jorgenson et al. (2013) show how to apply the Delta method for policy evaluation given the knowledge of the asymptotic covariance matrix of the parameters. It becomes also practicable to run project impact simulations using Monte Carlo methods (Taylor 2012; Taylor and Filipski 2014). Such simulation methods in an applied general equilibrium context may become a complement to randomized control trials (RCTs) and an effective tool for impact evaluation where RCTs are not feasible. When randomized control trials are not appropriate or cannot be implemented, impact evaluation at the local level can be performed comparing the same local economy before and after the program or with adjacent local economies "not treated" with the program.

This inferential feature of the micro-macro approach should become a standard feature of policy and project evaluation in an equilibrium framework, though it should be recognized that it is constrained more by data availability rather than modelling capabilities. It also opens the doors to the implementation of causal analysis with observational data, represented in our context by the micro integrated data base, along the lines traced by Heckman (2010), striving to reconcile structural and program evaluation policies by using LATE techniques or by studying causal models through Directed Acyclic Graphs (DAGs) or Bayesian Networks (Pearl 2013). Clearly, causal inference can be pursued provided that the applied framework is an econometrically estimated and calibrated general equilibrium model and is an important opportunity for policy analysis, because, as stressed by Imbens (2010: 401), questions concerning the causal effects of macroeconomic policies or involving general equilibrium effects can rarely be settled by randomized experiments.

So far, we have examined a circular micro-macro link that can be made operational either from the bottom up or indifferently top down exactly aggregating individuals to families, communities or societies or disaggregating society into their individual members. However, as it will be apparent in the next section, the micro-macro modelling approach is not fully integrated because the tax-impact component of the policy analysis is mainly omitted.

5.5 A Fully Integrated Micro-Macro Modelling Approach

A fully integrated micro-macro modelling approach refers to a modelling strategy that builds a formal communication flow between the macro equilibrium analysis, the micro behavioural analysis and the non-behavioural tax-benefit simulator. A tax-benefit microsimulation model calculates the effects of direct and indirect taxes and benefits on household incomes and work incentives for the population of a country or, in a comparable manner, for a set of countries. It executes a highly detailed and exhaustive set of policy rules, that must be updated on a yearly basis, using representative expenditure, income or standard of living surveys whose reported incomes are checked for consistency with available administrative data. The simulator is usually too large to be hosted in a general equilibrium model that normally incorporates a gross representation of a country tax system. This explains why it is rare to observe applications that integrate an analytical tax-benefit simulator with a less systematic and accurate macro representation of the fiscal system.[3] Previous discussions of

[3]Bourguignon and Spadaro (2006) explain that performing a microsimulation entails three basic inputs: the policy rules to be evaluated describing, for example, a tax reform, an appropriate behavioural model of individual response to policy and an informative micro dataset.

micro-macro modelling integration do not explicitly include a tax-benefit simulator (Savard 2003; Davies 2009; Ahmed and O'Donoughe 2007; Cockburn et al. 2010; Cockburn et al. 2014; Peichl 2016) in the communication flow.

The macro level is normally represented by a CGE with a representative agent or more agents mirroring the social class differentiation of the underlying SAM. To establish an exactly aggregable link, it is recommendable that the SAM information about household classes come from the same micro dataset used for the behavioural micro-simulation model. This feature would help retain the heterogeneity that CGE models alone normally do not account for and the micro-macro consistency of income and poverty simulations at the micro level by linking intra-group heterogeneity to a statistical relationship between averages and a measure of entropy dispersion.

Two main approaches can be identified to build a formal communication between the macro and micro layer of the analysis. The approach proposed by Savard (2003), Cockburn et al. (2010) and Cockburn et al. (2014) is an integrated method that incorporates all the "real" households in the micro data set directly in the general equilibrium model. It does so by simply extending the set of households treated in the model while ensuring the coherence between income and expenditure accounts in the household survey and in the SAM. This method uses traditional calibration techniques to recover demand parameters rather than importing directly more flexible and sophisticated consumption and labour supply models estimated before the execution of the micro-macro exercise. This evolution, along with the theoretical result in Magnani and Mercenier (2009), would eliminate both the need to reconcile incomes with expenditures and savings and the model aggregate account and would respond to the concern raised by Savard (2003) and Bourguignon et al. (2005) about the difficulties investigating policies involving discrete choices or regime switching behaviour.

The second main approach establishes a sequential top-down link, whereby the macro CGE model generates equilibrium prices that are passed on to a partial-equilibrium behavioural household micro-simulation model. This approach is usually performed with an iterative feed-back "top-down/bottom-up" process that ensures consistency between the behaviour of the aggregate classes in the CGE and that of individual households (or individuals if a collective approach is pursued) in the micro databases (Savard 2003; Bourguignon et al. 2005). As noticed also by Cockburn et al. (2014), the importance of the feedback effect critically depends on the aggregation error due to the lack of exact aggregation of the micro behavioural functions. If the exactness property is maintained, then the micro and macro models will be consistent and the iterative feedback process will be minimal.

The micro-macro dialogue depicted in Fig. 4 traditionally occurs between the macro models passing on information about equilibrium prices and incomes of the new post-reform economic situation, to the micro behavioural models. It simulates how consumers or workers respond to the change and the associated impact on poverty and income distribution, or between the tax-benefit non-behavioural simulators, thereby transferring information about post-reform incomes to the behavioural micro-simulation models. This clarifies what we mean by a fully integrated micro-macro modelling system that takes advantage from a circular communication flow

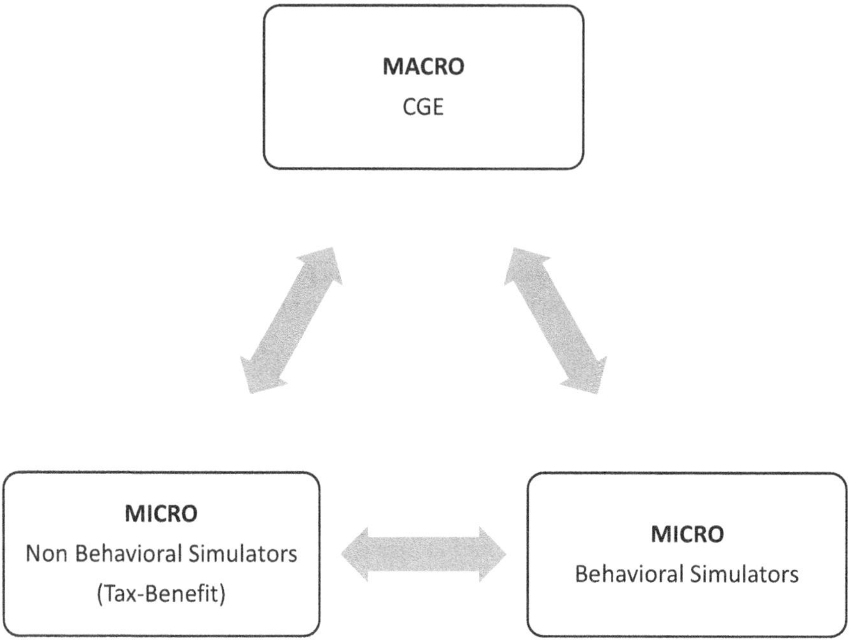

Fig. 4 A fully integrated micro-macro modelling approach

between the macro and both the behavioural and non-behavioural micro models, as illustrated in Fig. 4.

We now provide a formal representation of a fully integrated approach.[4] In the following expression, a tax-benefit simulator gets the records about households' gross income y_h from microdata obtained from information about the household *h* supply of working hours *l* at aggregate wage *w* and non-labour income as the rent *r* of capital *k*

$$y_h = rk_h + wl_h$$

and derives households' net income y_h^{net} by applying a household specific tax-benefit rule τ to gross incomes

$$\begin{aligned} y_h^{net} &= (1 - \tau_h)\, y_h + \mathrm{transf}_h = ((1 - \tau_h)\, rk_h + \mathrm{transf}_h) + (1 - \tau_h)\, wl_h \\ &= y_h^{exo} + (1 - \tau_h)\, wl_h, \end{aligned}$$

where net incomes are decomposed into an exogenous component y_h^{exo} and endogenous labour income.

[4]We thankfully acknowledge Jean Mercenier for this integrated representation and for many fruitful discussions on frontier topics of general equilibrium.

At the micro level, each member j of household h chooses her/his optimal bundle of consumption goods c and labour l by maximizing her/his own utility U subject to the individual budget constraint corresponding to a proportion $0 < \mu^j < 1$ of household net income

$$Max_{c_h^j, l_h^j} \left\{ U_h^j \left(c^j, 1 - l^j \right) | pc^j = \mu^j \cdot \left(\left(1 - \tau_h \right) l_h w_h + y_h^{exo} \right) \right\}$$

whose solution gives optimal collective consumption for each consumption aggregate (or sector) s such as food, housing, education, health, transportation, recreation, and others, and collective labour supply

$$\begin{cases} c_{(h|s)}^j = c_{(h|s)}^j \left(w, p, y_h^{exo}, \tau \right) \\ l_h^j = l_h^j \left(w, p, y_h^{exo}, \tau \right) \end{cases} \quad \forall j = \text{male, female, children and } \forall s \text{ sectors}$$

that feed the macro CGE level.

At the macro level, to exemplify we suppose that the economy reaches a Walrasian equilibrium that we summarize as follows

$$\begin{cases} \sum_h l_h = L^{den}(\frac{w}{p}, Q) \\ \sum_h k_h = K^{den}(\frac{w}{p}, Q) \Rightarrow (w^*, p^*, r_{numeraire}; Q^*) \\ \sum_h c_h = Q \end{cases}$$

where individual factor demands aggregate to the total factor endowments in the economy (L^{dem}, K^{dem}) and aggregate demand equals aggregate supply Q. The unique solution of the system gives equilibrium prices and output $(w^*, p^*, r_{numeraire}; Q^*)$ that may feed and feedback both the behavioural microsimulation models and, for a complete integration, the tax-benefit simulator

$$y_h^{net} = y_h^{exo} + (1 - \tau_h) \, w^* l_h^*.$$

Then, to maintain an exact micro-macro consistency, researchers would need to iterate until convergence, where two adjacent iterations give wages and prices differing for a small error of the 1×10^{-8} size. Incorporating labour supply and consumption schemes consists in replacing the first order conditions in the CGE by the estimated consumption and labour (collective) supply

$$\begin{cases} c_h^j = \hat{c}_h \left(w, p, y_h^{exo}, \tau \right) \\ l_h^j = \hat{l}_h \left(w, p, y_h^{exo}, \tau \right) \end{cases}.$$

It is relevant to note that failing to close the model at the macro level would be "distortionary partial". For example, focusing on labour supply alone, would break the utility optimization and $\hat{c}_h$ would be determined residually to satisfy the budget constraint $\mathrm{pc}_h = (1 - \tau_h)\, \mathrm{l}_h \mathrm{w}_h + \mathrm{y}_h^{exo}$.

The importance of a direct integration between CGEs and tax-benefit models should not be overlooked. For example, consider an aggregate shock, such an economic crisis or the exit of a European country from the Community. In this case the iterative use of the tax-benefit simulator and the CGE model would allow determining the endogenous policy rule $\hat{\tau}$ that neutralizes the shock.

This step is critical to place tax-benefit simulators as an integral component of the micro-macro behavioural modelling approach that is going to make more and more extensive use of big data warehouses. A relevant example is the Rhomolo model (Mercenier et al. 2016) that is a spatial computable general equilibrium model for EU regions and sectors. The model is calibrated using the SAMs of all regions in Europe, uses inter-regional trade information of both products and factors, and incorporates systematic information of the fiscal and legal systems specific to each region.

Similarly, there is a large investment in creating "big micro data" warehouses that are uniformed across European regions of interest so that it is practical to run the same behavioural (consumption and labour supply) models. In turn these would allow to support micro-econometric based inter-regional general equilibrium models that will be able to perform causal policy analysis. Such EU-level matched data bases, would allow a more thorough understanding of the policy impact at the micro level on the well-being of policy-relevant types of families, such as fragile and migrant families, and a more complete cross-country comparison of standards of living, possibly accounting for differences in prices and access to public services.

6 Conclusions

Computable general equilibrium (CGE) modelling is an evolving methodology of applied economics, that is fast becoming the main tool of policy analysis capable to link the aggregate level of the economy, to different disaggregate levels, such as, in particular, specific regions, population aggregates and investment projects. Although some of their theoretical foundations can be traced to the historical debate on the theory of market equilibria from Walras to Debreu and beyond, CGE models are pragmatic constructions based on available statistics and relatively simple estimates. They try to capture the essentials of the economic system using the mathematical and statistical structure of the Social Accounting Matrix, as a core element of their description of a transaction based economy. This implies not only that they are not an attempt to quantify the complicate relations described by the classical economists under the invisible hand hypothesis, but also that they do not address some of the big unresolved issues concerning equilibrium among stocks and flows and, in particular, by different capital theories.

Because of their reliance on SAMs, CGEs also present the peculiarity of being a consistent and complete set of accounts that represent a disaggregation of the national accounts and provide monetary indicators of value consistent with the UN methodologies and the gross domestic product (GDP) concept. They also allow to easily extend the same accounting principles to represent and measure values of non-market goods, externalities, environmental damages, and natural resources.

Linear homogeneity of the price system in the model makes equilibrium price variations and prices specified only in relative terms, with respect to a numeraire, which can be a particular or a composite commodity. A simple differential generalization of a CGE structure shows how to accommodate shadow prices and opportunity costs in policy and project evaluation.

By virtue of these characteristics, and in spite of its many limitations, both the theory and the application of CGE modelling has been growing fast, in part as the consequence of advancing knowledge in the fields of social accounting, mathematical programming and development of computational power. Even though the SAM transactions do reflect a fully circular economy and the CGE computations register the interdependencies across a plurality of autonomous agents, the results of the simulations tend to reveal well defined adjustment mechanisms and typically can be interpreted as the consequence of adding flexibility and price effects to a basic input-output structure. Model closures are useful to further avoid the "black box" effect characteristic of large models, by imposing explicit directions to the causal chains of the policy experiments and project evaluations. If we further consider that the SAMs are the aggregate accounting structure of microdata about firms and households transactions that can be used by micro econometric models that supply behavioural parameters and associated standard errors to the macro model, then policy and project evaluations techniques can be implemented in a causal framework.

Yet some important questions remain and suggest that many dimensions of a new frontier of CGE modelling are ready for exploration. Among these, the most important appears to be the reliance of the current models on aggregate agents, whose behavioural functions are assumed to be equivalent or similar to those of the individual, decentralized agents acting in the economic system. We know that this is not only a gross simplification, but because of the "everything goes" result, it is not consistent with economic theory. Any set of excess demand functions, in fact, may be capable of being solved for non-negative equilibrium prices, regardless of the properties of the underlying disaggregated structures that may have been used to generate them. This means that using well behaved demand and supply functions at aggregate level does not per se add any validity to the general equilibrium solutions found. In other words, more specific micro-foundations, such as those derived by the collective theory of the household, are not only useful, but necessary to fully legitimize the aggregate model and ensure the exact aggregation of the micro-macro link within a modelling approach that fully integrates the macro CGE environment with the tax-benefit analysis of fiscal reforms and the micro-behavioural impacts.

The above considerations also bring about the question of the causal nature of CGE models and their use in inference. As causal models, CGEs have to deal with three sets of tasks in this respect:

- Defining the set of hypotheticals or counterfactuals (a scientific theory),
- Identifying parameters (causal or otherwise) from hypothetical population data (mathematical analysis of point or set identification),
- Identifying parameters from real data (estimation and testing theory).

From the point of view of inference this implies that counterfactuals, even in dynamic settings, may face two different problems: (i) identifying causal models from idealized data of population distributions (infinite samples without any sampling variation), (ii) identifying causal models from actual data, where sampling variability is an issue. In case (i), the hypothetical populations may be subject to selection bias, attrition and the like, but all issues of sampling variability are irrelevant for this problem. In case (ii), the analysis must recognize the difference between empirical distributions based on sampled data and population distributions generating the data.

The tension that has been driving recent developments in the theory and applications of general equilibrium towards greater integration between policy and project analysis has led to major achievements in moving the frontier of knowledge forward. As this new stock of knowledge is increasingly transferred to institutions and practitioners, it is our hope that the quality of the policy process and project evaluation will be greatly enhanced.

References

Ahmed, V., & O'Donoghue, C. (2007). CGE-Microsimulation modelling: A survey. MPRA Paper No. 9307.

Arrow, K. J. (1951). An extension of the basic theorems of classical welfare economics. In J. Neyman (ed.), *Proceedings of the Second Berkeley Symposium on Mathematical Statistics and Probability*, .

Arrow, K. J., & Debreu, G. (1954). Existence of an equilibrium for a competitive economy. *Econometrica, 22*(3), 265–290.

Bourguignon, F., & Spadaro, A. (2006). Microsimulation as a tool for evaluating redistribution policies. *Journal of Economic Inequality, 4,* 77–106.

Bourguignon, F., Robilliard, A. S., & Robinson, S. (2005). Representative versus real households in the macroeconomic modelling of inequality. In T. J. Kehoe, T. N. Srinivasan, & J. Whalley (Eds.), *Frontiers in applied general equilibrium modelling*. Cambridge: Cambridge University Press.

Caiumi, A., & Perali, F. (1997). Female labor force participation: A comparison between urban and rural families. *American Journal of Agricultural Economics, 79,* 595–601.

Caiumi, A., & Perali, F. (2015). Who bears the full cost of children? Evidence from a collective demand system. *Empirical Economics*.

Chavas, J. P., Menon, M., Pagani, E., & Perali, F. (2017). Collective household welfare and intra-household inequality. Theoretical Economics (forthcoming).

Chiappori, P. A. (1992). Collective labor supply and welfare. *Journal of Political Economy, 100*(3), 437–467.

Chiappori, P. A., & Ekeland, I. (2011). New Developments in aggregation economics. *Annual Review of Economics, 3*, 631–668.

Chiappori, P. A., & Lewbel, A. (2015). Gary Becker's a theory of the allocation of time. *Economic Journal, 125*(583), 410–442.

Cockburn, J., Corong, E., & Cororaton, C. (2010). Integrated computable general equilibrium (CGE) micro-simulation approach. *International Journal of Microsimulation, 3*(1), 60–71.
Cockburn, J., Savard, L., & Tiberti, L. (2014). Macro-micro models. In C. O'Donoghue (ed.), Handbook of microsimulation modelling, Contributions to economic analysis (Vol. 293, pp. 275–304). Emerald Group Publishing Limited.
Colander, D. (1995). Marshallian general equilibrium analysis. *Eastern Economic Journal, 21*(3).
Dalla Chiara, E., Menon, M., & Perali, F. (2016). An integrated database to measure living standards. Working paper, Department of Economics, University of Verona.
Damania, R., & Scandizzo, P. L. (2016). The Serengeti ecosystem—Burden or bounty? *Journal of Policy Modeling*.
Davies, J. (2009). Combining microsimulation with CGE and macro modelling for distributional analysis in developing and transition countries. *International Journal of Microsimulation*, 49–65.
De Janvry, A., Fafchamps, M., & Sadoulet, E. (1991). Peasant household behavior with missing markets: some paradoxes explained. *Economic Journal, 101,* 1400–1417.
Debreu, G. (1951). The coefficient of resource utilization. *Econometrica*.
Debreu, G. (1974). Excess-demand functions. *Journal of Mathematical Economics, 1,* 15–21.
Devarajan, S., Go, D., Lewis, J., Robinson, S., & Sinko, P. (1997). Simple general equilibrium modeling. In J. Francois, & K. Reinert (Eds.), *Applied Methods for trade policy analysis*. Cambridge University Press.
Dixit, A. K., & Stiglitz, J. E. (1977). Monopolistic competition and optimum product diversity. *American Economic Review, 67*(3), 297–308.
Finizia, A., Magnani, R., & Perali, F. (2004). *A general equilibrium analysis of the mid term review of the CAP on the Italian Economy*. Rome: Franco Angeli.
Grosh, M., & Glewwe, P. (2000). Designing household survey questionnaires for developing countries: Lessons from 15 years of the living standards measurement study. World Bank.
Heckman, J. (2010). Building bridges between structural and program evaluation approaches to evaluating policy. *Journal of Economic Literature, 48*(2), 356–398.
Imbens, G. (2010). Better LATE than nothing: Some comments on deaton (2009) and Heckman and Urzua (2009). *Journal of Economic Literature, 48*(2), 399–423.
Jorgenson, D. W. (2007). 35 Sector KLEM'. http://hdl.handle.net/1902.1/10684.
Jorgenson, D. W., & Samuels, J. (2014). Long-term estimates of U.S. productivity and growth. In: *Prepared for Presentation at Third World KLEMS Conference Growth and Stagnation in the World Economy*, Tokyo, 19–20 May 2014.
Jorgenson, D. W., Lau, L. J., & Stoker, T. M. (1980). Welfare comparisons and exact aggregation. *American Economic Review, 70*(2), 268–272.
Jorgenson, D., Goettle, R., Ho, M., & Wilcoxen, P. (2013). *Double dividend. environmental taxes and fiscal reform in the united states*. Cambridge, MA: The MIT Press.
Löfgren, H., & Robinson, S. (1997). The mixed-complementarity approach to specifying agricultural supply in computable general equilibrium models. In *1997 Conference*, 10–16 August 1997, Sacramento, California, International Association of Agricultural Economists, No 197058.
Löfgren, H., & Robinson, S. (1999). To trade or not to trade: Non-separable farm household models in partial and general equilibrium. TMD Discussion Paper No. 37, IFPRI, Washington, D.C.
Magnani, R., & Mercenier, J. (2009). On linking microsimulation and computable general equilibrium models using exact aggregation of heterogeneous discrete-choice making agents. *Economic Modelling, 26,* 560–570.
Mantel, R. (1974). On the characterization of aggregate excess-demand. *Journal of Economic Theory, 7,* 348–353.
Matteazzi, E., Menon, M., & Perali, F. (2017). The collective farm-household model: Policy and welfare simulations. *Applied Economic Perspectives and Policy, 39*(1), 111–153.
McKenzie, L. W. (1959). On the existence of general equilibrium for a competitive economy. *Econometrica, 27*(1), 54–71.
Mercenier, J., Álvarez-Martínez, M., Brandsma, A., Di Comite, F., Diukanova, O., Kancs, A., Lecca, P., López-Cobo, M., Monfort, P., Persyn, D., Rillaers, A., Thissen, M., & Torfs, W. (2016).

Rhomolo v-2 model description. A spatial computable general equilibrium model for EU regions and sectors. JRC Technical reports, Valencia, Spain.
Norton, R., & Scandizzo, P. L. (1981). General equilibrium computations in activity analysis models. *Operations Research, 29*(2), 243–262.
Pearl, J. (2013). Reflections on Heckman and Pinto's 'causal analysis after Haavelmo. UCLA cognitive systems laboratory, Technical report (R-420), Working Paper.
Peichl, A. (2016). Linking microsimulation and CGE models. *International Journal of Microsimulation, 9*(1), 161–175.
Savard, L. (2003). Poverty and income distribution in a CGE-household micro-simulation model: Top down/bottom up approach. Laval University, CIRPEE Working Paper 03–43, CIRPEE, Université Laval: Quebec.
Scandizzo, P. L. (1995). A general equilibrium analysis of ownership, property rights and control in a transition economy, ESA Working Paper (FAO), no. 02-01/ FAO, Rome (Italy).
Scandizzo, P. L. (2014). Equilibrium theory. In *Encyclopaedia of Law and Economics*. Springer.
Scandizzo, P. L., & Ferrarese, C. (2015). Social accounting matrix, a New Estimation Methodology. *Journal of Policy Modeling*.
Scarf, H. E., & Hansen, T. (1973). The computation of economic equilibria. In *Cowles foundation for research in economics at Yale University*, Monograph No. 24, New Haven, CT and London, UK: Yale University Press.
Singh, I., Squire, L., & Strauss, J. (Eds.). (1986). *Agricultural household models, extensions, applications and policy*. Washington DC: The World Bank and The Johns Hopkins University Press.
Sonnenschein, H. (1972). Market excess-demand functions. *Econometrica (The Econometric Society), 40*(3), 549–563.
Sonnenschein, H. (1973). Do Walras' identity and continuity characterize the class of community excess-demand functions? *Journal of Economic Theory, 6*, 345–354.
Stone, R., & Brown, A. (1962). *A computable model for economic growth*. Cambridge, UK: Cambridge Growth Project.
Taylor, E. (2012). A methodology for local economy-wide impact evaluation (LEWIE) of cash transfers. Working Paper, 99, International Policy Centre for Inclusive Growth.
Taylor, Ed, & Adelman, M. (2006). *Village economies: The design, estimation, and Use of villagewide economic models*. Cambridge, MA: Cambridge University Press.
Taylor, Ed, & Filipski, M. (2014). *Beyond experiments in development economics. local economy-wide impact evaluation*. Oxford, UK: Oxford University Press.
The Wye Group. (2007). *Rural households' livelihood and well-being. statistics on rural development and agriculture*. NY: United Nations.
Walbroeck, J. L., & Ginsburg, V. A. (1986). *Activity analysis and general equilibrium modelling*. Amsterdam: North Holland Publishing Co. Webley, O.
Wolff, E. N., Zacharias, A., Masterson, T., Eren, S., Sharpe, A., & Hazell, E. (2012). A comparison of inequality and living standards in Canada and the United States using an expanded measure of economic well-being. Working paper n. 703, Levy Economics Institute.

Part II
Methodology and Estimation Issues

Demand-Driven Structural Change in Applied General Equilibrium Models

Roberto Roson and Dominique van der Mensbrugghe

Abstract This chapter analyzes the variations in industrial structure induced by income-sensitive patterns of final consumption, and how these changes can be captured by a multi-sector numerical model with a flexible demand system. We focus, in particular, on the estimation of parameters for an AIDADS (An Implicitly, Directly Additive Demand System) specification. We then test the latter by inserting it in the ENVISAGE global general equilibrium dynamic model, which is run under the SSP2 scenario from 2011 to 2050. It is found that time-varying income elasticity can generate sizable variations in the industrial structure. This finding has important practical implications, particularly when structural models are applied at a medium and long term horizon.

Keywords Demand systems · Structural change · Economic dynamics
Computable general equilibrium models

JEL Codes C33 · C68 · D58 · E21 · O11 · O41

1 Introduction

Structural change refers to the variations in the patterns of industrial output, consumption and trade flows inside an economic system. In the short run, this change is mainly determined by income and relative prices, but in the medium and long run other forces shape the economic structure in a more persistent way. Technological progress, modifications of production processes, shifts in aggregate consumption,

R. Roson (✉)
Department of Economic, Ca'Foscari University of Venice, Venice, Italy
e-mail: roson@unive.it

R. Roson
IEFE Bocconi University, Milan, Italy

D. van der Mensbrugghe
The Center for Global Trade Analysis, Purdue University, West Lafayette, IN, USA
e-mail: vandermd@purdue.edu

© Springer International Publishing AG, part of Springer Nature 2018
F. Perali and P. L. Scandizzo (eds.), *The New Generation of Computable General Equilibrium Models*, https://doi.org/10.1007/978-3-319-58533-8_2

possibly driven by demographic evolution, all contribute to long lasting structural change.

Understanding structural change, and its determinants, is clearly an interesting and relevant scientific topic in itself, with direct policy implications. It is also practically important when applied, multi-sector general equilibrium models are used for the assessment of policies and effects having impacts in the long run, like in the case of climate change. Indeed, whereas these models are usually characterized by a detailed account of the economic structure, which is often essential when dealing with impacts affecting specific sectors, they are also normally calibrated on the basis of some past data (e.g., input-output tables or their social accounting matrix (SAM) extensions), meaning that they mirror an economic structure quite different from the one we could possibly observe in the distant future.

Some of the factors affecting the long run structural change are clearly unpredictable. Most of the technological breakthroughs of the past, affecting various industries, appear to have occurred in a seemingly random fashion. Harberger (1998) points out that the whole dynamics of economic progress actually resembles the growth process of "mushrooms", rather than the steady rise of "yeast" (as neoclassical models of economic growth posit).

Some other factors, however, are quite predictable, in the sense that some of the forces which will affect the economic structure tomorrow are already active and observable today. Technology adoption and diffusion is under way. Catching up by fast growing developing economies is occurring. Demographic transitions are taking place, as well as mass migrations.

Broadly speaking, there are two classes of effects at work. There are supply side effects, affecting industrial productivity, either directly or indirectly, and there are demand side effects, involving variations in the structure of final demand. In this paper, we focus on the issue of modeling and numerical estimating changes in the pattern of aggregate household consumption, driven by varying (growing) levels of per capita income. Therefore, income levels are taken here as given, although in a full-fledged numerical model they could be determined endogenously, or obtained from an hypothetical scenario.

Modeling a time-varying and income-dependent structure of household consumption implies introducing a sufficiently sophisticated demand system, capable of capturing what Matsuyama (2016) terms "Generalized Engel Law": the fact that budget shares in consumption expenditure (and, more generally, industrial shares in terms of employment, value added or output) do not vary monotonically over time at progressively higher income levels. Therefore, in the next section, we briefly review what functional forms have been employed in the recent economics literature for this purpose. We focus, in particular, on the AIDADS (An Implicitly, Directly Additive Demand System; Rimmer and Powell 1992), presenting in Sect. 3 an exercise of parameters estimation for this demand system, based on the recently published Report of the 2011 International Comparison Program (ICP 2015). Section 4 illustrates how results obtainable from a dynamic, computable general equilibrium model may change when the AIDADS specification, instead of a simpler, conventional form is employed to model consumption demand. A final section draws some concluding remarks.

2 Long-Run Changes in Consumption Patterns

Several demand systems, utility and expenditure functions, all with differentiated income elasticity, have been proposed. Desirable properties for their utilization in applied economic models are: (1) relative simplicity and analytical tractability; (2) generation of well behaved demand curves; (3) easiness of parameters estimation. Of course, the choice should also depend on the characteristics of the underlying model and on its purpose, for instance:

- the model could focus either on relatively small variations in income or expenditure levels (e.g., a single country CGE for short run policy assessment), or on more substantial variations (long run scenarios or intercountry comparison);
- the model could primarily focus on changes in income, rather than changes in relative prices.

Assessing long run changes in the structure of consumption demand means considering significant changes in income, with variations in relative prices entering only as a second order effect. Therefore, the selection of a demand system should be restricted to functional forms that, at higher income levels but constant relative prices, simulate structural changes consistent with historical "stylized facts".

One interesting option is the Hierarchical Demand System (Matsuyama 2002; Buera et al. 2013). The idea behind the HDS is deceptively simple: goods and services are ranked from lowest to highest priority in terms of needs. All consumers spend their income in a sequential way, starting from basic needs and stepping up to the highest level they can afford with their income. Once a need is satisfied, the corresponding good or service provides no more marginal utility. This is broadly consistent with the observation that goods could be initially regarded as a luxury (e.g., air conditioning), and when they can be obtained they become a necessity. When associated with a given income distribution, HDS can produce some interesting dynamics, with goods/industries "taking off" at various stages of economic development, possibly generating "hump shaped" trajectories as well.

Generally, HDS works well for theoretical models (possibly to be validated econometrically), but its implementation in applied macro-economic models like the CGEs would require information about the distribution of income and how it could evolve over time. This may be quite problematic, especially when a large set of countries are considered, including data-poor developing countries.

Gohin (2005) illustrates how to implement any regular configuration of price and income effects through "latent separability". Latent separability can be seen within an intermediate production process, where goods are first used to produce commodities, which are the true arguments of the utility function and not the goods. Even if each intermediate utility function is homothetic, there is a wide spectrum of possible income and substitution effects for purchased goods generated from the combination of different groups to which each good belongs. The problem with this method here is that is assumes knowledge of income and substitution elasticities from the outset. Indeed, this information is used to infer a consistent latent separability structure, which is not observable.

A number of authors have recently work with some variants of CES functions, with industry-specific but time-constant income elasticities. In Fieler (2011) a single parameter plays the double role of substitution and income elasticity. Caron and Markusen (2014) set relative income elasticities equal to relative substitution elasticities, whereas Comin et al. (2015) use separate and independent parameters for the two good-specific elasticities.

In all cases, income elasticities are constant. This implies that the demand pattern does not stabilize over time and, actually, the good with the highest income elasticity would asymptotically cover 100% of the budget. Clearly, this is not an appealing property for a realistic assessment of long run changes in demand patterns.

A demand system for structural change simulation should be "sufficiently flexible" or, technically speaking, "full rank". Rank one demands, the most restrictive demand systems, are independent of income; rank two demand systems are less restrictive, allowing linear Engel curves not necessarily through the origin; while rank three (i.e., full rank) demand systems are least restrictive, allowing for non-linear Engel responses (Cranfield et al. 2003).

Among the many full-rank demand systems which have been proposed, AIDADS (An Implicitly, Directly Additive Demand System; Rimmer and Powell 1992) appears to be especially suited for implementation in multi-sector, applied general equilibrium models. Indeed, it was introduced by CGE modelers and it has already been applied in a number of CGE models (Yu et al. 2000, 2004; Golub and Hertel 2008).

The AIDADS can be seen as a generalization of the Linear Expenditure System (LES). The demand for good i is expressed as:

$$q_i = \gamma_i + \phi_i \frac{y - \sum_j p_j \gamma_j}{p_i} \tag{1}$$

where y is total income or expenditure, γ_i is a parameter and ϕ_i (which in a LES would itself be a fixed parameter) is given by:

$$\phi_i = \frac{\alpha_i + \beta_i e^u}{1 + e^u} \tag{2}$$

with α_i, β_i parameters and u being the *implicitly* defined, *cardinal* utility function. To understand how AIDADS behaves, notice that:

$$\lim_{u \to -\infty} \phi_i = \alpha_i \tag{3}$$

$$\lim_{u \to \infty} \phi_i = \beta_i \tag{4}$$

$$\alpha_i < \phi_i < \beta_i \tag{5}$$

$$\lim_{y \to \infty} \frac{p_i q_i}{y} = \phi_i = \beta_i \tag{6}$$

Expenditure shares therefore stabilize at the level ϕ_i in the long run, although at different "speeds". It is not possible to get a closed form solution for the utility level u, which must then be estimated numerically, alongside the parameters α_i, β_i and γ_i. A number of constraints must also be taken into account, to ensure regularity conditions for the system (Powell et al. 2002). Cranfield (1999) shows how to use maximum likelihood methods to this purpose, employing also bootstrapping techniques to get parameters statistics (e.g., confidence intervals) and maximum entropy for multiple demands, disaggregated in terms of per-capita income.

Furthermore, Cranfield et al. (2003) assesses the ability of five structural demand systems to predict demands when estimated with cross sectional data spanning countries with widely varying per capita expenditure levels. Results indicate demand systems with less restrictive income responses are superior to demand systems with more restrictive income effects. Among the least restrictive demand systems considered, the AIDADS and the Quadratic Almost Ideal Demand System (QUAIDS) seem roughly tied for best, while the Quadratic Expenditure System (QES) is a close second. They notice that an important advantage of the QUAIDS model over AIDADS is its ease of estimation. Yet, and despite the fact that AIDADS is not exactly aggregable, the latter has fewer price related parameters to estimate and is designed so that budget shares lie between zero and one at all expenditure levels. This property suggests a preference for AIDADS when expenditure (income) shows substantial variation (or when extrapolations would involve large changes in expenditure) but prices are anticipated to experience little change.

3 Estimation of an AIDADS Demand System

ICP (2015) provides data on real and nominal consumption expenditure for 180 countries at the year 2011, in 14 categories, which are further aggregated here in 11 consumption classes:

- Food and nonalcoholic beverages (**FOOD**)
- Alcoholic beverages, tobacco, and narcotics (**BEVTOB**)
- Clothing and footwear (**CLOTHING**)
- Housing, water, electricity, gas and other fuels + Furnishings, household equipment and maintenance (**HOUSE**)
- Health + Education (**HEAEDU**)

- Transport (**TRANSP**)
- Communication (**COMMUN**)
- Recreation and culture (**RECREAT**)
- Restaurants and hotels + Miscellaneous goods and services (**OTHER**)
- Machinery and equipment (**MACHINE**)
- Construction (**CONSTR**)

Ratios between real and nominal consumption readily give a set of country and sector specific price indexes. For the estimation of AIDADS parameters, we closely follow Cranfield (1999), by formulating the equations in terms of budget shares, and adding a stochastic error term:

$$w_{ir} = \frac{p_{ir}\gamma_i}{y_r} + \frac{\alpha_i + \beta_i \exp(u_r)}{1 + \exp(u_r)} \left(1 - \frac{\sum_i p_{ir}\gamma_i}{y_r}\right) + \epsilon_{ir} \tag{7}$$

where w_{ir} is the observed household budget for the item i in country r; y_r stands for total *per capita* expenditure (income) in country r; p_{ir} is the price index for the item i in country r; ϵ_{ir} is a normal multivariate error term, distributed independently across observation, with zero mean and finite covariance matrix, where the sum over all items in each country is zero. All remaining symbols, including the cardinal utility u_r, are parameters to be estimated.

The following restrictions apply:

$$\begin{aligned} \sum_i \alpha_i &= 1 \\ \sum_i \beta_i &= 1 \\ 0 \leq \alpha_i, \beta_i &\leq 1 \end{aligned} \tag{8}$$

The estimation is performed using a non-linear maximum likelihood procedure,[1] and gives the results shown in Table 1.

Figure 1 graphically displays how the budget shares evolve at constant prices, when annual per capita income (total consumption expenditure) varies from a minimum level of 8691 USD up to 168788 USD.

To interpret the meaning of the estimated parameters, consider that gamma (γ) expresses the fixed and unavoidable consumption, therefore the higher the value for this parameter, the more essential a certain good or service is seen, in terms of basic needs. On the other hand, beta (β) is the asymptotic budget share, for income levels going to infinity. The higher this share is, the more important a consumption item becomes, as we get very rich.

To make the AIDADS system functional for a numerical simulation model, an additional step is necessary. Indeed, the procedure illustrated above allows to estimate country specific values for the cardinal utility u, but that variable is not available in the destination model, so a link must be established between utility and income

[1]Technical details about the specific algorithm and software are available on request.

Table 1 Estimated parameter values

	Alpha	Beta	Gamma
FOOD	0.40	0.00	116
BEVTOB	0.02	0.02	16
CLOTHING	0.04	0.03	29
HOUSE	0.08	0.21	136
TRANSP	0.07	0.09	6
COMMUN	0.02	0.02	1
RECREAT	0.00	0.07	10
CONSTR	0.16	0.13	40
MACHINE	0.10	0.10	16
HEAEDU	0.08	0.14	98
OTHER	0.02	0.20	38

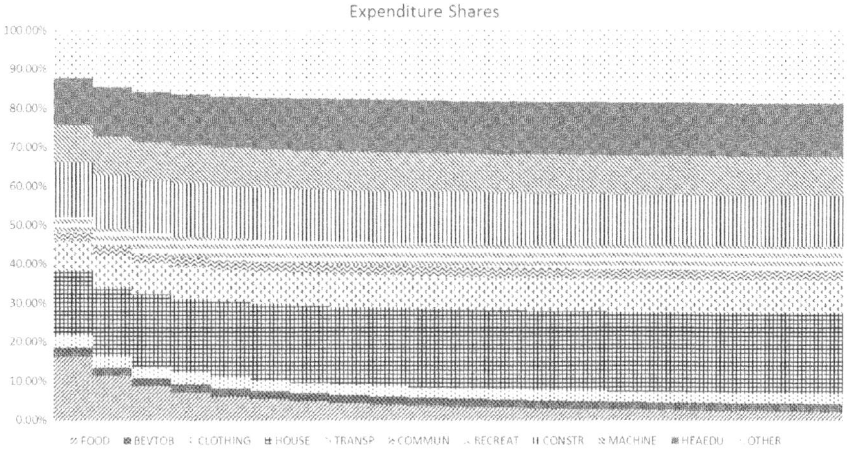

Fig. 1 Expenditure shares by income levels

levels. To this end, observe the plot contrasting income (vertical axis, logarithmic scale) with cardinal utility levels in Fig. 2.

The Figure suggests that the relationship is semi-logarithmic. Indeed, after trying several specifications of the functional form, the best regression results have been obtained with the following heteroskedasticity corrected OLS formulation, where u_r is regressed against $\ln(y_r)$:

```
Model 1: Heteroskedasticity-corrected, using observations 1-177
Dependent variable: u
           coefficient std.error t-ratio p-value
-----------------------------------------------------------
const      -7.17788    0.160788   -44.64 1.45e-097 ***
lnm         0.839040   0.0183408   45.75 2.80e-099 ***
Statistics based on the weighted data:
Sum squared resid 656.4597 S.E. of regression 1.936801
R-squared 0.922833 Adjusted R-squared 0.922392
F(1, 175) 2092.804 P-value(F) 2.80e-99
Log-likelihood -367.1501 Akaike criterion 738.3002
Schwarz criterion 744.6525 Hannan-Quinn 740.8764
Statistics based on the original data:
Mean dependent var -0.047430 S.D. dependent var 1.348156
Sum squared resid 28.07932 S.E. of regression 0.400566
```

When the estimated coefficients of the regression are plugged into the AIDADS demand (1), the latter becomes a function of income and prices only, as one would expect from a regular demand function:

$$q_i = \gamma_i + \left(\frac{\alpha_i + \beta_i K y^Z}{1 + K y^Z} \right) \cdot \frac{y - \sum_j p_j \gamma_j}{p_i} \tag{9}$$

where we have added the two constants $K = 0.000763284$ and $Z = 0.83904$.

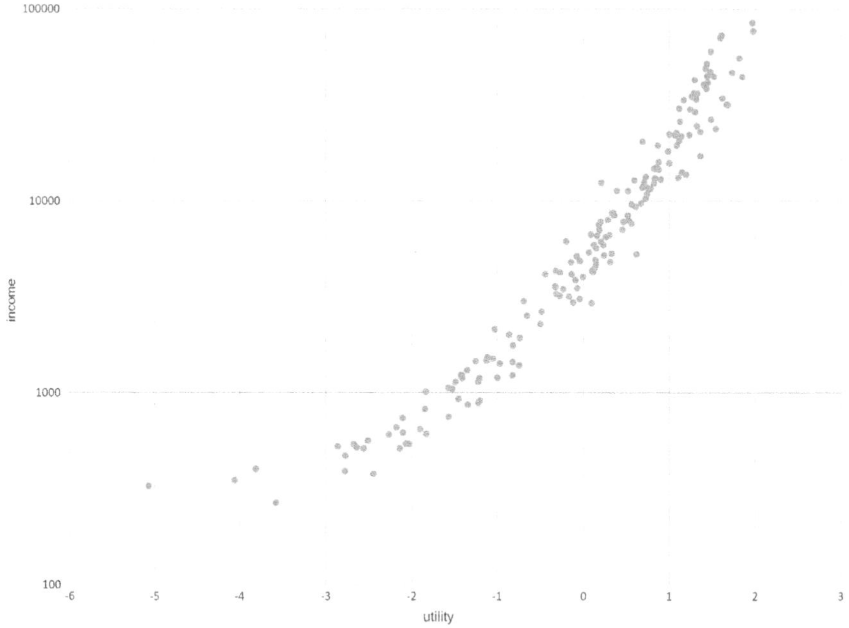

Fig. 2 Income versus cardinal utility levels

4 Introducing a Flexible Demand System into a Dynamic CGE Model

We have used the recursive dynamic global CGE model ENVISAGE (van der Mensbrugghe 2017) to assess how results may change in a multi-sector structural model, when a flexible demand system like the AIDADS is introduced. First, a baseline was built, by running the model with endogenous labor productivity[2] and exogenous GDP growth (using OECD projections) and population (using IIASA projections), according to the Shared Socio-Economic Scenario 2 (SSP2), from 2011 to 2050. In two subsequent rounds, labor productivity was kept fixed at its baseline level, but two alternative specifications for the final consumption demand were tested: a simple homothetic Cobb-Douglas and the more flexible AIDADS system.[3] The purpose is verifying how the model output could vary when income elasticity for households consumption is switched from constant and unitary values to non-constant and time-varying ones.

Figure 3 shows the differences in GDP produced by the AIDADS simulation with respect to the Cobb Douglas benchmark, for the following 14 macroregions in the ENVISAGE model:

- Low income East Asia & Pacific (LEAP)
- Middle income East Asia & Pacific (MEAP)
- South Asia (LSAS)
- Low income Europe & Central Asia (LECA)
- Middle income Europe & Central Asia (MECA)
- Low income Middle East & North Africa (LMNA)
- Middle income Middle East & North Africa (MMNA)
- Low income Sub-Saharan Africa (LSSA)
- Middle income Sub-Saharan Africa (MSSA)
- Low income Latin America & Carib. (LLAC)
- Middle income Latin America & Carib. (MLAC)
- European Union (EU28)
- United States (USAM)
- Rest of high-income countries (XHIC)

The different regions exhibit a differentiated response after the introduction of time-variant elasticities of substitution. Some regions get a higher growth, most notably Middle income East Asia & Pacific (MEAP), whereas other regions are characterized by lower growth, in particular Middle income Europe & Central Asia (MECA).

---

[2]Labor productivity growth is assumed to be smaller for Services and greater for Manufacturing industries.

[3]Parameters of the AIDADS system were adjusted to fit the different industrial classification in ENVISAGE.

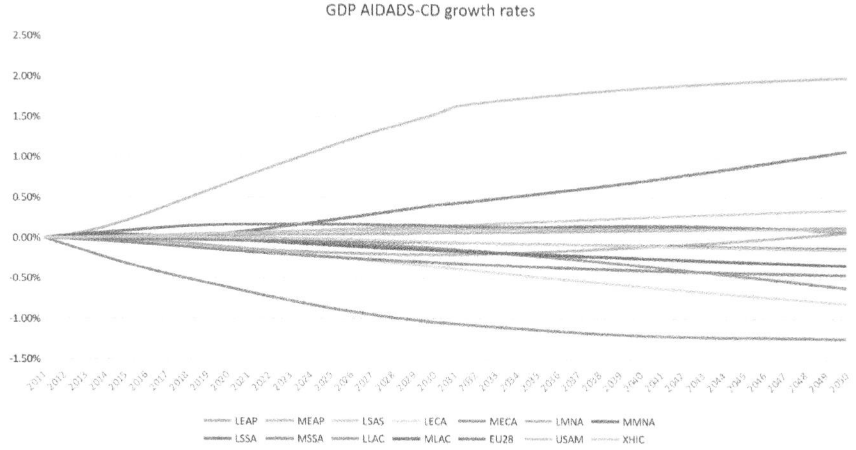

Fig. 3 Differences in GDP growth rates

To understand the reasons behind such divergence, we look at the composition of the gross industrial output in the two regions (Table 2), recalling that Manufacturing is the sector with the highest productivity growth in ENVISAGE, while Services is the slowest one. Notice that East Asia is characterized by a very large share of Manufacturing, whose component is significantly larger under the AIDADS specification. By contrast, middle income countries in Central Asia and Europe (outside the EU) are characterized by a much smaller Manufacturing sector, but a much larger share for Services (in particular, Transport and communications). As aggregate GDP growth can be seen as a weighted average of industrial growth rates, the different structure obtained under the AIDADS and CD formulations has direct implications for the national income increase. Regionswith relatively high shares of manufactures, relative to services, will see an accentuation of aggregate GDP growth when using the AIDADS specification relative to the C-D specification.

Table 3 presents the same industrial output composition, but for the whole world. With a unitary income elasticity (CD) all changes in the structure of final consumption must be due to variations in relative prices, so as to keep the shares in value terms constant. Here the drivers of variations in relative prices are differentiated productivity growth rates: since services are characterized by slower growth, their relative prices increases and real consumption first, then gross output decrease (in relative terms).

When the AIDADS formulation replaces the CD one, the effect of income elasticity overlaps to the relative price effect. For both Agriculture and Services industries, income and relative prices work to the opposite directions, and the industrial shares at 2050 do not differ very much from those of the 2011 base year (except Housing services). For Manufacturing, instead, the two effects reinforce each other, bringing about a share for "Other manufacturing" 3.81% larger than it was in 2011.

Table 2 Industrial output composition in MEAP and MECA regions

MEAP	2011 (%)	2050-CD (%)	2050-AID (%)
Cereals (CERL)	3.04	4.21	2.33
Livestock (LVST)	3.14	4.48	2.72
Processed food (PFUD)	3.80	5.22	3.23
Textile, apparel and leather goods (TXWP)	5.03	4.97	3.93
Other manufacturing (MANU)	40.76	38.62	40.73
Housing utilities (HUTL)	11.89	9.23	10.05
Wholesale and retail trade (TRAD)	5.30	5.74	6.85
Transport and communication (TRCM)	9.63	9.72	10.64
Financial services (FSRV)	7.68	7.54	8.90
Housing services (HSRV)	9.73	10.27	10.62
MECA	2011 (%)	2050-CD (%)	2050-AID (%)
Cereals (CERL)	2.15	2.21	1.29
Livestock (LVST)	4.08	4.40	2.74
Processed food (PFUD)	3.30	3.58	2.53
Textile, apparel and leather goods (TXWP)	2.56	2.36	1.72
Other manufacturing (MANU)	17.28	17.60	17.78
Housing utilities (HUTL)	15.27	15.98	16.50
Wholesale and retail trade (TRAD)	11.56	11.94	12.63
Transport and communication (TRCM)	22.56	22.34	22.46
Financial services (FSRV)	8.88	8.54	9.11
Housing services (HSRV)	12.35	11.05	13.24

Table 3 Industrial output composition—World

World	2011 (%)	2050-CD (%)	2050-AID (%)
Cereals (CERL)	2.03	3.64	2.04
Livestock (LVST)	2.47	3.44	2.54
Processed food (PFUD)	3.09	4.01	2.88
Textile, apparel and leather goods (TXWP)	2.12	2.80	2.25
Other manufacturing (MANU)	24.26	26.86	28.07
Housing utilities (HUTL)	11.50	11.35	11.90
Wholesale and retail trade (TRAD)	10.27	9.23	10.08
Transport and communication (TRCM)	12.64	12.72	13.21
Financial services (FSRV)	12.80	10.57	11.46
Housing services (HSRV)	18.81	15.38	15.57

5 Conclusions

Changes in the economic structure are due to variations in technology and preferences, but also to differentiated sectoral productivity growth and varying patterns of consumption, sensitive to income per capita levels. Whereas future technology and preferences are not observable, trends in productivity are, as well as the response of consumption patterns to different income levels.

This chapter has focused on the estimation of the latter effect, that is on the changes in the economic structure driven by a different composition of final consumption in the medium and long term. An empirical estimation of parameters for a flexible demand system has been presented, and the system was tested in a structural dynamic general equilibrium model. We found that time-varying income elasticity can generate sizable variations in the industrial structure.

This finding has important practical implications, because numerical structural models like CGE are increasingly been employed to assess long terms effecst of policies and other impacts (e.g., economic impacts of climate change), but structural parameters are still derived from past input-output and social accounting matrices.

More work is needed to understand and gauge how income effects interact with differentiated productivity growth rates. Different sectoral "speeds" have been assumed in the ENVISAGE model, but not empirically estimated. We leave this topic for future research.

References

Buera, F. J., Kaboski, J. P., & Zhao, M. Q. (2013). *The Rise of Services: the Role of Skills, Scale, and Female Labor Supply*. NBER Working Paper 19372.

Caron, J., & Markusen, J. R. (2014). International trade puzzles: A solution linking production and preferences. *The Quarterly Journal of Economics*, 1501–1552.

Comin, D. A., Lashkari, D., & Mestieri, M. (2015). *Structural change with long-run income and price effects*. National Bureau of Economic Research Working Paper Series No. 21595.

Cranfield, J. A. L. (1999). *Aggregating non-linear consumer demands: A maximum entropy approach*. Ph.D. thesis, Purdue University.

Cranfield, J. A. L., Eales, J. S., Hertel, T. W., & Preckel, P. V. (2003). Model selection when estimating and predicting consumer demands using international, cross section data. *Empirical Economics*, *28*, 353–364.

Fieler, A. C. (2011). Nonhomotheticity and Bilateral Trade: Evidence and a Quantitative Explanation. *Econometrica*, *79*(4), 1069–1101.

Gohin, A. (2005). The specification of price and income elasticities in computable general equilibrium models: An application of latent separability. *Economic Modelling*, *22*(5), 905–925.

Golub, A., & Hertel, T. W. (2008). Global Economic Integration and Land Use Change. *Journal of Economic Integration*, *23*(3), 463–488.

Harberger, A. C. (1998). A Vision of the Growth Process. *The American Economic Review*, *88*(1), 1–32.

ICP. (2015). *Purchasing Power Parities and Real Expenditures of World Economics: A Comprehensive Report of the 2011 International Comparison Program*. Washington, D.C.: World Bank.

Matsuyama, K. (2002). The Rise of Mass Consumption Societies. *Journal of Political Economy*, *110*(5), 1035–1070.

Matsuyama, K. (2016). *The Generalized Engel's Law*. In Search for A New Framework: Public lecture, Canon Institute for Global Studies.

Powell, A. A., Mclaren, K. R., Pearson, K. R., & Rimmer, M. T. (2002). *Cobb-Douglas Utility - Eventually*! CoPS Working Paper No. IP80.

Rimmer, M. T., & Powell, A. A. (1992). *Demand Patterns across the Development Spectrum: Estimates for the AIDADS System*. CoPS Working Paper No. OP75.

van der Mensbrugghe, D. (2017). *The environmental impact and sustainability applied general equilibrium (ENVISAGE) model: Version 10.01*. Technical report, The Center for Global Trade Analysis, Purdue University.

Yu, W., Hertel, T., Preckel, P., & Eales, J. (2004). Projecting world food demand using alternative demand systems. *Economic Modelling*, *18*(21), 205–236.

Yu, W., Hertel, T. W., Eales, J. S., Preckel, P. V. (2000). Integrating the AIDADS Demand System into the GTAP Model. In *3rd Annual Conference on Global Economic Analysis*.

Micro-Macro Simulation of Corporate Tax Reforms

Antonella Caiumi

Abstract Firm models are relatively rare in spite of the large number of models for households presented in the literature. The aim of this chapter is twofold. First, we illustrate the new microeconometric model on corporations currently used by Istat for revenue forecasting and policy analysis. Second, we discuss the advantages of combining microsimulation and computable general equilibrium models in simulating of corporate tax reforms.

Keywords Microsimulation · Taxation · General equilibrium · Ex-ante policy analysis

1 Introduction

The purpose of this chapter is to discuss how microsimulation and computable general equilibrium model (CGE) models can be effectively integrated in evaluating the impact of fundamental corporate tax reform proposals. While the use of microsimulation models is essential in modelling the distributive effects of corporate taxation and revenue forecast, it is limited, in the case of reforms involving changes in prices, wages and macro variables, by the inability of this kind of tools to model adjustments in several markets. Governments influence market outcomes by altering prices by means of taxes and subsidies and might exert significant impact on investment and the economic growth rate of various sectors of the economy. By contrast, CGE models—through their theoretical foundation in microeconomics—are powerful tools in the assessment of the impact of exogenous variables and policy measures (i.e., tax rates) on economic equilibria (i.e., prices and quantities) by the interaction of the demand and supply in goods and factor markets. However, since CGE models are based only on a few types of firms, they are unable to capture the full range of heterogeneity across firms. Henceforth, CGE models may fail to account for large part

A. Caiumi (✉)
ISTAT, Rome, Italy and CAPP, Università di Modena, Modena, Italy
e-mail: acaiumi@istat.it

© Springer International Publishing AG, part of Springer Nature 2018

F. Perali and P. L. Scandizzo (eds.), *The New Generation of Computable General Equilibrium Models*, https://doi.org/10.1007/978-3-319-58533-8_3

of both the distributional effects and the revenue impact associated to the economic adjustments.

In spite of the fact that firms are central units in economic decision-making, and information on the distributional and economic impact of business taxation is highly relevant for economic policy, to the best of my knowledge, no attempt has been made to combine microsimulation with CGE and macro modelling for the analysis of fundamental corporate tax reforms. For instance, the adoption of corporate tax reforms proposed in the literature to address the corporate debt bias was studied using several applied general equilibrium models (Keuschnigg and Diez 2007; Radulescu and Stimmerlmayr 2007; de Mooij and Devereux 2011) as well as in microsimulation analysis (see Finke et al. 2014; Caiumi and Di Biagio 2015), but the two methodologies were never combined.

Analogously, the proposal to adopt a single set of rules for companies operating within the EU was separately assessed through general equilibrium models (Betterndorf et al. 2009) and microsimulation studies (Fuest et al. 2007; Devereux and Loretz 2008). More recently, the re-launch of the Common Consolidated Corporate Tax Base (CCCTB) reform proposal was examined in the impact assessment using CORTAX, a CGE model describing 28 countries of the European Union, other relevant economies in the world, such as the US and Japan, and a tax haven. Macroeconomic results show that a fairer and more efficient corporate tax system will positively affect investment, employment, GDP and welfare in Europe. However, results vary across countries and the CORTAX model fail to provide forecast value for the tax revenue impact of the CCCTB for each Member State. For this purpose, the simulation exercise based on CORTAX should be complemented with a microsimulation approach. Alternative modelling approaches easier to implement, such as drawing a whole distribution of 'effective tax rates' based on the approach proposed by Eeger et al. (2009) and building models for 'average firms' in the subgroups of the whole population,[1] are not adequate to account for relevant impact of policy measures for heterogeneous populations.

More recently, Bhatterai et al. (2017) simulated the effects of corporate tax reform proposals in the United States, using a two-tier modelling design, with a large dynamic computable general equilibrium model to address the macroeconomic magnitudes. The dynamic CGE was also linked to a micro-simulation tax calculator model to measure the distributional effects on household income, while disregarding the impact on corporations.

Given this evolving background of converging macro and micro approaches, this chapter analyzes the advantages of combining microsimulation and computable general equilibrium models in simulating corporate tax reforms. To this aim, it also illustrates the analytical potential of the new micro-econometric model on corporations currently used by Istat for revenue forecasting and policy analysis. The chapter is organized as follows. Section 2 is devoted to the main tenets of microsimulation analysis of corporate tax reforms. Section 3 presents ISTAT-MATIS a new microsimulation model on corporations developed by the Italian National Institute of Statistics

[1] See for an example, Roggeman et al. (2014) based on the 'European Tax Analyzer' (Spengel 1995).

(ISTAT). Section 4 illustrates the ex-ante full distributional impact of the adoption of an allowance for corporate equity regime in Italy using MATIS. Section 5 concludes.

2 Microsimulation Modelling for Policy Analysis at the Firm Level

Corporate tax microsimulation models compute the net tax liabilities for individual firms and are used to forecast the revenue impact as well as the distributional consequences of tax reforms. These models are ultimately used to assess ex-ante whether policy initiatives had the intended or unintended effects on relevant targeted groups of the firm's population.

Compared to the expanding literature on households, microsimulation models for firms are relatively rare (for a survey see Ahmed 2006; Buslei et al. 2014). Firm models are more complex than household models both because firm behaviour involves inter-temporal aspects and tax rules are usually also more complex. In addition, access to firm data, especially tax, is more restricted compared to household data.

The starting point for tax microsimulation models is a (large) microdata set which provides comprehensive information on the determinants of individual tax liabilities. In principle, corporate tax models require the use of two complementary company level data sources—confidential corporate tax return data and accounting data —because usually corporate taxable income differs from economic income. Corporate tax returns allow researchers to precisely determine the tax position of corporations in each fiscal year as well as to recover information on the use of non-debt tax shields, like capital allowances, losses carry forwards and preferential tax treatments. Knowledge of loss offsetting and firms' ability to shift taxable profits over time are especially important for revenue forecasting. However, to completely identify heterogeneity in business activities other information are required. In particular, company accounts provide information of interest on the economic determinants of corporate profits.

Information from financial statements integrated with other sources of economic content at the corporate level is also valuable when the scope of the analysis requires to go beyond 'the dry run' (also called first-round effects) and estimates of empirical behavioural models may be welcome. Corporate tax reforms are sometimes targeted at affecting firm behaviour, such as investment, employment and financing decisions, as well as profit-shifting incentives. It follows that a key shortcoming of static modelling is neglect of behavioural responses to policy changes.

One viable solution to overcome such limitation is enriching a static microsimulation model with elements of behavioural responses as proposed by Chetty (2009), thus avoiding the need to develop a fully specified structural behavioural model. This has been done in a study of the impact of German 2008 corporate tax reform by Finke et al. (2013) by complementing non-behavioural computation with elasticities for several firm choice variables. Of course, there is a price to pay when the empirical measures of elasticities are taken from the related economic literature.

Another common shortcoming of firm's models is that they are usually static models—by definition—do not account for time, that is to say for the gradual entry into force of corporate tax reform. A notable exception is the ISTAT-MATIS microsimulation model which is based on a multi-period framework.

3 The ISTAT-MATIS Corporate Tax Model

ISTAT-MATIS is a corporate microsimulation model for Italy (for more details see Caiumi and Di Biagio 2015). MATIS simulates corporate tax liabilities according to fiscal rules and it is used on a regular basis by the Italian Central Institute of Statistics for revenue forecasting and policy analysis. It has two distinguishing features. First, it relies on the use of the largest complementary database in existence. To improve accuracy in revenue forecasting, the model relies on confidential corporate tax returns for all Italian corporations. Further, to assess ex-ante the full distributional impact of tax changes, the tax database is integrated with supplementary data.[2] The richness of the database allows to identify a broad range of category of firms in accordance with technological intensity, financing structure, profitability, size, age, location, export orientation, and ownership structure. Secondly, the model reproduces all the complexities of the corporate tax base through a multi-period framework.[3] This requires observations at the firm level for consecutive time periods (panel data). Currently, the integrated database covers the years 2005–2015.

Being based on the entire population of corporations, our results allows for conclusions on the distribution of the tax burden among taxpayers as well as on the revenue impact of tax changes. The model reproduces in detail the key features of the corporate tax in Italy, in particular the treatment of corporate losses, the consolidated taxation mechanism, the interest deductibility regulation, the local business tax and the allowance for corporate equity. At the current stage, the model does not account for behavioural responses by taxpayers to tax changes. Therefore, its analytical capacity is limited to first round effects.

The adopted model framework is particularly advantageous in the evaluation of tax reforms that are gradually introduced into force. Tax changes often provides advantages partially offset by restriction in other provisions and the sign of the net effect on tax liabilities may vary over time. More precisely, the model is aimed at monitoring the dynamic effects of an ACE-type regime that was integrated in the

[2]The sources involved in the integration process are the company accounts database, the ISTAT archive on national business groups, the statistical register of Italian active enterprises (acronym ASIA), information on spin-offs and mergers, and business structural surveys, in particular the survey on foreign trade (COE), the survey on Italian enterprises controlled by foreign firms (Fatsinward) and the survey on resident firms with foreign subsidiaries (Fats-outward).

[3]For example, interest deduction add-backs (carry forwards), losses carry forwards and tax allowances carry forwards.

Italian tax law ('Aiuto alla Crescita Economica') in 2011 with the aim to remove the favourable tax treatment of debt and stimulate companies' own capitalizations.[4] In the next section, we show the potential of MATIS in highlighting the effects of the Italian ACE regime on the corporate tax burden.[5]

4 The Distributional Effects of Introducing an ACE-Type Regime

Under the new ACE regime, a notional return on equity is deductible against corporate profits. The Italian ACE is applied on an incremental basis in order to minimize revenue losses. Starting from tax period 2011, taxable income is split into two components, ordinary and above-normal return. Ordinary income is exempt under ACE. The ordinary return is computed by applying a notional interest rate to new equity generated after 2010. Therefore, the increments of equity capital cut down the average tax rate of benefiting firms gradually over time. In practice, an incremental ACE-type reform induces a selective abatement of the average tax rate depending on the financial policy of firms with increasing effects in the long run.

In Caiumi and Di Biagio (2014) the model was used to analyze the revenue impact and the distributional effects of the newly introduced ACE regime both in the short and in the long run. One crucial aspects of an incremental ACE is that benefits are granted on the net increments of equity accumulated from a certain point in time. In the long run, however, new equity would have replaced old equity then the tax benefit will be granted to the entire capital stock.

Figure 1 shows how the implementation of the ACE affects the distribution of average effective tax rates (ETR) computed as the ratio of the company's tax-debt over before tax-profits for the whole population of Italian corporations. After only four years from its introduction, in 2014, the new ACE has provided significant advantages to beneficiary firms. After computing the ACE deduction, the ETRs for this type of firms drop below or at same level of the ETRs estimated for non-beneficiary firms (26.2% points at the median value) that likely adopt different tax shields, such as debt. The estimated cut in the average effective tax rates equals 2.3% points at the median value.

Focusing on companies benefiting from the reform, the full implementation of the ACE regime is simulated by considering companies' total equity as the ACE base.

---

[4]The ACE regime is a potential reform option that was originally proposed for the U.K by the Institute of Fiscal Studies (IFS 1991).

[5]A first scenario is obtained reproducing the legislation implemented in 2011 onwards over some consecutive periods (first year of simulation 2008). An alternative scenario ('Long-run ACE') is based on the assumption that ever since 2011 the ACE allowance were applied to the entire stock of equity. This simulation exercise (counterfactual scenario) allows investigating the impact of the *incremental* ACE in the long run, when companies would have accomplished a process of capitalization such that they will be granted a deduction against the taxable base for the entire stock of equity.

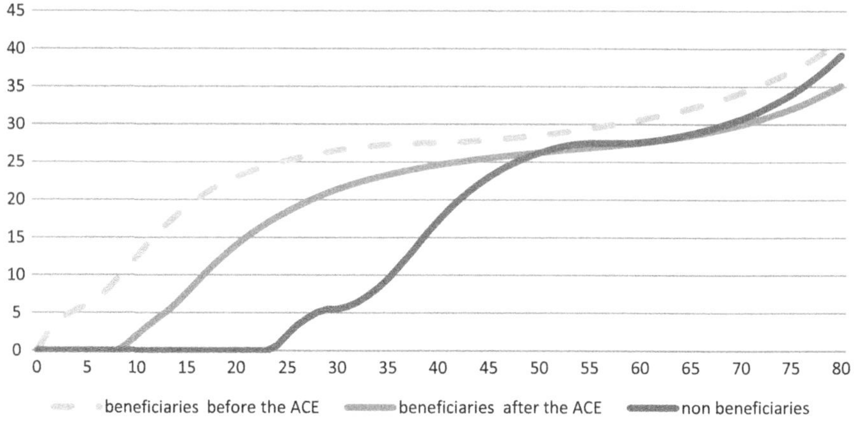

Fig. 1 The incremental ACE after 4 years from its implementation. Effective tax rates for ACE beneficiaries and non beneficiaries

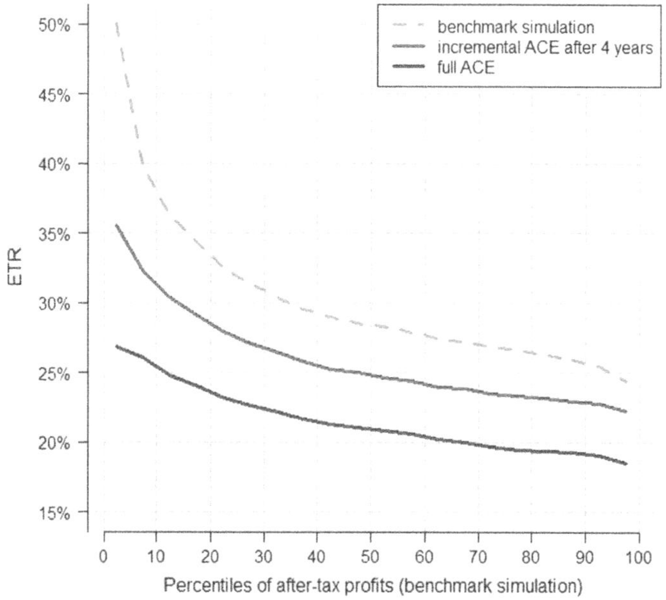

Fig. 2 The incremental ACE at work: short run and long run effects for beneficiary firms. Effective tax rates

Although ETRs are significantly further reduced for all companies, firms in the lower percentiles of the distribution of after-tax profits mostly benefit by the progressive convergence towards the full ACE regime (Fig. 2).

In Table 1 (ISTAT Annual Report (2014), Chap. 5) tax savings for beneficiary firms are measured in terms of the reduction of the statutory tax rate (27.5%), the average tax rate (ATR) being computed as the ratio of the tax debt over the taxable base before the ACE deduction. In 2012, around 20% of the industrial and commercial companies benefited of the allowance mechanism with an average tax saving lower than 1% point (0.8%). After five years (in 2015) the share of beneficiaries will raise to 31.4% and the tax cut will increase to 2.9% points. In the long-run the entire stock of equity will be qualifying for ACE and almost half of the taxpayers will be granted a tax cut equal to 8.5% points of the statutory tax rate.[6]

Looking inside the distribution of the ACE benefits, we see that the ACE mechanism is more frequently used by manufacturing firms, especially those characterized by high and medium-high technological intensity. The share of beneficiaries also increases with firm size. However, the intensity of the tax benefits decreases with firm size, as the abatement of the statutory tax rate granted by the incremental ACE is higher for small firms in comparison to larger ones. This holds true both in the second year of implementation (real data) and after five years (simulation results). Indeed, the distributional effects increases over time, likely because the cumulative framework of the allowance mechanism. In 2012 the tax bonus ranges from 1.7% points for smaller firms (turnover less than 500,000 euro) to 0.6% points for larger ones (with turnover higher than 50 million of euro). After five years, in 2015, the tax discount ranges from almost 6% points for firms in the first turnover class to less than 2% points for larger firms. In the long run the latter effect vanishes, all firms are granted a *full* ACE deduction and the tax cut turns out to be less affected by firm size.

Figure 3 shows the differentials in the speed of convergence towards the *full* ACE. After five years the base qualifying for the ACE is approximately equal to 50% of the entire equity stock (*long-run* ACE base) for beneficiary firms with turnover lower than 500.000 euro, whereas it remains below 20% for corporations with turnover above 50 million euro. Therefore, smaller firms reach the tax exemption of the ordinary return of equity faster than larger companies. In contrast to the allegation that the ACE is mainly a tax relief for profitable and large firms, our microsimulation analysis shows that an ACE-type regime can be very beneficial for smaller companies and innovative firms that usually suffer from restrictions of their outside financing capacity. This is also confirmed by a recent econometric analysis on the effect to the Italian ACE on debt choices of companies (Branzoli and Caiumi 2017), showing that the reaction of SMEs to the changes in the tax incentives to equity financing has been even stronger than large companies.

Since firm size is usually identified at the core of different aspects explaining the poor performance of Italian firms, the ACE-type reform can proved to be a valuable policy option. To our knowledge this important result has not been highlighted in previous studies. In particular, this analysis can be compared in the literature with the study by Finke et al. (2014), which focus on the consequences of introducing an ACE regime in Germany using the behavioural microsimulation model ZEW

[6]In this exercise, the ATR for year 2012 is computed directly from the tax returns data filed by corporations and fiscal groups ("UnicoSC" form and "CNM" form). The MATIS model was used to estimate the two alternative scenarios as described in footnote 4.

Table 1 The ACE at work: share of beneficiaries and average tax rate (2012 and 2015)

	Standalone companies and fiscal group	2012		2015		*'Long-run ACE'*	
		Beneficiaries (%)	Average tax rate (%)	Beneficiaries (%)	Average tax rate (%)	Beneficiaries (%)	Average tax rate (%)
Total	842,184	20.1	26.7	31.4	24.6	49.0	19.0
Economic sector							
Mining and manufacturing	128,411	26.0	26.7	36.4	24.8	54.6	19.6
Water, electricity, gas supply and sewerage	14,130	26.8	26.8	35.9	25.7	49.2	19.3
Construction	151,465	17.4	26.3	28.4	24.0	44.2	19.5
Trade	184,334	20.9	26.7	31.5	24.8	48.8	20.4
Other services	363,844	18.5	26.6	30.7	24.4	49.1	17.9
Technology and knowledge							
Manufacturing—technological intensity:							
High	3,881	30.0	26.9	39.2	25.7	57.7	20.7
Medium-high	23,338	32.0	26.8	42.6	25.3	61.5	19.8
Medium-low	50,304	26.3	26.6	37.2	23.9	55.6	18.2
Low	48,914	22.7	26.7	32.8	24.5	50.2	19.0
Services—knowledge intensity							
High	97,440	18.8	26.8	32.6	25.9	51.8	18.9
Low	450,738	19.4	26.5	30.6	23.4	48.4	18.4
Other	167,569	18.2	26.6	29.1	25.0	44.7	20.2

(continued)

Table 1 (continued)

	Standalone companies and fiscal group	2012		2015		*'Long-run ACE'*	
		Beneficiaries (%)	Average tax rate (%)	Beneficiaries (%)	Average tax rate (%)	Beneficiaries (%)	Average tax rate (%)
Turnover							
<1	37,504	1.9	25.3	3.4	21.2	6.7	18.5
1–500.000	519,269	13.7	25.8	25.6	21.6	43.2	17.6
500.000–2 mln	178,061	28.2	26.4	41.4	23.3	61.6	19.5
2–10 mln	83,931	42.2	26.5	51.9	24.1	69.9	19.9
10–50 mln	19,204	51.1	26.7	56.7	24.5	73.4	19.4
>50 mln	4,215	54.2	26.9	56.4	25.6	73.7	18.7
Employees							
0	213,310	12.1	25.8	21.5	21.8	35.5	17.7
1–9	484,733	19.4	26.4	32.3	23.5	51.4	19.6
10–19	84,244	31.9	26.6	42.4	24.5	60.8	20.6
20–49	41,201	36.2	26.6	44.5	24.3	61.0	20.3
50–249	16,214	40.0	26.8	45.5	24.7	60.9	19.3
250–499	1,414	38.2	26.6	42.1	24.6	56.5	18.1
>=500	1,068	44.6	26.9	47.4	26.0	62.8	18.0
Location							
North West	243,762	25.1	26.8	38.0	24.8	56.0	19.2
North East	173,100	24.5	26.6	36.8	24.3	54.7	18.0
Centre	211,983	17.8	26.6	29.0	24.9	46.9	19.6
South	213,339	13.2	26.5	22.0	23.9	38.5	19.3

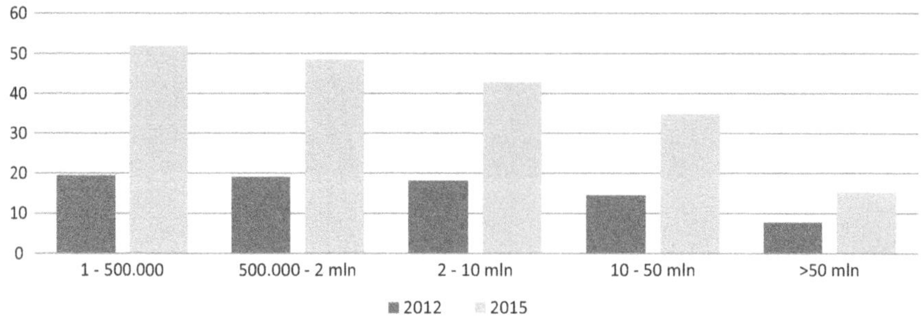

Fig. 3 Speed of convergence towards the *full* ACE by turnover class: ratio between net increments of equity over company's net worth (*percentage value*) *Source* ISTAT Annual Report, Chap. 5 (2014)

TaxCoMM. Their analysis relies however on clustering the sample of corporations in Germany based on structural parameters from the financial statements, instead of more direct firm characteristics, therefore restricting the possibility to precisely identify the reform "winners".

5 Concluding Remarks

The ISTAT-MATIS model provides many interesting insights on the dynamic role of an ACE-type regime on the tax burden distribution across the population of firms and over time. By relating the reform effects with the firm characteristics of policy interest, our analysis shows that the ACE relief is particularly favorable for smaller and innovative firms that are the backbone of our economy.

Nevertheless important developments of the toolbox remains to be accomplished in two key areas. The first relates to incorporating behavioural responses to tax reforms in our simulation framework. As the ACE, currently integrated into the Italian tax code in constancy of the statutory tax rate, entails a reduction of the incentive for indebtedness, not accounting for the debt/equity substitution effects implies that the revenue impact of the incentive mechanism is somewhat overestimated. Indeed, recent results (Branzoli and Caiumi 2017) suggests that the Italian ACE, although limited on capital increases, works effectively as a substitute for interest deductions in lowering the effective marginal tax rate for corporations. Of course, there are also other decision margins that may be affected by the reform. In principle, the ACE is designed not only to address the debt bias but also to promote investments. By decreasing the cost of capital, the allowance is expected to boost investment, leading to increased employment and growth all other things being equal. Location decisions and profit shifting, although relevant *per se*, are not of major concern in the current policy context in Italy, characterized by decreasing statutory tax rates. The second area for improvement and for development is therefore to consider feedback effects between the micro and macro level.

A fully integrated micro-macro model is potentially a powerful tool to go beyond the partial equilibrium framework in which microsimulation models operate, if the objective is to disentangle the reform effects at the macro level and to explore the implications on welfare. The investment function estimated at the micro level can be aggregated and incorporated in the macro model which can be used to assess the overall impact of the reform—i.e. the introduction of the ACE (or its repeal if it is already implemented) for the context of interest here, on the economy as a whole and on tax collection, considering, in addition to corporate tax revenues, variation in personal income tax and consumption tax revenues triggered by the simulated policy changes.

It should be stressed that it would be relevant both for policy decision makers and the corporate community to have the opportunity to use the information stemming from the micro-macro simulation of tax reforms, incorporating institutional and economic changes in real time, through web accessible extension services targeted to all potential beneficiaries. The information is available, but its economic and social value has not been fully exploited yet.

This study may be interpreted as an initial endeavour towards a greater effort to routinely incorporate firms behaviour and their response to tax reforms in microsimulation models—not only at the national level but also at the EU level—and to consistently link them to macro analysis.

References

Ahmed, S. (2006). Corporate Tax Models: A Review. *SBP Working Paper Series*, (13).

Bhattarai, K., Haughton, J., Head, M., & Tuerck, D. G. (2017). Simulating corporate income tax reform proposals with a dynamic CGE model. *International Journal of Economics and Finance, 9*(5), 20–35.

Branzoli, N., & Caiumi, A. (2017). *Tax incentives and firm capital structure: evidence from the Italian ACE regime*. Taxation Papers, European Commission, forthcoming.

Buslei, H., Bach, S., & Simmler, M. (2014). Firm level models. *Handbook of microsimulation modelling,* Contributions to Economic Analysis (Vol. 293, pp. 479–503). Emerald Group Publishing Limited.

Caiumi, A., & Di Biagio, L. (2014). The redistributive effects of an ACE Tax: The case of Italy, *Paper presented at the 70th Annual Congress of the International Institute of Public Finance*, August 20–23, 2014. Lugano, Switzerland. www.iipf.org.

Caiumi, A., & Di Biagio, L. (2015). Corporate effective taxation in Italy using a new microsimulation model (Istat–MATIS). Istat Working Papers, 13.

Chetty, R. (2009). Sufficient statistics for welfare analysis: a bridge between structural and reduced-form methods. *Annual Review of Economics*, *1*, 451–488.

De Mooij, R., & Devereux, M. P. (2011). An applied analysis of ACE and CBIY reforms in the EU. *International Tax and Public Finance, 18,* 93–120.

Devereux, M. P., & Loretz, S. (2008). The Effects of EU Formula Apportionment on Corporate Tax Revenues. *Fiscal Studies*, *29*(1), 1–33.

Eeger, P., Loretz, S., Pfaffermayr, M., & Winner, H. (2009). Bilateral effective tax rates and foreign direct investment. *International Tax and Public Finance*, *16*, 822–849.

Finke, K., Heckemeyer, J., Reister, T., & Spengel, C. (2013). Impact of tax-rate cut cum base-broadening reforms on heterogeneous firms: Learning from the German tax reform of 2008. *Finanzarchiv: Public Finance Analysis*, *69*(1), 72–114.

Finke, K., Heckemeyer, J., & Spengel, C. (2014). *Assessing the impact of introducing an ACE regime* (pp. 14–033). ZEW Discussion Paper: A Behavioural Corporate Microsimulation Analysis for Germany.

Fuest, C., Hemmelgarn, T., & Ramb, F. (2007). How would the introduction of an eu-wide formula apportionment affect the distribution and size of the corporate tax base? an analysis based on german multinationals. *International Tax and Public Finance*, *14*(5), 605–626.

Institute for Fiscal Studies (1991). Equity for Companies: A corporation tax for the 1990's. A report of the IFS capital taxes group. London: The Institute for Fiscal Studies. Commentary 26.

ISTAT (2014), "The state of the nation, Annual report, Chapter 5, Rome.

Keuschnigg, C., & Diez, M. D. (2007). A growth oriented dual income tax. *International Tax and Public Finance*, *14*(2), 191–221.

Radulescu, D. M., & Stimmerlmayr, M. (2007). ACE versus CBIT: Which is better for investment and welfare? *CESifo Economic Studies*, *53*, 294–328.

Roggeman, A., Verleyen, I., Van Cauwenberge, P., & Coppens, C. (2014). Impact of a common corporate tax base on the effective tax burden in Belgium. *Journal of Business Economics and Management*, *15*(3), 530–543.

Spengel, C. (1995). *Europäische Steuerbelastungsvergleiche*, Düsseldorf.

Estimating an Energy-Social Accounting Matrix for Italy

Marco Rao, Umberto Ciorba, Giovanni Trovato, Carmela Notaro and Cataldo Ferrarese

Abstract This chapter describes an application of the Social Accounting Matrix (SAM) to the analysis of the economic impact of energy policies in Italy. An Energy Social Accounting Matrix (ESAM) is estimated for the year 2010 using as an input a general purpose technology-oriented model (of the MARKAL—TIMES family) representing the evolution of the Italian energy environment and a variant of the Wolsky procedure.

Keywords Social accounting matrix · Wolsky procedure · Economic impact of energy policies · Input-output analysis

1 Introduction

This study proposes a methodology for the estimation of the economic flows related to the energy sector and their inclusion in the scheme of the Social Accounting Matrix (SAM). Since the SAM as well as representing an accounting scheme is also a tool of economic analysis, the expansion of the matrix has as main objective the evaluation of the economic impact on the energy sector of energy and macroeconomic policies.

M. Rao
ENEA FSN-TECFIS-APAM, Rome, Italy

U. Ciorba
ENEA UVER SECAM, Rome, Italy

G. Trovato
CEIS, University of Rome "Tor Vergata", Rome, Italy

C. Notaro
Invitalia, Rome, Italy

C. Ferrarese (✉)
OpenEconomics srl, Rome, Italy
e-mail: dino.ferrarese@openeconomics.eu

© Springer International Publishing AG, part of Springer Nature 2018

F. Perali and P. L. Scandizzo (eds.), *The New Generation of Computable General Equilibrium Models*, https://doi.org/10.1007/978-3-319-58533-8_4

In the absence of systematic and detailed surveys, the enlargement of the SAM to the energy sector is carried out through estimation procedures based on the use of partial information. First, it was necessary to construct a scheme of reconciliation between the national energy balance (considered as an organic framework of the existing energy statistics) and the SAM. Secondly, we estimated the missing coefficients of the extended SAM through a method of distribution of economic flows based largely, but not entirely, on a procedure identified by Wolsky (1984).

This chapter is divided into four sections: the first illustrates the theme of the economic classification of the Italian energy system; the second explains the main features of a social accounting matrix (SAM); the third describes the national energy balance, and elaborates patterns of connection with the SAM together with procedures of disaggregation for the energy sector of the SAM. Some applications of the disaggregated SAM are described in Sect. 4, where we highlight the effects on employment and gross domestic product from strengthening the upstream energy sector in Italy.

This work is the end product of a collaboration agreement between the National Agency for the new Technologies, Energy and Sustainable Economic Development (ENEA) and the Center for Economic, International Studies (CEIS) of the University of Rome "Tor Vergata" and OpenEconomics. The authors would like to thank Maria Gaeta and Bruno Baldissara (both at ENEA) for their support in the analysis of the national energy system. Heartfelt thanks go also to Professor Pasquale Lucio Scandizzo, Gian Carlo Tosato, Maria Rosa Virdis and Daniela Palma for their scientific supervision.

2 The Economic Classification of the Italian Energy System

The need to link the statistical information related to the energy sector with the national accounts scheme arose during the seventies both nationally and internationally. In 1975, the statistical office of the European Community, the precursor of Eurostat, entrusted the Institute of Studies for Informatics and Systems (ISIS), under the direction of Prof. Vera Cao Pinna, with a research project aimed to estimate a system of relations between the various sources of data related to the energy sector and the Italian national accounts. The purpose of the project was to include the energy system into the input-output table of the Italian economy, in order to improve forecasting of energy requirements. The object of the research was to quantify both in physical units and monetary value the individual energy sources supplied by the energy system and used as intermediate or final services in the rest of the economy. At that time the data for the energy sector were scattered and poorly organized, as they were based on various surveys carried out by different organizations. It was hoped that the same statistical data, properly organized, would help to illustrate the functioning of the energy system in a detailed and reliable way.

ISIS subsequent research work helped to improve both the national accounts and the energy surveys by establishing meaningful connections and consistency requirements. Today the whole system of surveys is coordinated nationally through SISTAN and statistics on transactions of energy commodities, in physical and monetary units, appear to be more solid than in the past.

Despite this, the energy sector is still represented in a rather aggregate way, while some analyses, such as those focused primarily on the *green recovery* (Pollin et al. 2008), would require a more detailed representation of the energy sector. With a less ambitious level of detail, the purpose and objectives of this research are the same of the work carried out by ISIS: facilitate the comparison between data on supply and use of energy sources as they are represented in the National Energy Balance (NEB) and data on energy commodities contained in the Social Accounting Matrix (SAM), which is derived from the input-output tables and has the same classification of the ISTAT tables for resources and uses (NACE Rev. 2).[1]

The ISTAT classification presents a fairly detailed breakdown of the productive sectors (58 economic branches) but its representation of flows within the energy system is limited to four main branches: *Hard coal, Petroleum and natural gas, ancillary services to the extraction of oil and gas*, for the extraction of energy minerals, *Coking and refineries* for the transformation of these minerals and a sector called *electricity, gas, steam,* which includes the different ways of producing electricity together with the distribution of gas and energy. In order to improve the flow representation and provide at the same time a basis for a SAM including energy at some interesting level of disaggregation, our study has proceeded through four consecutive stages:

- reclassification of NEB items in the SAM scheme;
- collection of data from different sources;
- disaggregation of the energy sectors represented in the SAM;
- estimation of the missing coefficients (the non-energy input needed by the new energy sectors arising from the disaggregation).

The first two stages are the same of the ISIS work, while the third and fourth constitute its natural complement: rebuild in an input-output framework the structure of the most relevant energy sub-sectors.

The national energy balance is considered, for this work, as the statistical scheme that organically describes the functioning of the national energy system. It does not represent the maximum disaggregation available with respect to information on energy sources (which are presented with a degree of aggregation in the NEB scheme) but, with minor modifications, can provide a comprehensive framework of use and supply of energy goods as well as of technologies and methods of transformation. The national energy balance can in fact be easily connected to a purely economic accounting scheme represented by the SAM. This statistical framework presents a complete overview of the national economy and is often used for the calibration of general equilibrium models. Connecting the two accounting schemes facilitates

[1]ISTAT uses the Ateco classification, i.e. the national classification derived from NACE (European nomenclature of economic activities). The Nace is the European reference for the production and dissemination of statistical data on economic activity. Since it was drafted in 1970, the EU Member States have relied on it (or on national classifications derivatives), by using it in the European statistical system for the production of internationally comparable data.

detailed analysis of the impact of the development of the energy sector on the whole economy and the impact on the structure of the energy sector generated by the evolution of energy consumption.

3 The Social Accounting Matrix

The Social Accounting Matrix (SAM) is an accounting scheme that represents the structure of the economy and highlights the circularity of relations among economic agents and activities of the market system. The matrix can also be used as a tool to address and support public investment policies by analyzing the relations of interdependence between the agents of an economic system. The SAM has been widely utilized in the macroeconomic and sector analysis of both developing and advanced countries. Starting from the SAM, in fact, it is possible to integrate the distribution of income within the input-output representation of the economic process, and analyze how the redistribution process interacts with other phenomena.

While SAM origins can be traced back (Pyatt and Round 1979) to the experiences of Norway and the Netherlands, respectively, in the years 1930 and 1940, the first official SAM was built for the British economy in 1960 by Richard Stone, as part of the Cambridge Growth Project, in response to the need to develop a national accounting system taking into account the process of income formation both from the production and the consumption side, as well as its distribution across institutions. Previously, the detailed information about the economic system had taken the form of Input-Output tables for the part regarding the productive sectors, while the accounts related to the distribution and redistribution of income and their connection with households, government, savings and trade accounts had been considered in summary and incomplete ways.

In the early 50' the need arose to create a common information system, characterized by standardized rules for collecting, coding and presenting data. In this context, in 1953, the UN published the System of National Accounts and Supporting Tables (SNA53) adopted by countries with a market economy, and the Material Product System (MPS) adopted by countries with collectivist economies. In 1968, thanks to the idea of Stone to present accounts in matrix form, the UN revised the previous SNA53 replacing it with the SNA68 and highlighting the accounting aspect of the matrix. In 1968, the Statistical Office of the European Communities (SOEC), elaborated the European System of Integrated Economic Accounts (ESA), subsequently adopted by the member countries in 1970. The SNA68 underwent further changes, becoming SNA93, laying the foundations for additional revisions to the European System of Integrated Economic Accounts and finally resulting into the current accounting system, ESA95, which came into force in 1999.

The SAM accounts report the existing transactions between the key aggregates that constitute an economic system, including production, income formation and income redistribution. To integrate the last two accounts with the input-output matrix, two additional elements and their connections are included: (i) the factors of production (labor and capital) and, (ii) the so-called institutional sectors (households, businesses, government, capital formation and rest of the world) to which the factors (land, capital and labor) distribute the income received from production activities.

The SAM is typically a square matrix (it has the same headers for rows and columns) and is formed by a series of sub-matrices (the block in the upper left in the table is Leontief input-output matrix), where for each aggregate the sum of all transactions of a sector by column equates the sum for the same sector by row. Each row of the matrix reports the sources of revenue of each institution or productive sector, while each column shows how revenue is used to purchase goods and services or to make money transfers to the other sectors and institutions.

The SAM is not only an accounting scheme but also a tool for economic analysis. While the monetary flows represented in the matrix typically reproduce the equilibrium between demand and supply of an economic system in a given year, it is possible to change the levels of demand of one or more accounts (represented by the transactions reported in their corresponding columns) and evaluate the changes needed in the other sectors to meet these changes and reconstitute equilibrium. This means that the accounts of the matrix can be considered either endogenous or exogenous, and the variation of expenditure attributed to the exogenous accounts constitutes the shock vector that impacts the economic system.

The simplest way to evaluate the effect of exogenous shocks on the endogenous accounts consists in using the system of multipliers derived from the so called Leontief input-output open equilibrium model.

$$X = (I - A)^{-1} * f$$

A	Matrix of the expenditure coefficients ($a_{ik} = X_{ik}/X_k$)
$(I - A)^{-1}$	Matrix of global multipliers for endogenous accounts
f	Shock vector (exogenous)

In this model, the changes in production (and therefore in employment and income) are estimated by multiplying the multiplier matrix for the vector of exogenous accounts, thereby incorporating two distinct multiplicative circuits. The first circuit connects the increase in final demand of an exogenous sector, the increase in production in the sectors directly connected as providers of goods and services, and, through the supply chain of intermediate goods, the sectors indirectly activated. The second circuit connects the increase of production to an increase in factor incomes, their distribution to the institutions, and the ensuing increase in the final demand for goods and services.

3.1 *The SAM Estimation Procedure*

The SAM used in this study was estimated by using official ISTAT (the National Institute of Statistics) statistical sources (national accounts data, "supply and use" matrices, survey on household consumption), suitably re-aggregated and organized in order to create a detailed picture of the national economy for the base year 2010. The estimated SAM has the following structure:

- 58 productive sectors (25 services, 29 industries, 1 building construction, 3 agriculture),
- 2 factors of production (labor and capital),
- 4 institutions (Families, Companies, Government, Investments)
- Rest of the world.

The estimation procedure begins with the collection and organization of data available from institutional sources. The first part of the SAM that is estimated is the input-output matrix. The matrix of intermediate flows using the supply (resources) and use (uses) tables published by ISTAT. Starting from these tables and following the procedures outlined by ISTAT a symmetric input-output is finally obtained. In the estimation process of the symmetric table, estimates of the input-output relationships can rely on two alternative assumptions, product technology or branch technology. Depending on the assumption adopted, both the outputs and the inputs of the secondary productions are reallocated using different mathematical methods. With the first assumption, it is assumed that each product is produced with the same technology, no matter where it is produced, hence the input structure of the technology that produces a given good is the same everywhere. Under the technology branch hypothesis, it is assumed that the inputs are consumed in the same proportions in each production activity carried out by a branch. This means that main products and by-products are all manufactured using the same technology, hence having the same structure of inputs. In the estimation procedure adopted for this study, product technology was assumed, thus implying that there is only one technique for producing each product and each product has, therefore, its typical input-output structure.

The model also required the estimation of the revenues and expenditures of the institutional sectors: households, companies, government, capital formation, the rest of the world. For these national account figures from ISTAT were used, except for the households' accounts, for which we used the time series analysis of the surveys on household spending conducted by ISTAT and the surveys on household budgets conducted by the Bank of Italy.

3.2 The National Energy Balance

The National Energy Balance (NEB), aggregates the energy statistics of a country in a matrix scheme that represents the functioning of the energy system, including supply, transformation and energy demand. The data contained in the NEB, processed according to standards encoded by international statistics authorities such as the International Energy Agency (IEA) and Eurostat, are the most suitable statistics for international comparisons (data from energy questionnaires, which comply with the same standards, are not published, except through the NEB). The column headings of the NEB matrix report the energy sources while the row headings describe the stages of production, transformation and consumption of commodities used in the system itself. The outline of the NEB is a rectangular array organized into three sections.[2] (supply, transformation and end-use), whose order and whose disaggregation depend on conventions adopted by the organization that publishes the balance. For example, the NEB published by the Ministry of Economic Development aggregates the data of the final energy consumption, the residential sector and services into a single "Civil Sector", while EUROSTAT and IEA present the two sector data separately.

The transformation section shows the flow of commodities as inputs to transformation plants (for example cooking plants, refineries, power stations) and the outputs of these plants. This section records the production of secondary energy and shows the different flows of commodities (primary and secondary) that feed the final demand, also reporting the energy consumption of the same generation plants and losses along the transmission and distribution network.

The end-use section shows the energy consumption of the productive sectors (in economic terms these constitute intermediate consumption), energy consumption of the residential sector (imputable to households' final consumption) and consumption of the transport sector (part of consumption for road transport is households' final consumption, the rest is classified as intermediate consumption of the sector that produces transport services).

The Energy Balance gives details of final consumption of non-energy sectors (3 main groups of productive sectors grouped according to the classification ISIC Rev4, the residential sector and non-energy use). The productive sectors include 13 industrial sectors of which five are Energy intensive (Iron and steel, non-ferrous metals, non-metallic Minerals, Chemical, Petrochemical, Pulp and paper), one is a sector for trade and services and six are modes of transport. The classification is consistent with the OECD guidelines for the homogenization of the input-output matrices.

3.3 The Economic Classification of the National Energy Balance

The SAM (base-year 2010), estimated according with the procedure detailed in Sect. 3.1, was disaggregated in the energy sectors included in the scheme of energy balance adopted by the IEA.[3] This disaggregation was designed through the construction of

[2] See, for example the structure of Eurostat Energy Balances: http://epp.eurostat.ec.europa.eu/portal/page/portal/product_details/publication?p_product_code=KS-EN-13-001.

[3] The TIMES-Italy model (http://www.enea.it/it/seguici/le-parole-dellenergia/glossario/parole/times) could provide even more disaggregated data (the level of activity and cost of each single technology). It has been chosen to adopt the National Balance (NEB) aggregation scheme for two main reasons: because its a reference framework universally accepted by the experts of the energy sector and cause it minimizes the problems in estimating missing data through a complete explication of the TIMES-Italy technologies.

a so-called *bridge* matrix, estimated to allow greater accuracy in the evaluation of the economic impact of energy scenarios and, ultimately, to use energy balances arising from technological scenarios developed by the TIMES-Italy model.

The integration of the SAM with technological bottom-up models of the MARKAL-TIMES[4] family allows to evaluate the time evolution of the structure of the economy (the coefficients of the SAM) as a consequence of national energy policy and to gain analytical insights to support energy policy.

The basic SAM estimated disaggregates the Energy sector into 4 sub-sectors:

1. Coal;
2. Oil and natural gas, ancillary services to the extraction of oil and gas;
3. Coke and refined petroleum products;
4. Electricity, gas and steam.

The *bridge* matrix, with a greater disaggregation of the energy sector, allows the connection between the SAM and the NEB (that can be the statistical framework of a past year as well as an output of the TIMES model regarding future years).

After the division of the sector Electricity, gas and steam, the resulting breakdown is:

- Coal
- Crude Oil and Natural Gas and auxiliary activities
- Coke and refineries
- Natural Gas Distribution
- Thermal Power Generation
- Thermal Power Generation (Biomass only)
- CHP
- Hydro
- RES for electricity generation (PV CSP WIND)
- RES for electricity generation (Geothermal)
- RES (heat generation)
- Biomass
- Electricity transmission and distribution

The estimated flows of the *bridge* matrix include the quantitative data of the national energy balance, while estimation of energy prices was performed using different statistical sources (mainly ISTAT data and IEA).

The *bridge* matrix reports value transactions within the disaggregated energy sectors, and between each energy sector and the rest of the economy, for a level of disaggregation higher than that required by the SAM originally estimated. In practice, through the bridge matrix several energy sectors can be added to the original SAM and for the sectors added it is possible to estimate the transaction flows related to

---

[4]The MARKAL/TIMES family are energy/economic/environmental models supporting a rich technology detail. MARKAL/TIMES were developed in a collaborative effort under the auspices of the International Energy Agency's Energy Technology Systems Analysis Program, which started in 1978. See: http://www.energyplan.eu/othertools/national/markaltimes/.

energy sales through the NEB data. The information for the purchase of energy sectors from the other non-energy production sectors, however, is not available and to estimate them it is necessary to use other statistics, and where necessary, numerical simulation methods.

The general criteria used to project the NEB flows estimates into a SAM are as follows:

- two sectors were added for distribution to users of gas and electricity;
- fossil fuels and renewables (appropriately aggregated) indicated in the column headings of the NEB, became areas of mining or energy sectors of primary energy production in the *bridge* matrix;
- transformation processes in the transformation section of the NEB became new production sectors in the *bridge* matrix;
- the energy flows in the transformation section of the NEB, were considered intermediate purchases of energy sub-sectors;
- consumption of final energy shown in the lower part of NEB was considered as intermediate purchase from productive sectors by energy transformation industries or by the distribution networks;
- in the *bridge* matrix, the gas extraction industry was assumed to sell only to the gas distribution sector and the latter to the intermediate and final users;
- in the *bridge* matrix, sectors involved in electricity generation buy inputs from extractive sectors, from sectors of production of primary energy, from the foreign sector, and from other sectors of transformation (refineries);
- the sector "transmission and distribution of electricity" purchases electricity from electricity generation sectors and sells to the intermediate and final users;
- imports and exports in the NEB scheme are treated in a similar manner in the bridge matrix;
- energy consumption of households in the bridge matrix derive from the NEB final consumption in the residential sector and consumption in road transport.

3.4 *Disaggregation of the SAM*

A computer program has been designed to perform what has been described at the theoretical level. The program takes as inputs the energy data from a NEB, the social accounting matrix and a set of prices for Energy commodities and uses them according to the algorithm detailed in Fig. 1.

While the process of estimation of the bridge matrix is described in detail in Rao and Ciorba (2010), Figs. 2 and 3 provide a synthetic description of the procedures implemented.

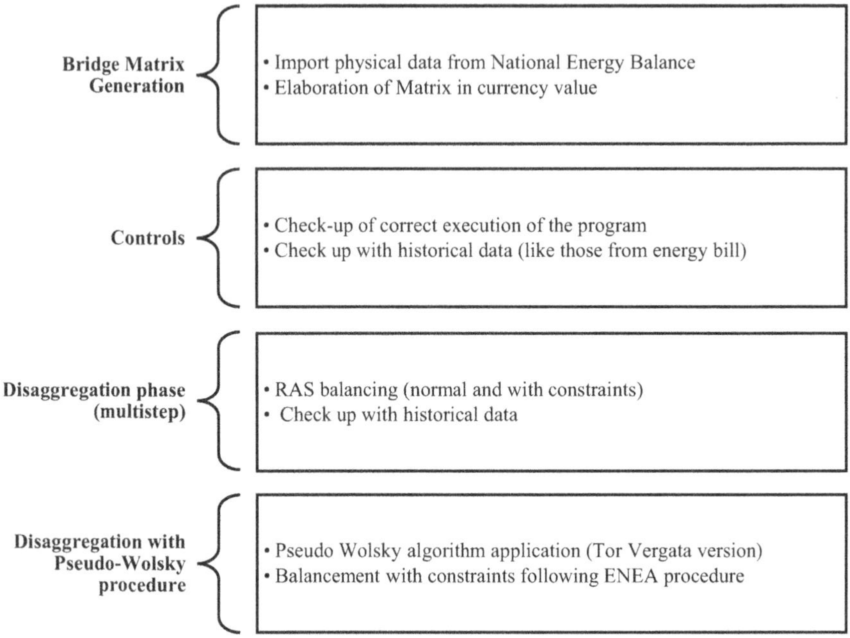

Fig. 1 E-SAM construction algorithm. *Source* Authors' elaboration

3.5 SAM Disaggregation Based on the Pseudo-Wolsky Algorithm

3.5.1 The Wolsky Method

A standard methodology can be used to build a disaggregated SAM matrix (Table 1) using the information known about some of the subsectors of disaggregation. The solution is exact and explicit, see Wolsky (1984) for detail. In our study, the Wolsky (1984) method for matrix disaggregation was implemented using a dedicated routine developed in MATLAB environment. In particular, the aggregate "Energy" column of the original SAM is disaggregated into new component sectors using the weights estimated for of the energy subsectors, using additional information when available (e.g. the ISTAT data for the rest of the world accounts on the national energy bill).

Table 1 E-SAM—Italy 2010

	LAV	CAP	FAM	IMPR	Agr	Fish	Mining	Food &	Text	Wood	Paper	Chem	Non m m
LAV	0	0	0	0	10029	703	941	13150	17309	9725	8069	17679	7777
CAP	0	0	0	0	20876	1146	1294	13453	11524	8894	5982	10027	6338
FAM	617411	399634	273816	331281	0	0	0	0	0	0	0	0	0
IMPR	0	326167	0	0	0	0	0	0	0	0	0	0	0
Agr	0	0	20895	55	4271	0	0	18406	874	545	245	366	16
Fish	0	0	1675	7	2	8	0	169	3	22	1	16	1
Mining &	0	0	0	0	3	0	17	15	3	40	86	402	1716
Food &	0	0	66468	163	3048	38	0	13793	1588	17	64	1174	13
Text	0	0	44639	123	94	28	1	37	26340	947	233	620	70
Wood	0	0	15892	55	18	10	6	347	126	13784	123	249	245
Paper	0	0	11451	50	8	4	6	1380	647	727	9486	1619	428
Chem	0	0	8741	74	1197	14	20	2341	4616	1808	2156	41102	1239
Non m min	0	0	1238	7	124	3	65	1126	51	323	3	1404	5263
Iron &	0	0	1173	8	169	25	54	632	610	4278	547	1702	918
Mach	0	0	12347	33	23	12	31	681	753	773	451	2245	470
Tr equip	0	0	0	34	156	28	3	23	79	136	61	627	38
Constr	0	0	5565	27	146	1	15	284	515	152	371	323	213
Comm	0	0	738528	19453	2749	84	151	21744	22710	12020	13191	22818	9838
Roa & R	0	0	33343	302	614	23	23	3001	1645	1030	1061	2419	1772
Dom nav	0	0	9628	278	0	0	0	0	15	0	0	76	14
N&i Aviat	0	0	3760	41	3	5	2	124	529	64	63	431	69
GOV	0	0	159336	57445	937	26	116	13478	1513	978	714	1950	733
FCAP	0	0	415897	200186	0	0	0	0	0	0	0	0	0
RDM	0	0	93375	0	6601	589	1883	16318	18203	7232	4717	37018	2256
Carbon	0	0	1	0	0	0	0	1	5	2	0	58	37
Petrolio	0	0	0	0	0	0	0	0	0	3	0	2	0
Coke	0	0	14173	82	1251	72	532	295	239	165	96	1391	831
Dgas	0	0	8157	58	57	0	735	618	505	0	341	822	332
Termoel	0	0	0	0	0	0	0	0	0	0	0	0	0
Termo bio	0	0	0	0	0	0	0	0	0	0	0	0	0
CHP	0	0	87	0	0	0	27	83	97	18	194	595	0
Idro	0	0	0	0	0	0	0	0	0	0	0	0	0
FER PV CSP	0	0	0	0	0	0	0	0	0	0	0	0	0
FER Geoter	0	0	0	0	0	0	0	0	0	0	0	0	0
FER ter	0	0	56	0	0	0	0	0	0	0	0	0	0
Biom	0	0	945	0	71	0	13	0	0	0	0	0	0
Elettr	0	0	8551	58	734	50	1426	2371	1934	875	1300	3105	2750
Totale	617411	725802	1949687	609822	53185	2867	7361	123870	112370	64557	49553	150236	43374

	Iron &	Mach	Tr equip	Constr	Comm	Roa & R	Dom na	N&I Av	GOV	FCAP	RDM	Carbo	Petrolio	Coke
LAV	27284	42254	11529	36779	379610	19644	5793	2147	0	0	0	2	647	1253
CAP	20080	24525	2985	52803	493665	24288	3124	1158	0	0	0	9	2647	3663
FAM	0	0	0	0	0	0	0	0	291058	0	36486	0	0	0
IMPR	0	0	0	0	0	0	0	0	30860	0	252795	0	0	0
Agr	162	74	11	42	3339	24	89	3	0	234	3514	2	0	0
Fish	1	1	1	0	642	1	109	0	0	12	198	0	0	0
Mining &	3415	40	2	723	143	1	0	0	0	18	693	0	0	2
Food &	30	14	17	2	22102	1	204	19	0	249	14845	0	0	23
Text	357	185	824	266	2468	105	14	18	0	610	34268	0	1	56
Wood	968	1082	590	3516	4072	232	277	294	0	7765	14629	0	1	14
Paper	788	1454	162	593	13162	340	33	68	0	405	6487	0	1	24
Chem	3590	7826	3739	3259	13848	2508	47	88	0	578	49966	0	40	772
Non m min	1917	1275	699	17980	1208	9	45	16	0	948	9408	0	6	40
Iron &	37939	29852	10113	12865	3525	299	112	25	0	16413	32525	1	3	42
Mach	2951	29530	6785	4662	13512	1666	128	86	0	65908	92486	5	9	229
Tr equip	1699	2038	13250	240	8362	3284	1044	1864	0	22093	40956	0	2	4
Constr	562	711	96	13311	14321	372	18	30	0	156743	0	0	5	0
Comm	33945	39316	9031	32040	461940	26998	6249	7652	56761	15637	882	0	193	1789
Roa & R	2350	2501	1827	4697	28301	18029	256	410	2586	0	0	0	26	66
Dom nav	31	0	16	0	7793	16	875	55	747	0	0	0	0	0
N&i Aviat	268	788	30	1060	7827	209	316	240	292	0	0	0	1	16
GOV	2880	2957	830	7176	56993	1274	425	370	0	44601	0	3	116	22263
FCAP	0	0	0	0	0	0	0	0	2033	0	110938	0	0	0
RDM	6570	46555	32229	0	1319	0	0	0	0	395064	0	1373	37514	4201
Carbon	438	1	0	0	13	0	0	0	0	0	4	0	0	44
Petrolio	940	3	0	0	10	0	0	0	0	1777	948	0	0	19880
Coke	1469	286	56	1688	8552	6546	329	1636	0	0	12072	0	2	1453
Dgas	625	668	0	0	3483	99	35	2	0	0	0	2	17	174
Termoel	0	0	0	0	0	0	0	0	0	0	0	0	0	0
Termo bio	0	0	0	0	0	0	0	0	0	0	0	0	0	0
CHP	18	9	137	0	18	0	0	0	0	0	0	0	0	0
Idro	0	0	0	0	0	0	0	0	0	0	0	0	0	0
FER PV CSP	0	0	0	0	0	0	0	0	0	0	0	0	0	0
FER Geoter	0	0	0	0	0	0	0	0	0	0	0	0	0	0
FER ter	0	0	0	0	0	0	0	0	0	0	0	0	0	0
Biom	0	0	0	0	0	459	0	0	0	0	23	0	0	0
Elettr	8207	3385	1072	554	13304	205	35	2	0	0	224	2	17	174
Totale	154486	237280	96030	194257	1563531	106608	19558	#####	384336	729055	714348	1399	41246	56182

	Dgas	Termoe	Termi	CHP	Idro	FER P	FER G	FER	Biom	Elettr	Totale
LAV	1589	108	2	126	101	44	11	5	136	2967	617411
CAP	5411	368	6	428	343	149	39	16	462	10102	725802
FAM	0	0	0	0	0	0	0	0	0	0	1949687
IMPR	0	0	0	0	0	0	0	0	0	0	609822
Agr	4	0	0	0	0	0	0	0	0	8	53183
Fish	0	0	0	0	0	0	0	0	0	0	2867
Mining &	13	1	0	1	1	0	0	0	1	24	7361
Food &	0	0	0	0	0	0	0	0	0	0	123870
Text	20	1	0	2	1	1	0	0	2	38	112370
Wood	82	6	0	6	5	2	1	0	7	153	64557
Paper	73	5	0	6	5	2	1	0	6	136	49553
Chem	208	14	0	16	13	6	1	1	18	388	150236
Non m min	67	5	0	5	4	2	0	0	6	125	43374
Iron &	205	14	0	16	13	6	1	1	18	383	154486
Mach	471	32	1	37	30	13	3	1	40	879	237280
Tr equip	3	0	0	0	0	0	0	0	0	6	96030
Constr	149	10	0	12	9	4	1	0	13	278	194257
Comm	2440	166	3	193	155	67	17	7	208	4556	1563531
Roa & R	102	7	0	8	6	3	1	0	9	190	106608
Dom nav	4	0	0	0	0	0	0	0	0	8	19558
N&i Aviat	14	1	0	1	1	0	0	0	1	27	16183
GOV	2257	154	2	179	143	62	16	6	193	4213	384336
FCAP	0	0	0	0	0	0	0	0	0	0	729055
RDM	1009	0	0	0	0	0	0	0	97	224	714348
Carbon	248	17	0	20	16	7	2	1	21	464	1399
Petrolio	5523	376	6	437	350	152	39	16	471	10311	41246
Coke	926	63	1	73	59	25	7	3	79	1728	56182
Dgas	0	1623	0	2469	0	0	0	0	0	0	20820
Termoel	0	0	0	0	0	0	0	0	0	2972	2972
Termo bio	0	0	0	0	0	0	0	0	0	196	196
CHP	0	0	0	0	0	0	0	0	0	2967	4138
Idro	0	0	0	0	0	0	0	0	0	1256	1256
FER PV CSP	0	0	0	0	0	0	0	0	0	544	544
FER Geoter	0	0	0	0	0	0	0	0	0	141	141
FER ter	0	0	0	0	0	0	0	0	0	0	57
Biom	0	0	174	103	0	0	0	0	0	0	1788
Elettr	0	0	0	0	0	0	0	0	0	0	45283
Totale	20820	2972	196	4138	1256	544	141	57	1788	45284	

64,752 mln euro

64.752 mln euro

12.444 mln euro

Source Authors' elaboration

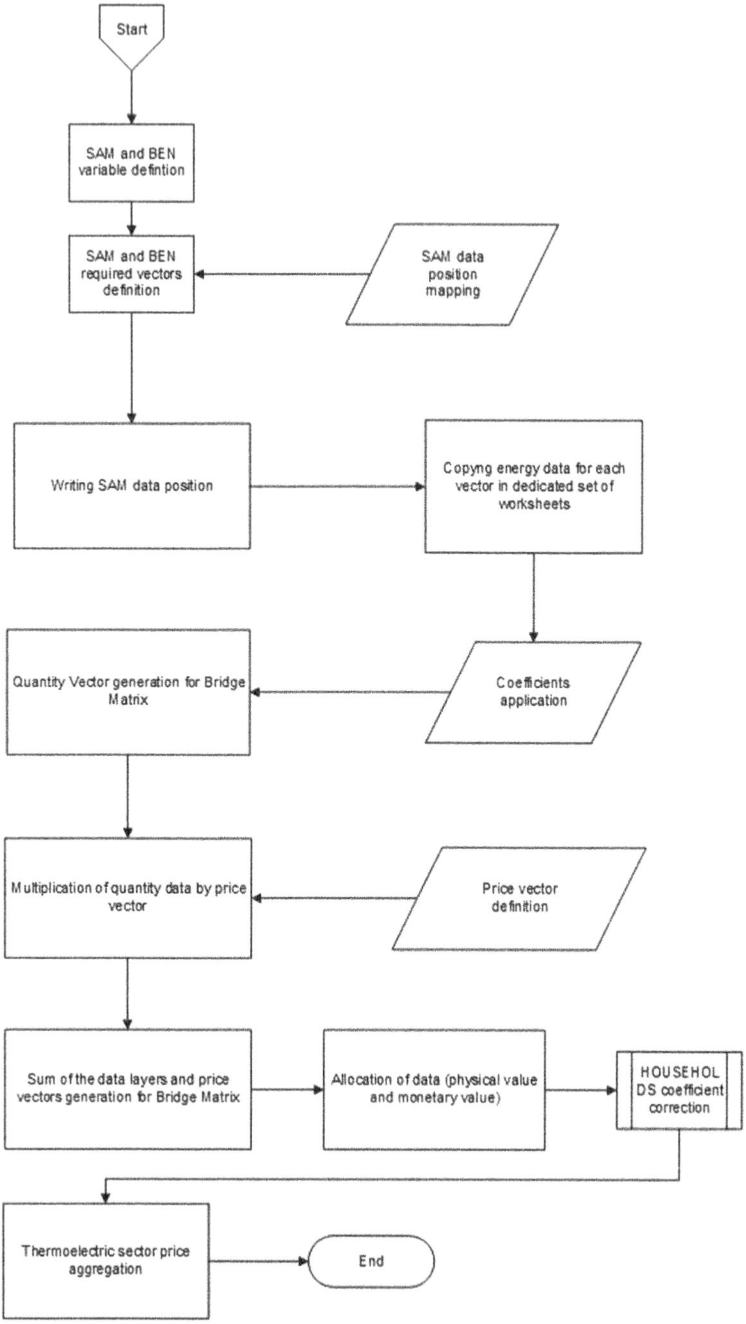

Fig. 2 *Bridge* matrix algorithm. *Source* Authors' elaboration

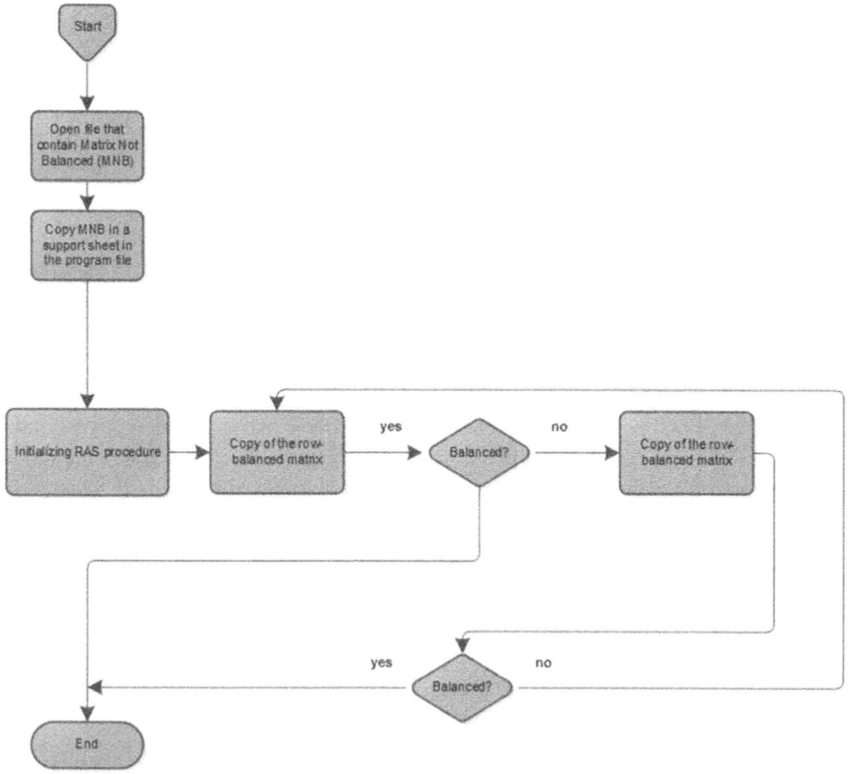

Fig. 3 RAS balancing algorithm. *Source* Authors' elaboration

4 A Case Study: The Development of the Upstream Sector in Italy

Italy is highly dependent on energy imports, with significant implications on the foreign energy bill and the security of supply. Despite this situation, the country has unexploited reserves of oil and gas. The aim of this section is to quantify the impact, in terms of job creation and economic growth, of investment in the upstream sector.[5]

---

[5]Such an investment would likely produce negative effects on the environment and on human health. However, the evaluation of these effects goes beyond the scope of our study.

Moving from the above premises, the simulation performed[6] was based on the targets reported by the Italian National Energy Strategy (SEN)[7]:

- increase the current domestic production of about 24 million barrels of oil equivalent (BOE) of gas and 57 million BOE of oil, moving from ~7 to ~14% in terms of contribution to the total primary energy consumption.
- invest for ~€15 billion, with a saving in energy bill of about €5 billion for year, due to the reduction of fossil fuel imports.

The simulation performed gives two types of results, concerning, respectively, the "Construction period" and the "Operating period" of the projects implemented.

For the development of the Italian upstream sector, the National Energy Strategy forecasts an amount of 15 billion € of Investment in the period (2012–2020) i.e. approximately 1.67 billion € each year. An independent survey, carried out in 2010, identified the activities needed for this objective.[8] These activities are reported in the left column of Table 2 and have been regrouped by sector of supply. The percentages have been used to estimate the vector of increased demand of investment goods (the "investment vector") of the simulations performed.

Using the expenditure vector in Table 2 to simulate the impact of investment in the energy upstream activities, we obtain an estimate of a series of impacts on the economy and, in particular, of the investment cumulative impact on value added, production and employment due to the following effects:

- direct effects (related only to the investment),
- indirect effects (related to income and final consumption increases),
- induced effect (related to supply chain activation).

These effects are estimated to cause a cumulative growth of value added of 20.8 Billion € (i.e. 2.3 Billon € per year or 0.1% of annual GDP). Such a growth and related factor's remuneration boosts tax revenues that increase by 9.6 Billion € (i.e. 46.2% of GDP growth) over the whole construction period. Overall production grows by 43.6 Billon € with manufactures (especially iron and steel) increasing by 7.8 Billion €, construction by 5.4 Billion €, services by 19.2 Billion €, and the upstream sector by 7.5 Billion €. The production of other energy sectors grows by nearly 700 M€ for the production and distribution of electricity, by more than 500 M€ for the refining sector, and by 256 M€ for the gas distribution sector.

Production growth, essentially demand driven, shows a contained impact on the rest of the energy sector, if compared with other sectors that provide intermediate and final goods or services. Investment is estimated to create almost 30,000 full time

---

[6]We distinguish between impact analysis for the "construction period" (CP) and for the "operating period" (OP). In the CP the "Investment" sector is exogenous and the effect of each additional investment is evaluated. In the OP, the sector that became the "owner" of the investment is considered exogenous and the impact of the change on current expenditure in the sector is assessed.

[7]Strategia Energetica Nazionale, in Italian language: see point 4.6 of the approved document, page 110.

[8]Nomisma Energia—"Idrocarburi e occupazione" presentation at the Assomineraria Conference 30-3-2010.

Table 2 Upstream investments in Italy. Type of activities by productive sector

Logistic/Vessel/Helycopters/Control	0.4%	
Transport (onshore)	0.0%	
Pipelay on shore	1.5%	
Pipelay off shore	7.0%	Crude oil, natural gas and auxiliary activities
offshore installation	7.0%	
Compression	13.0%	
Drilling plants	16.7%	
Geophysics, seismic and field studies	1.9%	
Other	2.8%	
Construction/Installation	11.5%	Construction
Civil works (onshore + storage)	3.3%	
Facilities/Equipment/Central	14.8%	
Sealine (offshore), pipelines (onshore)	3.3%	Iron and steel
Casting/Tubing/Chisels/Range	3.9%	
Sludge/Cements	0.9%	Non-metallic minerals
Engineering and project management	4.6%	
Logging/Testing/Auxiliaries	7.4%	Comm. and public services

Source Authors' elaboration

equivalent (FTE) new jobs per year (of which only 6,732 direct jobs). Direct jobs are created in the construction (3,674 FTE), the tertiary sector (1,750 FTE), oil and gas extraction (454 FTE), iron and steel production (764 FTE) and in the non metallic minerals production (99 FTE). Indirect and induced job creation accounts for the majority of the impact (23266 FTE, 78% of which in the tertiary sector); a relevant impact is also expected in manufacturing industries (3892 FTE), and Construction (494 FTE). The detail of the energy sector shows that expected employment grows by 151 FTE in the production and distribution of electricity, by 46 FTE in the refining sector, and by 69 FTE in the gas distribution sector (Figs. 4 and 5).

The National Energy Strategy hypothesizes an operating period of 30–40 years for the new plants. It is estimated that the production of about 80 million of oil barrel equivalents with an average oil price of about 60 €/barrel) would avoid imports of crude oil and natural gas for an amount of 5 billion € per year. This is assumed to be the value of production and is distributed across the production inputs of the oil and gas sector, according to the column coefficients of the SAM (excluding import). In order to provide a range of results for the impact analysis, an alternative average international price of 30 €/bbl is also assumed. The value of production is thus halved and the difference is subtracted entirely from the profits of the sector. During the operating period, we estimate a yearly growth of the value added in the range 9.1–18.2 Billion € (0.1% of annual GDP for the highest estimate).

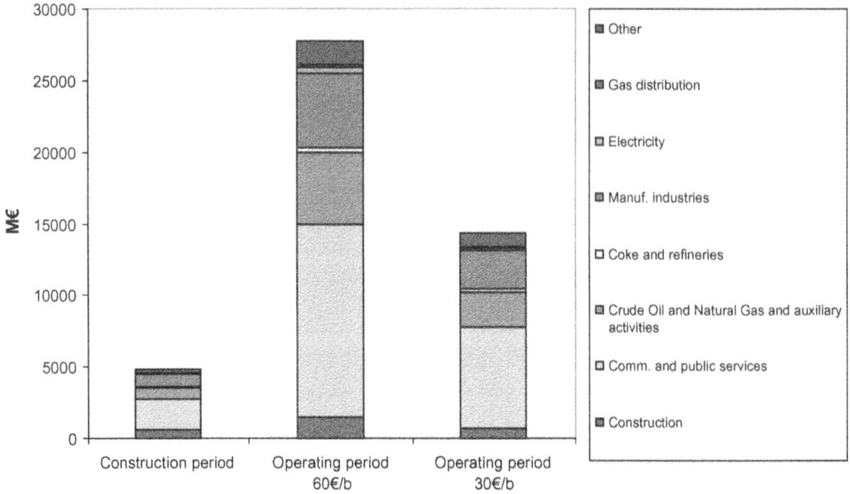

Fig. 4 Annual impact on production by sector (M€). *Source* Authors' elaboration

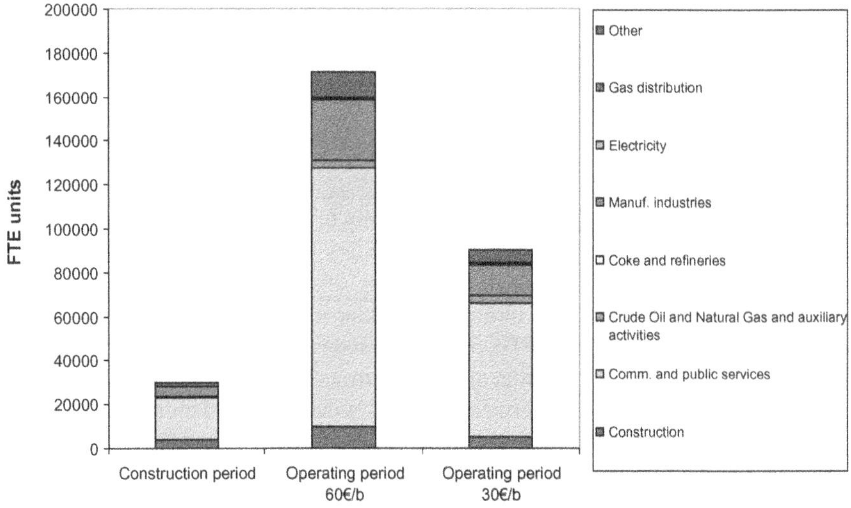

Fig. 5 Annual impact on employment by sector. *Source* Authors' elaboration

The simulation results suggest that in the operating period overall production would grow by 14.4–28 Billion € per year. Manufactures would post an increase of 2.7–5.2 Billion €, the tertiary sector of 7.6–14.7 Billion € and construction of 0.7–1.5 Billion €. Energy sectors production would grow by 204–396 M€ for the generation and distribution of electricity, by 188–362 M€ for the refining sector, and by 124–202 M€ for the gas distribution sector. From the employment point of view, the long term investment effects would consist in 90,400–171,500 full time equivalent

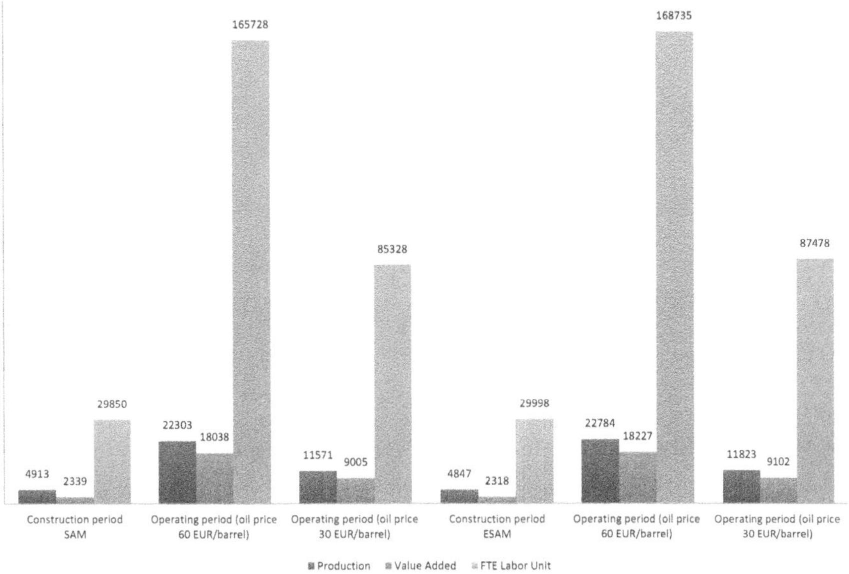

Fig. 6 Main simulation results of SAM and ESAM—millions of Euros. *Source* Authors' elaboration

Table 3 Percentage differences between SAM and E-SAM—%

	Construction period (%)	Operating period (oil price 60 €/barrel) (%)	Operating period (oil price 30 €/barrel) (%)
Production	1.4	−2.1	−2.1
Value added	0.9	−1.0	−1.1
FTE labor unit	−0.5	−1.8	−2.5

Source Authors' elaboration

(FTE) new jobs per year (of which 3,136 direct jobs), with a relevant impact in manufactures (14,000–27,300 FTE), services (65,800–126,400 FTE), construction (4,921–9,942 FTE). Tax revenues would increase by 4.3–8.4 Billion € (i.e. nearly 47% of GDP growth) each year. Employment in the energy sector would grow by 457–887 FTE in the production and distribution of electricity, by 151–290 FTE in the refining sector, and by 341–555 FTE in the gas distribution sector (Fig. 6).

The results show a substantial matching between the simulations performed by the SAM and the E-SAM.

The mean percentage difference among the main aggregated results (Production, Value Added and FTE Labor Unit) is less than 1.0%: the disaggregate estimates of the energy system provided by ESAM thus don't appear create significant distortions in the direction and intensity of the effects measured by the base matrix (Table 3).

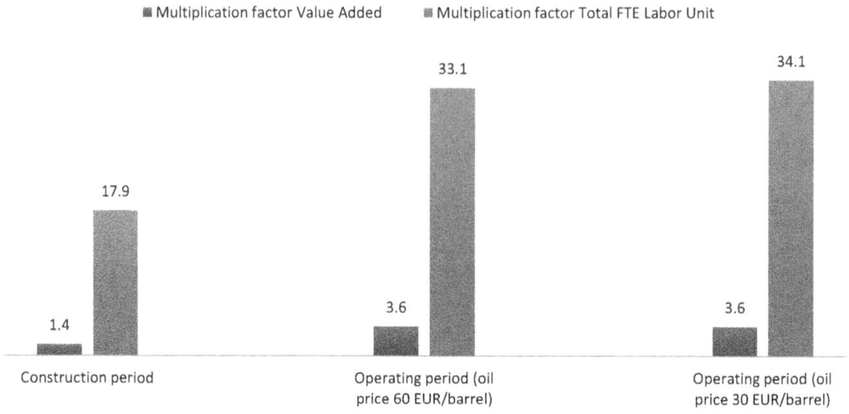

Fig. 7 Main simulation results by E-SAM. *Source* Authors' elaboration

As previously explained in detail, the simulation results are significantly positive both in monetary terms and in terms of FTE Labor Units. Given the substantial equivalence between SAM and ESAM results, we can summarize the effects using the ESAM as shown in Fig. 7.

References

Batten, D. F., & Boyce, D. E. (1986). Spatial interaction, transportation and interregional commodity flow models. In P. Nijkamp (Ed.), *Handbook of regional and urban economics* (Vol. 1, pp. 357–406). Amsterdam: North-Holland.

Batten, D. F. (1983). *Spatial analysis of interacting economics*. In Boston et al. (Eds.), Kluwer-Nijhoff.

Cao Pinna, V. (Ed.). 1980, Il sistema energetico italiano nel quadro dell'economia nazionale. Anno 1975. *Special Issue of the Journal ENERGIA*.

Caputo, M. R., & Paris, Q. (2000). *Comparative statics of the generalised maximum entropy estimator of the general linear model*. Department of Agricultural and Resource Economics, University of California, Davis: Working Paper.

Commissione Europea. (1999). Regolamento Comunitario N.1260/99, 21 June 1999.

Commissione per la garanzia dell'informazione statistica. (2001). *I Conti Economici Regionali: Alcune Possibili Estensioni*. Presidenza del consiglio dei ministri: Rapporto di Ricerca.

Deming, W. E., Stephan, F. F. (1940). On a least squares adjustment of a sampled frequency table when the expected marginal totals are known. *The Annals of Mathematical Statistics, 11*(4), 427–444. Institute of Mathematical Statistics.

Fossati, A. (1991). *Equilibrio generale e simulazioni, Aspetti teorici e applicazioni al caso Italia: IVA-IRPEF-oneri sociali*. Milano: Franco Angeli.

Gaeta, M., & Baldissara, B. (2011). *Il modello energetico TIMES-Italia*. ENEA: Struttura e dati.

Golan, A., Judge, G. G., & Miller, D. (1996). *Maximum entropy econometrics: Robust estimation with limited data*. Hoboken: Wiley.

Hartman, R. S., Wheeler, D., Singh, M. (1997). The cost of air pollution abatement. *Applied Economics 29*, 759–774.

ISTAT. (2006). Le tavole delle Risorse e degli impieghi e la loro trasformazione in tavole simmetriche—Nota metodologica, ISTAT Roma.
Kadas, S. A., & Klafszky, E. (1976). Estimation of the parameters in the gravity model for trip distribution: A new method and solution algorithm. *Regional Science and Urban Economics, 6,* 439–457.
Kolmogorov, A. N. (1968). Logical basis for information theory and probability theory. *IEEE Transactions on Information Theory, 14,* 662–664.
Kullback, S. (1968). Probability densities with given marginals. *The Annals of Mathematical Statistics, 39*(4), 1236–1243. Institute of Mathematical Statistics.
Leontief, W. (1965). The economic impact—Industrial and regional—of the arms cut. *Review of Economics and Statistics 43.*
Loulou, R., et al. (2004). Documentation for the Markal family of models. IEA-ETSAP. http://www.iea-etsap.org/web/Documentation.asp.
Loulou, R., et al. (2005). Documentation for the TIMES model part I and part II, IEA-ETSAP. http://www.iea-etsap.org/web/Documentation.asp.
McDougall, R. (1999). Entropy theory and RAS are friends. *GTAP working papers* 300. Center for Global Trade Analysis, Department of Agricultural Economics, Purdue University.
Norton, R., Scandizzo, P. L. (1981). Market equilibrium computations in activity analysis models. *Operations Research, 29*(2).
OECD/IEA. (2004). *Energy statistics manual*. Paris, OECD/IEA.
Pennisi, G., Scandizzo, P. L. (2003). *Valutare l'Incertezza, L'Analisi Costi Benefici nel 21 Secolo.* Giappichelli, Torino.
Pollin, R. et al. (2008). Green recovery: A program to create good jobs & start building a low-carbon economy. Political Economy Research Institute, University of Massachusetts at Amherst.
Pyatt, G., Round J. I. (1979). Social accounting matrix for development planning. *The Review of Income and Wealth, 23*(24).
Rao, M., Ciorba, U., Trovato, G., Notaro, C., Ferrarese, C. (2014). Costruzione di una matrice di contabilità sociale allargata al settore energetico—Un'applicazione della procedura di Wolsky alla SAM italiana del 2010. RT ENEA/14/2012.
Round, J. I. (1985). Decomposition multipliers for economic systems involving regional and world trade. *Economic Journal, 95,* 383–399.
Scandizzo, P. L., Ferrarese, C., Vezzani, A. (2010). La Matrice Di Contabilità Sociale: Una nuova metodologia di stima. *Il Risparmio Review.*
Scandizzo, P. L. (1990). The estimation of input-output coefficients: Methods and problems. *Ricerche Economiche, 44*(4), 455–474.
Shannon, C. E. (1948). A mathematical theory of communication. *Bell System Technical Journal, 27,* 379–423, 623–656.
Sinkhorn, R. (1967). Diagonal equivalence of matrices with prescribed row and column sums. *The American Mathematical Monthly, 74,* 402–405.
Snickars, F., & Weibull, J. W. (1977). A minimum information principle: Theory and practice. *Regional Science and Urban Economics, 7,* 137–168.
Stone, R. & Brown, A. (1962). *A computable model of economic growth*. University of Cambridge. Dept. of Applied Economics, London:Chapman & Hall.
Theil, H. (1967). *Economics and information theory*, Amsterdam, North Holland.
Thorbecke, E. (2001). The social accounting matrix: Deterministic or stochastic concept?, Cornell University.
Wauschkuhn, U. (1982). *Anpassung von Stichproben und n-dimensionalen Tabellen an Randbedingungen*. München, Wien: R. Oldenbourg.
Wolsky, A. M. (1984). Disaggregating input-output models. *The Review of Economics and Statistics, 66*(2), 283–291. The MIT Press.
Xie, J. (2000). An environmentally extended social accounting matrix. *Environmental and Resource Economics 16*, 391–406. Kluwer Academic Publishers.

Analysis of Local Economic Impacts Using a Village Social Accounting Matrix: The Case of Oaxaca

Cataldo Ferrarese and Enrico Mazzoli

Abstract This chapter describes a methodology to estimate a local economic model based on a social accounting matrix (SAM), for the district of Villa Alta within the Sierra Norte region of the Mexican State of Oaxaca. The estimates combine secondary statistics with data obtained through a direct survey. A SAM based model is then used to assess the impact of a rural development program on the local (village) and regional (state) economy.

Keywords Social accounting matrix · Local economy · Economic impact Rural development

1 Introduction

The applied economic research literature over the last years has collected studies focusing on project economic impacts using different evaluation techniques. Some of the limits of this kind of analysis are related to the fact that only selected beneficiaries and variables directly linked to projects/programs are accounted for, and is therefore not possible to capture the main impacts on the overall economy. Computable General Equilibrium (CGE) and Social Accounting Matrix (SAM) models can be used to fill this gap.

The aim of the present investigation is to assess the economic impact of a rural development programme[1] on a local (village) and regional (state) economy using a SAM model. More specifically, this research shows whether, and under which

[1] While acknowledging the distinction between programme and project, we opted to use the two terms interchangeably for the purpose of this research.

C. Ferrarese (✉)
OpenEconomics srl, Rome, Italy
e-mail: dino.ferrarese@openeconomics.eu

E. Mazzoli
Independent Consultant, Rome, Italy
e-mail: e.mazzoli@ifad.org

© Springer International Publishing AG, part of Springer Nature 2018
F. Perali and P. L. Scandizzo (eds.), *The New Generation of Computable General Equilibrium Models*, https://doi.org/10.1007/978-3-319-58533-8_5

conditions, a rural development project that includes Climate Change Adaptation strategies (such as the Adaptation for Smallholder Agriculture Programme, ASAP) would trigger economic growth in a region or in a group of villages within a specific geographical area.

The study applies an innovative methodology for estimating village-wide SAMs, which builds on the most recent literature on the subject, and integrates estimation methods by the World Bank and the Italian Government (Scandizzo et al. 2010, 2015). The SAM methodology identifies investment impacts on: (i) household income; (ii) household welfare; (iii) productive sectors; and (iv) local administration. We expand this method to make a direct comparison among three project scenarios, controlling for different types of investment projects and relative total costs. This approach allows us to quantify the extent to which a climate adaptation intervention would enable the local economy to achieve more stable paths of economic growth. Furthermore, our methodology distinguishes between short-medium term and long term expected impacts on the different economic and social sectors, thus increasing the informative capacity of the analysis performed.

We test the methodology in a small geographic area, the district of Villa Alta within the Sierra Norte region in Oaxaca State (Mexico). As a result of the direct comparison between three different project scenarios implemented over a period of five years, we identify which of the socio-economic sectors are benefitting from each programme intervention. Through this research, we contribute to generate more evidence-based decisions on investments project designs that aim to stimulate rural economic development and include Climate Change Adaptation strategies.

The plan of the Chapter is as follows. Section 2 begins with a theoretical discussion on SAM methodologies. Section 3 discusses additional practical implication for the SAM estimation at regional and local level. Section 4 presents the expected outcomes for the scenarios envisaged. Section 5 concludes.

2 The SAM: Theoretical and Methodological Aspects

The Social Accounting Matrix (SAM) is considered an extension of the traditional Input—Output[2] (I/O) model proposed by Leontief (1966), which records in monetary terms the exchange flows occurred within an economic system, during a specific period of time (usually an year). The Matrix allows to consider the entire structure of relations characterizing an economic system through the different phases of the production, distribution, utilization and income accumulation process.

As shown in Fig. 1, any economic system can be described by the circular income circuit where economic agents, productive sectors and institutions are connected to one another through real transactions. For example, households' incomes are related to remuneration of capital and labor, government assistance in the form of social

[2]The Input—Output accounts provide detailed industry and commodity accounts and show the supply and demand flows in a specific economy.

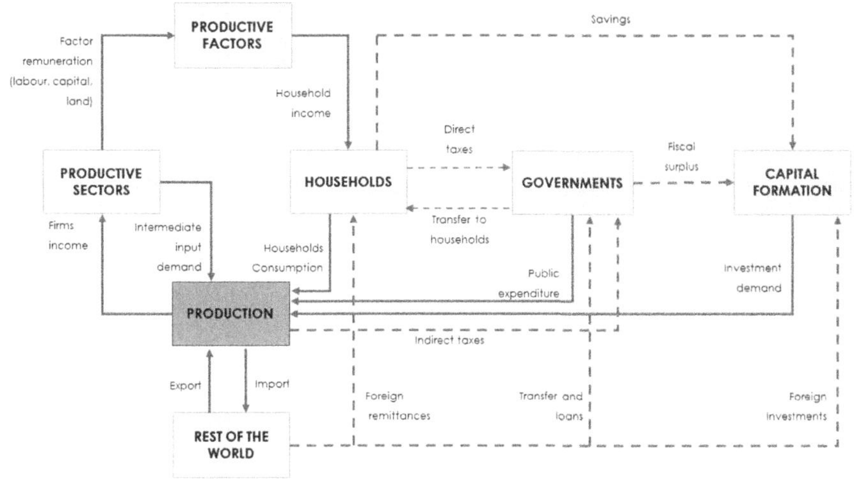

Fig. 1 The income circuit

transfers, and foreign remittances from the Rest of the World.[3] Conversely, families decide to allocate their wealth on both consumption and savings following their preferences, once taxes—both direct and indirect—are paid. In such a comprehensive framework, each actors' outflow becomes someone else's inflow and, considering that all transactions between people and institutions are monitored and quantified, the system does not present leakages.

A SAM thus consists of a set of interrelated subsystems that, on the one hand, provide the analytical framework of the economy studied by tracking monetary flows occurring between sectors and, on the other hand, measure the structural changes within the economy (injections and multiplying effects in the system), as a result of policy changes or a project interventions.

The information is compiled in a double-entry table (the matrix), describing the structure of the economic system through disaggregation in key blocks (actors, productive factors and activities), assumed as origins and destinations of the transaction flows. Each key block is further disaggregated into accounts headed to the institutional sectors (e.g. type of households, specific commodities, production sectors) depending on the detailed data availability. The economic system is typically disaggregated into the following blocks:

i. Primary production factors (Labour and Capital);
ii. Households (eventually disaggregated by income or income source);
iii. Government (Public Administration);
iv. Production sectors and Commodities (Agriculture, Industry, Services and their disaggregation);

[3]The Rest of the World can be defined as another Country/State, Region or geographical area depending whether the scale of the analysis is National, Regional or Local-wide respectively

	Productive Factors	Household	Government	Production sectors and Commodities	Savings and investment	Rest of the World	Total inflows
Productive Factors		Domestic Employment	Government Employment	Value-added		Payments from abroad	Total factor income
Household	Labour incomes and profits	Inter-household transfers	Social transfers			Foreign remittances	Total household income
Government	Taxes on labour and profits	Direct taxes		Indirect taxes	Taxes from capital account	Foreign grants and loans	Total Government income
Production sectors and Commodities	Domestic supply	Private consumption	Recurrent spending	I/O Matrix (intermediate demand)	Investment and stock	Export payments	Total demand and activity income
Saving and Investment		Private savings	Fiscal surplus			Current account balance	Total savings
Rest of the World	Factors payments abroad	Household transfer	Government transfers	Imports payment			Total imports
Total outflows	Total factors spending	Total Household expenditure	Total Government Expenditure	Total Gross output	Total investment spending	Total Export	

Fig. 2 A simplified SAM scheme (*Source* Own elaboration)

v. Savings and Investment (Public and Private gross fixed investments);
vi. Rest of the Country (ROC) or Rest of the World (ROW).

In a typical SAM structure, columns represent the outflows of the different economic agents that is, the expenditure of any aggregate with respect to the others, while rows represent the inflows, namely the income formation. Since total incomes equal total expenditures and material balances between demand and supply mist also hold,[4] a SAM is a square and balanced[5] matrix. A simplified scheme of the SAM is presented in Fig. 2.

An interesting evaluation in the context of developing countries relates to the simulation of structural changes of the economy in response to policy changes. Some exemplary questions to which this analysis could respond are: What would happen to the economy if technical change in agricultural production were brought in? How would the economy change after a shift in import? What would be the trickle-down effect due to the establishment of a new production activity?

All these interventions cannot be simply studied as the effects of an increase in households' disposable income, since changes in the economy have potential important effects on the structure of the SAM in terms of coefficients and multipliers. For instance, long lasting impulses in the Agricultural sector (as in the form of ODA interventions) would generate an increase in rural household income that would trigger a rise in goods and services demand. Thereafter, a likely increase in goods and services supply would generate a structural change within the local economy.

For this reason, we can base ours simulation on a variation of the linear Input—Output model, according to the equation (Scandizzo and Ferrarese 2015):

[4]Surplus or deficit in the balance of trade are compensated within the "rest of the world" account.

[5]A square matrix contains an equal number of rows and columns while a Matrix is balanced when the sum (total value) of each row is equal to the sum (total value) of each column for each of sectors included.

$$\Delta X = (I - A^{*})^{-1}[(\Delta A)X + \Delta Y]$$

where A and A* are the SAM matrices, respectively, with and without the Adaption for Smallholder Agriculture Programme (ASAP) component, and ΔY is the vector of exogenous changes in receipts or expenditure of the capital account (Project intervention or exogenous investment). In our specific case, $\Delta Y = 0$, since the policy examined consists only in the selective change in the sector coefficients interested by the project intervention.

In the case of Oaxaca the evaluation consists of a two-step process. The first step relates to the evaluation of an investment programme at regional scale, with and without the ASAP component. Using the Oaxaca SAM, we evaluate the short-term effects of the project in the five investment years, and the effects of the expected mid-long term structural change of the local economy, in response to the project and the related climate change adaptation measures.

The second step consists in assessing project effects on the targeted economy. In order to perform this evaluation we need to scale down the project at village level using the local SAM presented above. As in the analysis at regional level, the estimation procedure will consider short and long term effects on the local targeted area, differentiating impacts related to the ASAP inclusion or exclusion.

2.1 The Local SAM

CGE modelling and SAM-based research require the use of the most recent economic data available in a coherent framework. However, these data generally come from quite diverse sources of information such as Input—Output tables; national accounting data; households, firms and enterprise surveys; Sector-wide census; labour market surveys; government and international trade accounts. One of the main issues when constructing a SAM both at national and local level is how to combine and incorporate information, harmonizing both primary and secondary dataset, derived from previous periods.

While the original idea was based on the articulation of national accounts, the structure of a SAM appears particularly appealing to represent the interconnections of a smaller economy, such as a region, a town, a village or a group of villages, particularly in the process of investigating the aspects of the mutual relationships of obligation and exchange that characterize local communities. In this respect, a SAM can be used with a twofold aim to:

(i) focus on the local detail of the linkages among disaggregated production and consumption activities (agricultural production, rural works, personal services, etc.), and,
(ii) quantify the monetary and non-monetary transactions within and between the households and the formal and informal community groupings.

Because of its characteristics of a balanced network of exchanges among a variety of producers and users, a local SAM can also capture some of the more subtle linkages that characterize social cohesion, cooperative behaviour and institutional strength in a small community. These linkages may lead to estimates of multipliers and indicators of growth capacity that depend on the relational structure of the community, rather than merely on its resource endowments and performance indicators. In addition, the same linkages may shed lights on the phenomenon of development as a result of complex interactions between competitive and cooperative interrelations in a local context, and on the importance of network closure—dense connections between network participants—in maintaining trusting relationships and building up social capital (on this, see, for example Coleman 1988).

Depending on the degree of integration with external markets, villages are characterized by stronger or weaker market interactions amid village households. Following a U-shaped relationship, market interactions tend to be stronger in case all goods are non-tradable between villages, while they are weaker in economies perfectly integrated with external markets. The villages in our study could be depicted as in Fig. 3.

Compared to the more aggregate SAMs, local SAMs try to capture the complexities of a closely integrated, but small socio-economic systems. In fact, a village SAM is based on a representation of a local economy which has considerable more breadth and depth and, as such, demands a closer investigation of the elements of modularity and interconnection characterizing the structure of village life. Because of social capital, a local SAM spans a broader set of functions and non-monetary transactions, for example, payments in kind, reciprocal exchanges, management of the commons, social rewards and sanctions and a variety of social rites and customary activities.

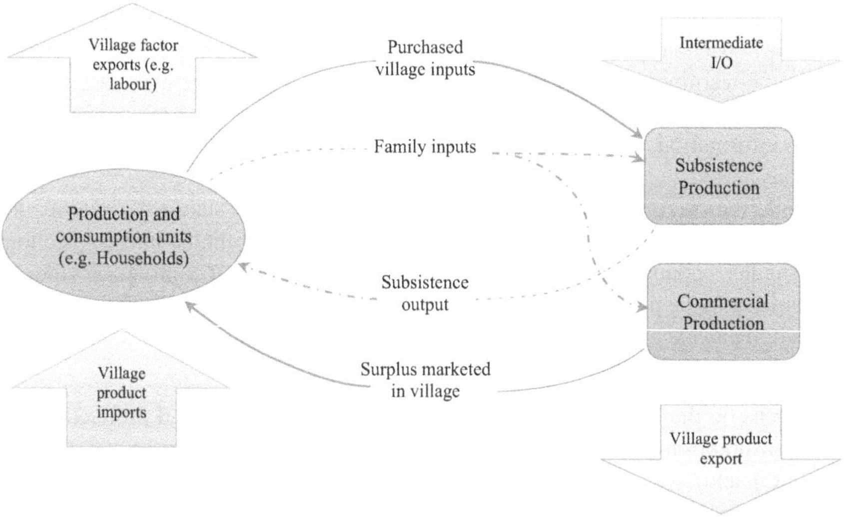

Fig. 3 Economic flows in a village with intermediate degree of interaction with external markets (*Source* Own elaboration)

Furthermore, owing to of the higher disaggregation level of economic activities, a local SAM may contain a deeper analysis of the productive relations, with a finer detail of agricultural activities, rural industries, small business and personal services (Taylor et al. 2006).

From the point of view of the target group, or the nodes of the social network, a village SAM may also include stakeholders other than the classical groupings defined in national accounts in order to capture, for example, exchanges within the extended family and repeated interactions, such as those occurring between farmers to govern the distribution for irrigation water. While households and firms may be disaggregated into finer categories, village level institutions may also be included in a local SAM as important nodes of interdependencies within the local community.

The integration of specific primary data information coming from the household, the business and the community questionnaires into a unique dataset, allowed tracking down the exchange relations between the sectors characterizing the economic system.

2.2 *Literature Review of Local SAM*

One of the first studies on local level SAM has been implemented by Bell, Hazell and Slade (1982) who analysed the effects on paddy land of an irrigation project for the Muda River basin. The authors focused on the evolution of some key variables (output, income, wage and rent) to estimate direct impacts of the project by means of a linear programming model. Indirect effects have been analysed developing a Social Accounting Matrix model at regional level. The regional SAM was disaggregated into forty-five accounts. Results of the analysis showed an increase in the regional value added but no changes in the distribution of income within the region.

Years later, Adelman et al. (1988) were the first to undertake a study and construct a village level SAM. The authors constructed a SAM using household data collected from a major migrant-sending village in Central Mexico in 1982 and focused on the economic structure of such economy. The study highlighted the importance on internal and international migration in the village economy. Findings showed that national and international migration has a central role in the village economy and that stressed the vulnerability of the village economy to external shock resulting from U.S immigration policy reform. Further, it showed how anti-poverty policies are crucial in addressing the problem of landless households.

Subramanian and Sadoulet's study (1990) elaborated a village-wide SAM for the village of Kanzara in Western India. The SAM was used to estimate the effects of an irrigation investment program in this cotton-producing and rain-fed area. Given the agricultural nature of the village, fewer commodity—producing activity sectors were considered in the SAM which instead provided greater details on services, labour flows, transfers, and income distribution.

An interesting town-based analysis through a SAM was carried out by Lewis and Thorbecke (1992). The analysis focused on a Kenyan town Kutus, comprising both the town population—of around 5,000 inhabitants—and the 8 km zone around it (hinterland) with a population of 42,000 people. The SAM was used to test the governmental assumption of agriculturally-driven regional economies and to evaluate non-agricultural production sector activities in the Kutus region. According to the authors, agricultural activities were indeed very important for stimulating regional output and income.

A vast and diverse set of issues have been analysed through Village-SAMs—from the impact of remittances from Mexican workers abroad or in urban centres (Adelman et al. 1988) to the impact of decentralized rural industrialization on employment, incomes and modernization trends within the village (e.g. Parikh and Thorbecke 1996); or the nutritional consequences of different exogenous policies (e.g. Ralston 1992).

Extensive application of village-SAM analysis is done by Taylor and Adelman (1996), which they applied to India, Indonesia, Kenya, Mexico, and Senegal. In their book—Village Economies—the authors present a general framework for modelling village economies based on computable general-equilibrium techniques. They estimate models for villages and a village-town and conduct a series of comparative experiments. In addition, they built a complementary CGEs calibrated and designed to capture the impact of policy, market and environmental changes on the different village economies.

Taylor et al. (2006) extend village SAMs to include household groups as well as separate components of a rural economy. In this type of model each "household level SAM" or rural group is integrated into a rural sector "mega-SAM". The SAM provides the data input into the micro economy-wide, CGE model.

A different microeconomic modelling approach is used by Subramanian and Qaim (2009) used to analyse welfare and distribution effects in a typical village economy in India. Unlike previous SAMs, which were based on sample surveys, their SAM was built on a village census and considered 156 agricultural and non-agricultural activities. Cotton production and numerous other crops are included within the Agricultural activities accounts. Non-agricultural activities included other village-based production (e.g. construction) and agriculture services (e.g. hiring out machinery), retail trade, private services (for example, doctor, barber etc.), government services (for example, ration shop, post office) and transportation.

2.3 A General Framework for Village-SAM Analysis

A typical village SAM can be described as essentially a scaled-down version of a national or regional SAM. In particular, the following sectors can be considered for the village-SAM structure:

Production activities: Production sectors normally included in the SAM are: (1) crop production—coffee, cocoa, wheat, maize, other pulses, oilseeds, cotton, fruits and vegetables, and other crops (cultivation of these crops could be divided for irrigated and rain-fed areas, but in SAM we can have only one column for each crop); (2)

animal husbandry—milk and milk products, wool and meat, cow dung manure, and bullocks; (3) construction; (4) service providers and the self-employed—small shops, grocery, fruit and vegetable vendor, cloth shop, general shops, transport, carpenter, and other services; (5) manufacturing—cotton ginning factory; and (6) services—government services (education, welfare) and private services.

Factors of production: Factors of production included in the village SAMs are tipically: (1) Labour—divided by sex; and (2) Capital, measured as income from managing one's enterprise—in various forms, including mixed income from the self-employed.

Institutions: Institutions considered in village SAMs are normally: (1) households divided by family size and by occupation—small, medium, large farmers, labour, self-employed in non-agriculture, service, and other households; and (2) government at various levels depending on the depth and breadth of the analysis (local, district, provincial).

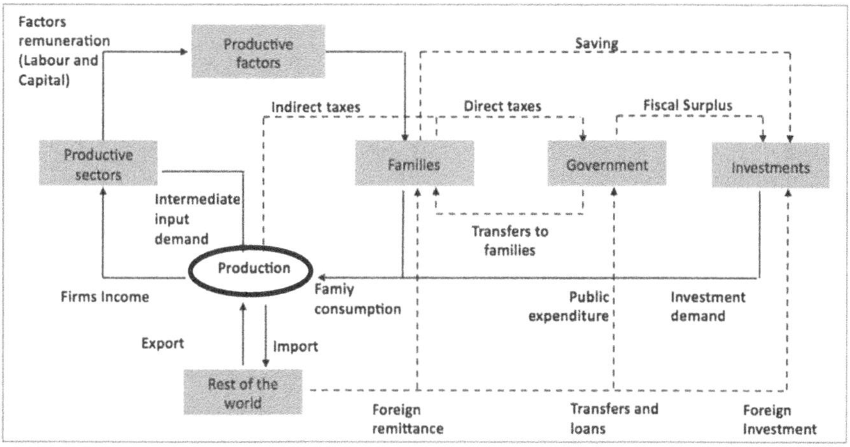

The construction of village-level SAMs can be a challenging task, considering the possibility to consider and to investigate both monetary and non-monetary transactions within a small community, and the need to collect primary data and household census data to represent these transactions. A typical description of local SAM would include: (*i*) *Primary Production Factors;* (*ii*) *Natural resources;* (*iii*) *Stakeholders,* (*iv*) *Production sectors;* (*v*) *Capital formation,* (*vi*) *Rest of the world.*

Transactions between the village and rest of the world are recorded in the Rest of World accounts. Depending on the geographical area of the analysis, The Rest of the World account can be further disaggregated into three different components including Rest of the Area, Rest of the Country and Rest of the World to describe domestic and international trade.

3 The Regional SAM of Oaxaca

3.1 Estimating the Oaxaca SAM

While no recent estimation of the Social Accounting Matrix for the Oaxaca region appears to be available, we were able to use Input—Output estimates made by Bautista (2008) and Martinez Jimenez (2012) integrated with economic data 2004/2010 collected by the Research Team of the Global Trade Analysis Project (GTAP) and INEGI. We thus estimated using a computational algorithm (Scandizzo and Ferrarese 2015) a regional SAM consisting of 4 agriculture economic sectors, 11 industrial sectors, 4 services sectors, 2 production factors, 2 institutions (Household and Government), Capital Formation and The Rest of the World and rest of the Mexico (see Table 1: SAM sectors).

Table 1 SAM sectors for the Oaxaca region

Productive sectors	Production factor
Agriculture	Labour
Animal	Capital
Forestry	
Fishing and hunting	
Mining	
Food, beverage and tobacco	
Textiles and textile products, leather and leather products	
Wood and wood products	**Institutions**
Paper and paper products, publishing and printing companies	Households
Manufacture of coke and refined petroleum products	Government
Mineral	
Metal product	
Manufacturing	
Construction	
Electricity, gas, steam and water supply	
Wholesale, restaurant and hotels	**Other sectors**
Transport storage and communications	Capital formation
Finance and real estate	Rest of the Mexico
Social and personal services	Rest of the world

3.2 Villages Profiles

In order to develop the estimates of the Village SAM for Oaxaca we conducted a statistical survey of two municipalities within the Villa Alta district, precisely in: (i) San Ildefonso Villa Alta and (ii) San Cristóbal Lachirioag.

The two communities are located in the northern eastern part of Oaxaca in the centre of the Sierra Norte region at about 140 km to Oaxaca de Juarez at an altitude of 1200 mt (3939 ft.). San Cristobal Lachirioag total area is of about 24.24 km^2 which represent the 0.03% of Oaxaca state while San Ildefonso Villa Alta covers a larger total area of 136 km^2 (0.14% of Oaxaca State).

The Villa Alta municipality includes, among others, the villages of San Juan Yalahui, San Francisco Yatee and San Jaun Tagui which have been part of the study. The total population of two municipalities is of 4,708 peoples (INEGI 2012). The first production activity within the target area is agriculture and in the observed villages there is only one exporting industry (Mezcaleria).

3.3 Survey Descriptive Statistics

To estimate the local SAM and analyse relevant sectors of the village matrix, we collected data through households and business-activities surveys in each of the above mentioned communities. In detail, we have gathered values on several variables such as output of crops and other activities; inputs of land, labour, capital, and purchased inputs, food and non-food consumption expenditures and pattern over time, public and private transfers, saving and remittances flows, economic shocks, climate change and adaptation strategies. Preliminary meetings with local authorities were held in each of the communities visited so as to being officially introduced to the inhabitants and get a better understanding of both the local government spending and the village formal and informal markets.

For household data we opted for a Random sampling technique[6] with the intention of reducing the likelihood of bias favouring, wherever possible, women's interviews since they are considered a more accurate and reliable source of information. The household sample consisted of 520 people (335 females and 285 males) representing 104 households. Seven local enumerators helped the team during data collection.

The data collected show that 20% of the population does not carry out any agricultural activity—despite the fact that minimal production for household consumption is generally present—while over 24% of population live exclusively on agriculture (hereafter defined as Farmers). More than half of the population (54%) relies both on agriculture and other activities. Figure 4 describes how agriculture production contributes to poor households' incomes.

[6]Random sampling is a sampling technique where we select a group of subjects (a sample) for study from a larger group (a population). Each individual is chosen entirely by chance and each member of the population has a known, but possibly non-equal, chance of being included in the sample.

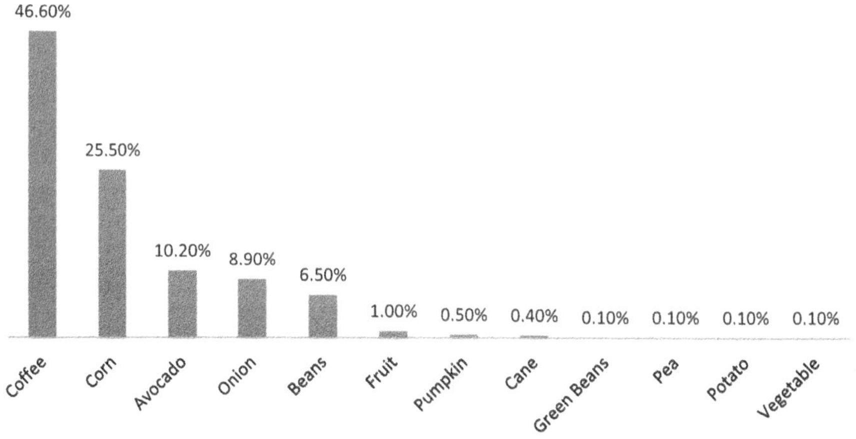

Fig. 4 Crop production value of poor households

The "non-poor" households, mostly with a double activity, are well integrated the local economy and also interact with neighbouring communities. The main activities carried out by this categories are: (i) Food store; (ii) Restaurant; (iii) Hardware; (iv) Blacksmith; (v) Internet Point; (vi) Household goods store; (vii) Bakery; (viii) Taxi; (ix) Other store.

In order to include the business section in the matrix we have surveyed 50 different shops in the various communities covering at least one shop for each business sector.[7] Even for this data collection process we opted for a random sampling technique, while considering as well spatial aspects such as proximity to the main road, visibility and ease of access. To the extent possible, we tried to cover the majority of villages' shops including those not immediately accessible. Table 2 summarizes the mean values of costs, revenues and profits and presents a breakdown for revenues' composition.

3.4 *Estimating the Village SAM*

Thanks to the information collected through the survey we identified 30 socio-economic sectors relevant in the local economy:

Some of these sectors represent the typical services produced and consumed by rural households and other productive sectors in the target area, while others pertain to goods and services consumed in the area but produced in a different region/community.

From the coefficient matrix we then estimate the ***Multiplier matrix***. The latter describes the effects of an exogenous expenditure on the economic system. Similarly

[7]Despite Mezcal production is a common practice in the communities we visited, only one person was formally running a prolific business activity on it.

Table 2 Business activities

	Revenue (pesos)	Costs (pesos)	Profit (pesos)	Labour costs (pesos)	Other costs (pesos)	Internal sales (business) (%)	Internal sales households (%)	Sales to institutions (%)	Sales to other communities (%)	Sales to the rest of Oaxaca	Sales to the rest of Mexico (%)	Sales to the rest of the world (%)
Food and beverage	20,911	14,444	6,467	2,000	11,518	–	80	–	20	–	–	–
Public phone	19,000	17,500	1,500	–	–	50	50	–	–	–	–	–
Restaurant	37,000	30,400	6,600	–	27,000	–	17	–	67	–	16	–
Transport	20,000	19,000	1,000	–	–	–	95	–	5	–	–	–
Carpentry	82,500	55,100	27,400	–	–	10	7	–	83	–	–	–
Hard industry	150,000	120,000	30,000	8,000	–	–	20	10	70	–	–	–
Construction	57,400	–	–	–	–	–	50	–	50	–	–	–
Internet point	72,500	30,300	42,200	–	–	25	50	5	20	–	–	–
Beauty shop	8,160	1,560	6,600	–	–	–	50	–	50	–	–	–
Oil	32,400	–	–	–	–	–	80	–	20	–	–	–
Mezcaleria	1,200,000	960,000	240,000	360,000	–	1	2	–	2	–	1	94
Boutique	80,000	56,000	24,000	–	–	–	100	–	–	–	–	–

Productive sector	Value added
Agriculture	Capital
Coffee	
Maize	
Avocado	
Spring onion	
Rest of agriculture	Labour
Mezcaleria	
Oil	
Energy	
Telecommunication	
Construction	
Food and beverage	
Accomodation and restaurants	**Institutions**
Transport	Farmers HHs
Carpentry	No farmers HH
Metalurgy	Government
Hardware	
Internet point	
Beauty shop	
Gas station	
Clothing shops	**Other sector**
Other shops	Capital formation
Repair services	Rest of Mexico
Instruction and public services	Rest of the world

to the Keynesian Multiplier, an initial expenditure of one MU in a specific sector will generate impacts equal to the multiplier factors to the respective interlinked sectors.

Starting from the Multiplier matrix we can generate the Restricted[8] Multiplier (Forward and Backward multipliers). Forward multipliers express the increase in the activity level of a particular sector in response to an equi-proportional increase in all sectors. They thus measure the importance of the sector considered as a supplier of goods and services and, in a broader sense, the capacity of a sector to participate to overall growth. Sectors possessing low forward multipliers indicate that these industries sell their output mostly to final demand and depend mostly on intermediate flows. Backward multipliers, on the other hand, measure the extent to which a sector autonomous rise in activity level spills over all the other sectors. Therefore they measure the importance of a sector as a centre of demand for the rest of the economy,

[8]The multipliers are defined "restricted" because the balance of payment is assumed to be constrained by the base year conditions, so that exports are prevented from growing to match the increase in imports.

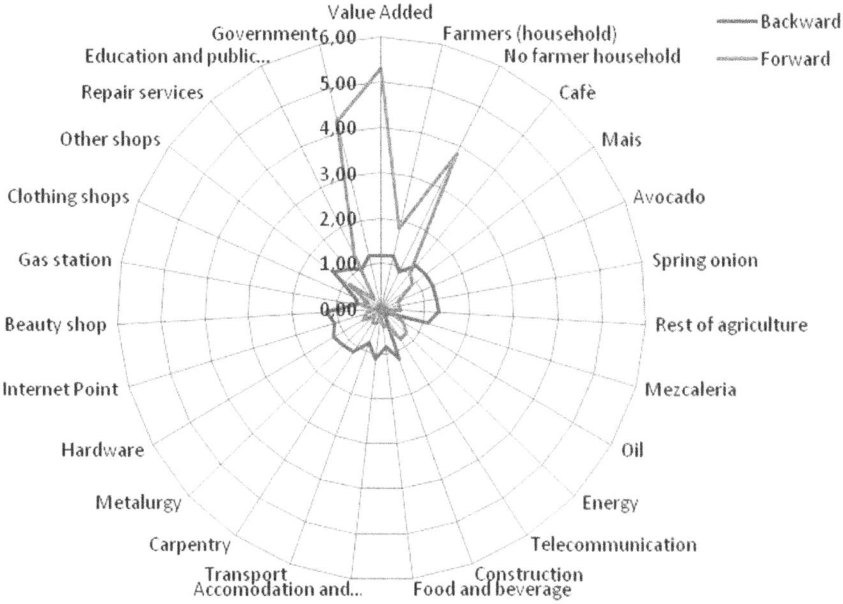

Fig. 5 Backward and forward multipliers

and can be considered as an index of the positive externalities generated by the network structure, which relates to the capacity to propagate a shock from one to other sectors. Those sectors characterized by low backward multipliers indicate that their dependence on other sectors for their inputs is comparatively very low, i.e., their principal inputs are provided mainly by imports.

In conclusion, forward linkages determine the relationship between the activity in one sector and its sales to others. Backward linkages display the relationship between the activity in a sector and its purchases from the others. In the case of the municipalities analysed, the sector with highest multiplier mean value are Coffee, Maize, Avocado, Spring Onions (cash agriculture) and public services. The key results of our estimation on the Local SAM restricted multipliers are summarized in Table 3 and Fig. 5.

Using the data of foreign expenses in the community we can estimate the multiplier effect for each Peso spent in the village. As it could be expected, given the socio-economic context of the area, Households and Services appear to be the most sensitive sectors. These results can be certainly justified observing that these two sectors are the most connected, therefore those with the highest capacity to spread the initial shock over the rest of the economy. The multiplier for value added is equal to 1.254 which means that each peso injected in the target area creates 1.254 pesos of value added in the village economy. Table 5 shows the results (Table 4).

To evaluate the multiplier effect in the study area we can simulate different investment scenarios. The following tables and figures depict the effects of alternative

Table 3 Local SAM, restricted multipliers

	Backward		Forward	
	Total	Mean	Index	Mean
Value added	1.17	0.26	5.30	1.16
Farmers (household)	1.22	0.27	1.83	0.40
No farmer household	0.91	0.20	3.82	0.84
Cafè	1.25	0.27	1.02	0.22
Maize	1.27	0.28	0.90	0.20
Avocado	1.27	0.28	0.42	0.09
Spring onion	1.27	0.28	0.41	0.09
Rest of agriculture	1.33	0.29	0.43	0.09
Mezcaleria	1.10	0.24	0.17	0.04
Oil	0.17	0.04	0.57	0.13
Energy	0.17	0.04	0.78	0.17
Telecommunication	0.17	0.04	0.81	0.18
Construction	1.19	0.26	0.39	0.09
Food and beverage	0.85	0.19	0.41	0.09
Accomodation and restaurants	1.12	0.24	0.21	0.05
Transport	0.81	0.18	0.33	0.07
Carpentry	1.13	0.25	0.37	0.08
Metalurgy	1.18	0.26	0.23	0.05
Hardware	1.23	0.27	0.49	0.11
Internet point	1.08	0.24	0.36	0.08
Beauty shop	1.24	0.27	0.18	0.04
Gas station	0.53	0.12	0.51	0.11
Clothing shops	0.66	0.15	0.33	0.07
Other shops	1.35	0.30	0.88	0.19
Repair services	1.14	0.25	0.27	0.06
Education and public services	0.98	0.21	1.34	0.29
Government	1.20	0.26	4.25	0.93

investments in cash agriculture, transport, services or government transfers to households. The results show a larger than unity effect on local value added in the case of agriculture investment, while the biggest impact on local industry and services are provided in the case of investment in transport services.

Another interesting simulation concerns the likely impact generated by remittances flows towards the area. Remittances represent 25% of farmers' total income and 40% of the income for other households. Remittance flows show a leverage capacity on value added of 1, 8 (11%) and 4, 7 (28%) million pesos respectively for farmers and other households. In the case of farmers the sectors with the larger effect

Table 4 Local SAM, restricted multiplier

	Multiplier
Value Added	1.254
Farmers (household)	0.394
No farmer household	0.865
Cafè	0.180
Maize	0.155
Avocado	0.047
Spring onions	0.045
Rest of agriculture	0.056
Mezcaleria	0.000
Oil	0.077
Energy	0.133
Telecommunication	0.139
Construction	1.188
Food and beverage	0.048
Accomodation and restaurants	0.009
Transport	0.115
Carpentry	0.101
Metalurgy	0.014
Hardware	0.295
Internet Point	0.037
Beauty shop	0.002
Gas station	0.097
Clothing shops	0.035
Other shops	0.159
Repair services	0.024
Education and public services	0.317
Government	1.256

are agriculture and in the case of other households cash agriculture and services. In general the remittances contribute for over 40% of total GDP creation for the Villa Alta area (Fig. 6).

4 Project Description

In order to assess project's effect on the targeted areas we hypothesized a typical IFAD investment programme of 20 USD million of which 7 million relates to ASAP

Table 5 Local SAM, alternative investment allocation

	Cafè	Maize	Avocado	Spring onion	Transport	Education and public services	Farmers (household)	No farmer household
Value added	1.776	1.806	1.806	1.815	0.455	1.334	1.417	0.880
Farmers (household)	0.556	0.565	0.565	0.568	0.143	0.417	1.444	0.276
No farmer household	1.223	1.244	1.244	1.250	0.314	0.919	0.977	1.607
Cafè	1.255	0.259	0.259	0.260	0.066	0.191	0.484	0.207
Maize	0.219	1.223	0.223	0.224	0.057	0.165	0.438	0.169
Avocado	0.067	0.068	1.068	0.069	0.017	0.050	0.091	0.071
Spring onion	0.064	0.065	0.065	1.065	0.016	0.048	0.075	0.073
Rest of agriculture	0.079	0.080	0.080	0.081	0.020	0.059	0.178	0.051
Mezcaleria	0.001	0.001	0.001	0.001	0.000	0.001	0.001	0.001
Oil	0.041	0.040	0.040	0.036	0.642	0.022	0.055	0.023
Energy	0.170	0.173	0.173	0.173	0.053	0.127	0.229	0.180
Telecommunication	0.168	0.171	0.171	0.171	0.066	0.235	0.144	0.206
Construction	0.022	0.023	0.023	0.023	0.012	0.024	0.019	0.025
Food and beverage	0.068	0.069	0.069	0.069	0.017	0.051	0.098	0.069

(continued)

Table 5 (continued)

	Cafè	Maize	Avocado	Spring onion	Transport	Education and public services	Farmers (household)	No farmer household
Accomodation and restaurants	0.013	0.013	0.013	0.013	0.003	0.009	0.010	0.016
Transport	0.027	0.027	0.027	0.027	1.087	0.020	0.058	0.017
Carpentry	0.044	0.045	0.045	0.045	0.020	0.032	0.039	0.053
Metalurgy	0.017	0.017	0.017	0.017	0.006	0.012	0.014	0.020
Hardware	0.057	0.058	0.058	0.059	0.018	0.035	0.047	0.053
Internet point	0.047	0.047	0.047	0.048	0.022	0.034	0.047	0.054
Beauty shop	0.003	0.003	0.003	0.003	0.001	0.002	0.002	0.004
Gas station	0.052	0.050	0.050	0.045	0.813	0.028	0.070	0.029
Clothing shops	0.049	0.050	0.050	0.050	0.013	0.037	0.039	0.064
Other shops	0.224	0.228	0.228	0.229	0.058	0.168	0.179	0.295
Repair services	0.030	0.031	0.031	0.031	0.048	0.022	0.044	0.029
Education and public services	0.269	0.273	0.273	0.273	0.155	1.285	0.240	0.251
Government	0.870	0.879	0.879	0.880	0.660	0.454	0.751	0.658

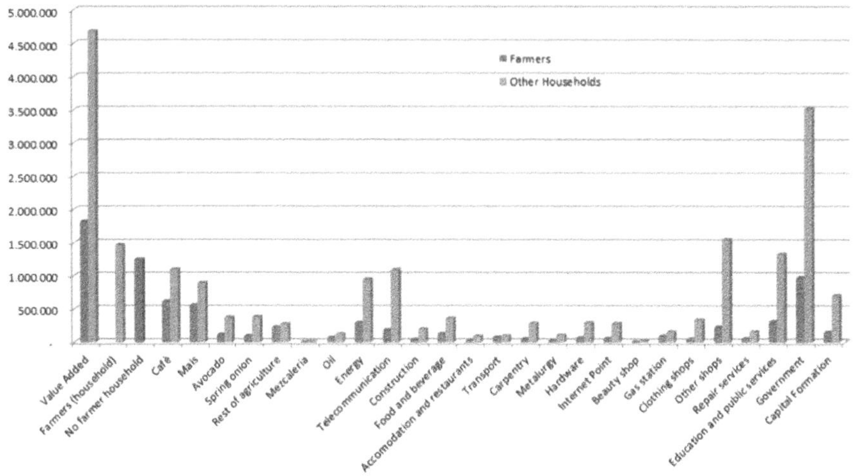

Fig. 6 Local impact of remittances

Table 6 Typical activities of a project with an ASAP component

Strengthening adaptive capacity of local institutions	Natural Buffer zones against climate extremes
Improving water resources	Livelihood diversification
Soil rehabilitation and protection	Improved processing and storage systems
Protection of communal infrastructure	Diversifying energy source
Climate information systems	Climate risk financing and transfer

contribution.[9] Programme investment and recurrent costs represent the expenditure vector[10] thanks to which we can estimate the short term impacts at both region and local level. The investment and recurrent costs considered in the analysis are: (i) Civil works; (ii) Goods and Supplies; (iii) Vehicles; (iv) Technical assistance; (v) Capacity building; (vi) Knowledge management; (vi) Salaries and Allowances.

The ASAP programme long term objectives represent the drivers upon which we have estimated the structural changes accrued to the targeted areas over a 10 years' time period after project implementation, vìs-á-vìs a traditional investment program lacking the ASAP component.

A typical project with an ASAP component consists of different activities and actions (Table 6).

In our investment project ASAP activities pertain to: (i) Strengthening adaptive capacity of local institutions; (ii) Improving water resources; (iii) Soil rehabilitation

[9]We calculated this amount as the average ASAP contribution to the IFAD investment portfolio in the year 2013/2014.

[10]Each expenditure item is classified accordingly to the SAM accounts and the NACE sector classification. The items are then reconciled and grouped as a vector.

Table 7 Investment and recurrent costs of alternative scenarios

	WP	WOP	WOP+
Labour	6.42	3.70	5.87
Agriculture	2.97	1.86	2.95
Forestry	4.95	4.00	6.35
Manufacturing	0.04	0.04	0.06
Construction	1.98	2.00	3.17
Social and personal services	3.65	1.00	1.59
Total	**20.00**	**12.60**	**20.00**

and protection; (iv) Natural Buffer zones against climate extremes; (v) Livelihood diversification.

Table 7 presents the investment and recurrent costs vectors related to three different scenarios:

1. ASAP (WP) for a total of 20 USD million;
2. Without (WOP) ASAP component for a total 12.60 USD million;
3. Without ASAP (WOP+) component for a total of 20 USD million.

Through these scenarios we would like to pursue a twofold objective of measuring short term incremental expected impacts on the economy, as the difference generated by two alternative projects (with and without ASAP), and simultaneously, to prove that the expected changes are not exclusively driven by budget amounts (Table 7).

4.1 *Short Term Effects on the Oaxaca Region*

4.1.1 A Direct Comparison Between WP and WOP

Estimates of the short term effects of the investment project on the Oaxaca economy are presented in Fig. 7. In the ASAP project scenario the results show an impact on value added equal to 50 USD million over an investment period of 5 years. In the WOP the value added impact is 30 USD million. In the production sectors the highest effect occurs in the Services account with a 43 USD million impact in the ASAP project and 25 USD million in the WOP (Fig. 8).

We can further analyse the different impacts on the productive sectors by dividing them into direct (expenditure) and indirect (multiplier) effects. In this specific investment scenario the sectors characterized by higher direct effects are associated with lower indirect effects. For instance, while on Agriculture and Forestry more than 40% of investment costs are spent, this initial spending accounts for only 19% of the total project impact (Fig. 10).

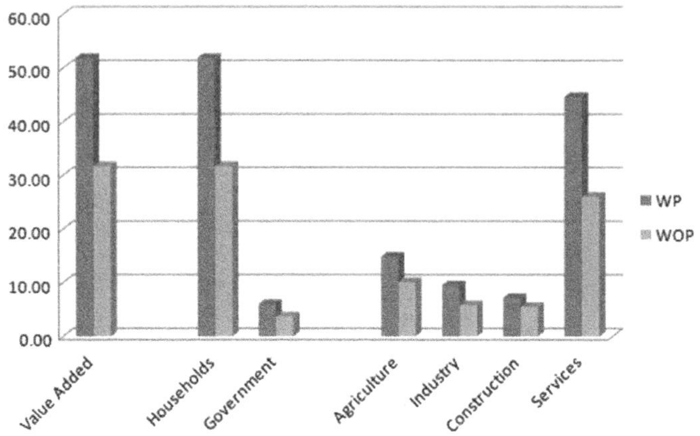

Fig. 7 Short term impact during investment period

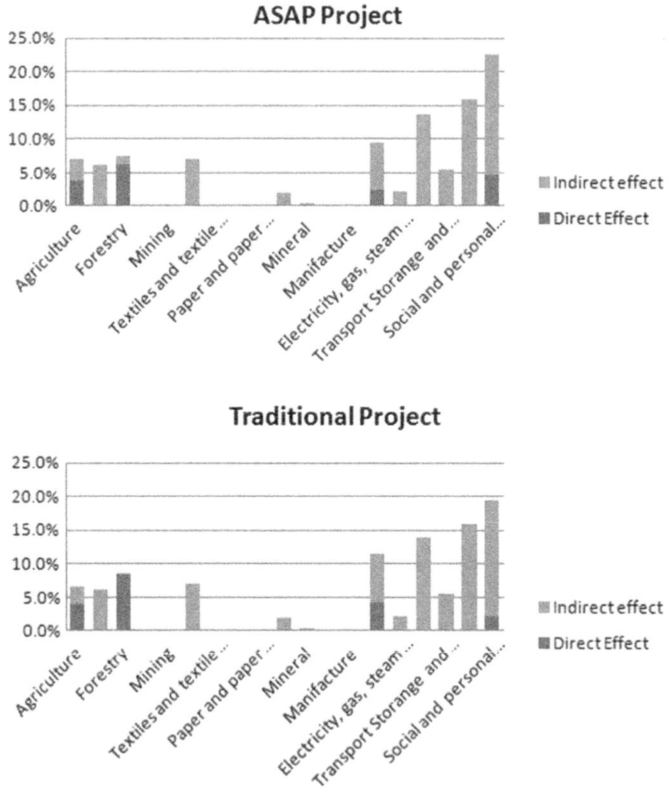

Fig. 8 Effects during the investment period in productive sectors

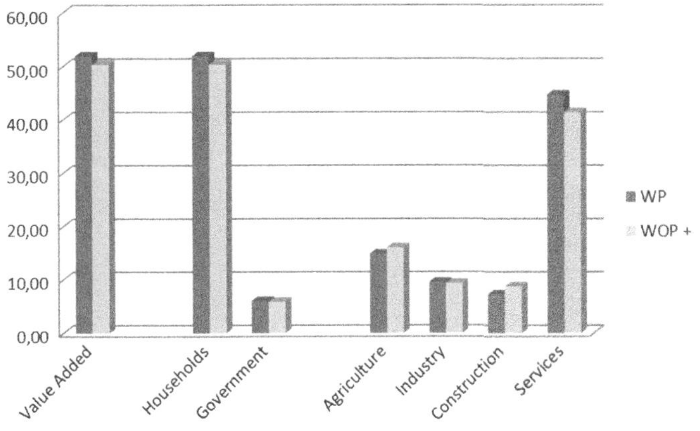

Fig. 9 Short term impact during investment period

4.1.2 A Direct Comparison Between WP and WOP+

Figure 9 shows the comparison between ASAP project (WP) and non-ASAP project (WOP+) for the same amount of resources.

In the ASAP project scenario the results indicate an impact on value added equal to 50 USD million over an investment period of 5 years. In the WOP+ scenario the value added impact is of about 49 USD million. In the productive sector the highest effect occurs in the Services account with a 43 USD million impact in the ASAP project and 41 USD million in WOP+.

In the midterm perspective, we consider production changes occurring in the sectors mainly affected by the programmes. The estimation is carried out over 10 years assuming an adoption timespan for the proposed interventions in line with what expected from the preliminary study of the project.

The likely effects on the Oaxaca State are measured as the difference between the development trends triggered after completion of the ASAP and non-ASAP project. In order to factor in the externalities related to climate change, we revised SAM's coefficients and multipliers, according with the Intergovernmental Panel on Climate Change (IPCC) long-term scenarios for the region[11] and the medium and long term OECD scenarios[12]. Table 8 summarizes the long term projections for Mexico.

The mid-term net effect, which is calculated as a cumulative difference of the two projects' trends, presents a growth pattern in Value Added and Natural sectors 15% higher for the ASAP *vis-à-vis* the non-ASAP, with a net gain for the Government of about 12% (Fig. 10).

[11]The IPCC scenario for Latin America are available at: http://www.ipcc.ch/ipccreports/tar/wg2/index.php?idp=45.

[12]OECD Economic Outlook 2014.

Table 8 Mexico long term scenario (%)

	2014–2030	2031–2060
Potential GDP	2.9	3.2
Potential GDP per capita	2.0	2.9
Trend productivity	0.9	2.4
Potential employment ratio	1.0	0.5

Source Author elaboration on public data OECD 2014

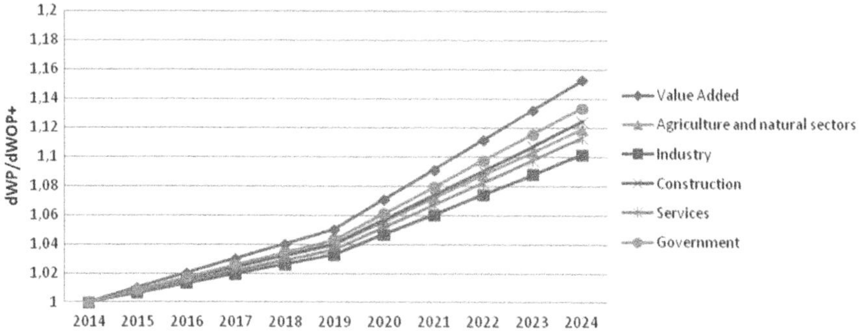

Fig. 10 Midterm growth difference in Oaxaca (WP—WOP+)

4.2 Impacts on the Local Economy

In order to downscale the analysis to the local level we reduced the expenditure vectors of the proposed projects, so as to estimate the share of project costs for each of the communities. Therefore, we assumed that 15% of total investment cost would be spent in the local economy. As shown in Fig. 11, in the short term the big bulk of the effects are concentrated in value added and agriculture sector. As mentioned in the previous section, the village rural economies in Oaxaca presented low level multipliers and the results on the short-term impact analysis confirms this characteristic.

During the five investment years, the ASAP project would generate and increase sector value added of about 5 USD million, 0.06 USD million more than the traditional project. In the productive sectors the impact would reach 8.17 USD million and 8.11 USD million respectively for the ASAP and the traditional project. The overall ASAP project impacts on the different sector determine a 31% increase on the local GDP.

The likely effects in the Oaxaca State are measured as the difference between the development trends triggered after completion of the ASAP and non-ASAP project. In order to factor in the externalities related to climate change, we revised SAM's coefficients and multipliers, according with the Intergovernmental Panel on Climate Change (IPCC) long-term scenarios for the region and the medium and long term

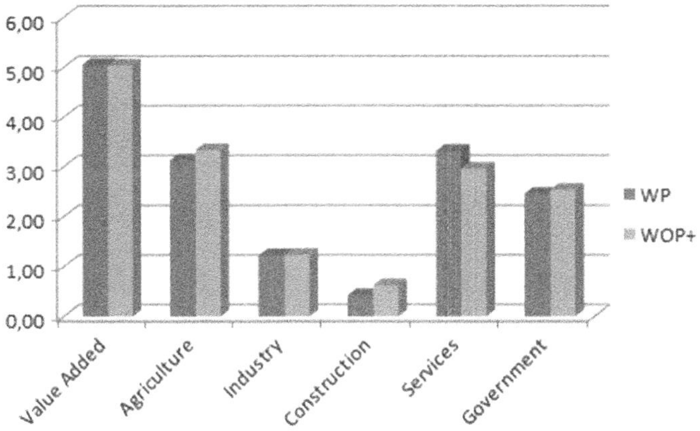

Fig. 11 Short term effect on local economy

OECD scenarios. For a more correct evaluation, we considered the different scenarios created after ASAP and non-ASAP implementation, within a nineteen year timeframe. The results show that in the standard project, production value of agriculture would increase of about 8% per year while the ASAP intervention would result in an increase of 12% per year.

These results notwithstanding, the most relevant results are foreseen in term of incomes of rural households. In fact, in the ASAP project their income would increase of 50% thanks to the knowledge acquired through the project on how to adapt to climate change. The following figures shows in summary the effect on Value Added, Households, Agriculture production, Industry, Construction, Services sectors and Government. The graphs depict the growth rates for each of the sectors with respect to the base year (Fig. 12).

5 Conclusions

The main objective of this study was to gain insights on whether, and under which conditions, a rural development project which includes Climate Change Adaptation strategies (as in the case of an ASAP investment) would trigger the economic growth in a region or in a group of villages.

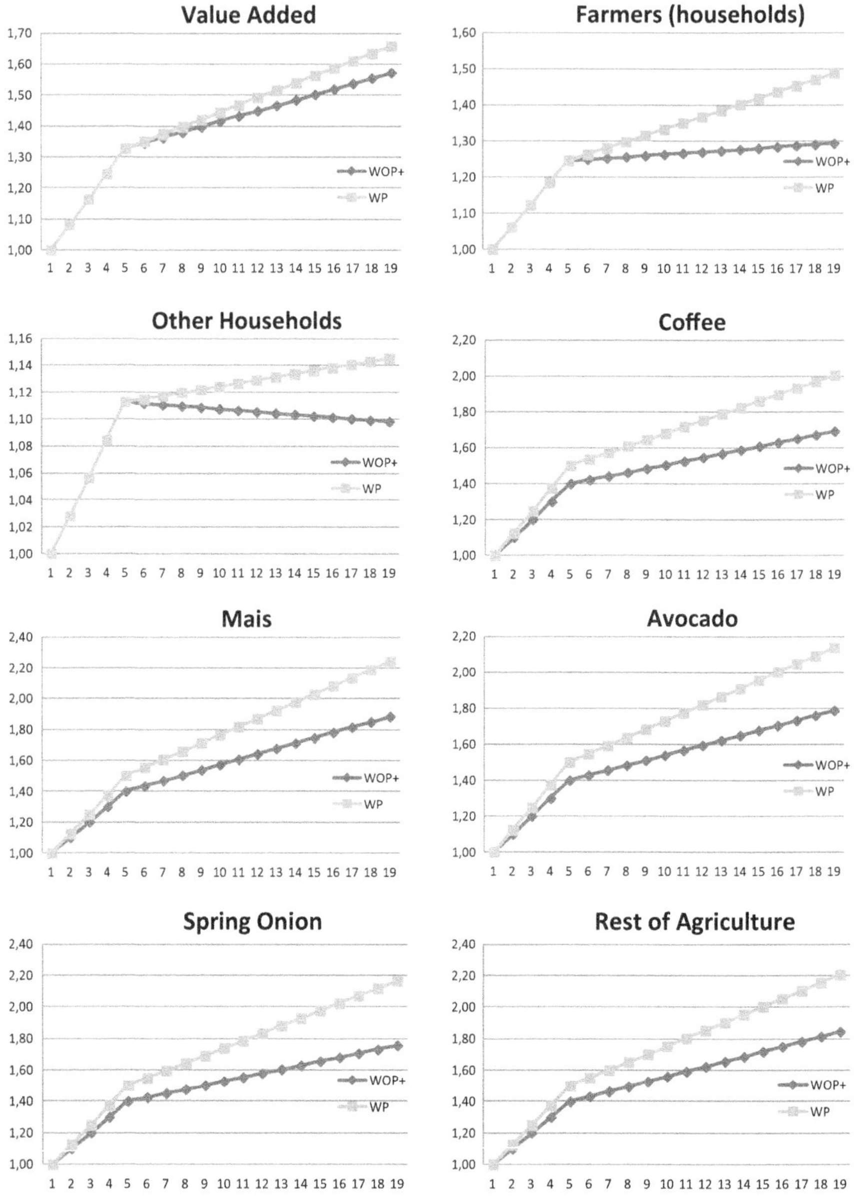

Fig. 12 Mid-term effect in local economy

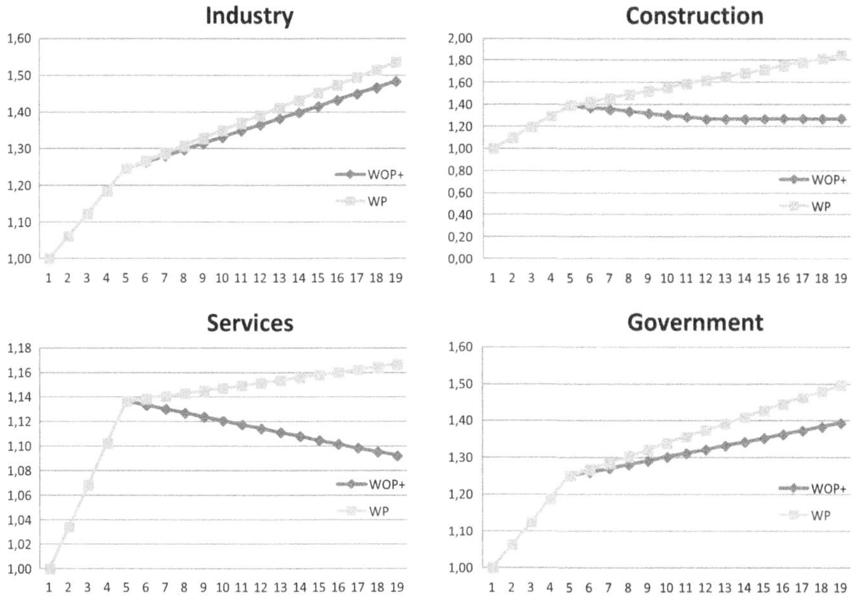

Fig. 12 (continued)

In particular, we applied an innovative methodology for estimating village-wide SAMs to make a direct comparison among three investment project scenarios (traditional investment project, ASAP project, and traditional investment project with total costs as ASAP project of reference). We therefore measure the extent to which a climate adaptation intervention would enable more stable paths of economic growth.

The geographic area under analysis is the district of Villa Alta in the region of the Sierra Norte, Oaxaca State (Mexico). In a first step of the analysis, we estimated the expected outcomes of the programmes in Oaxaca both in the short and medium term. In a second step, we develop a Village SAM to analyse the impacts of the three project scenarios at local level. Finally, we include a long term IPCC scenario to enhance the predictive capacity of our model over the medium–long range by factoring in climate change hazards for the region.

We believe that our results can usefully contribute evidence-based decisions on investments that aim to stimulate rural economic development and help develop strategies of adaptation to climate change. In the short term, we find some evidence of differences in impact between an ASAP and a traditional project, both regionally and locally. Differences however are smaller when we control for total project costs. Conversely, in the medium and long-term, the differences in impact between the scenarios are more evident, and could be explained in the light of the specific design features and components of a typical ASAP project and simulated through the changes in the SAM coefficients. ASAP projects in fact generally invest in strengthening relevant capacities and skills among the rural population, and thus can guarantee sustained growth even in the face of climate change phenomena. Indeed, thanks to the new knowledge acquired during the implementation of the project, farmers may apply new farming techniques, which in turn induce adaptive changes in the production structure of the local and regional agricultural sector. As results, farmers and the local, regional economies are better positioned to cope with climate change in the future.

Annex 1: Proposed Estimation Methodology for Village SAM (Scandizzo and Ferrarese 2015)

We propose to estimate the Village SAM with the methodology applied to the estimation of the system of regional social accounting matrices for Italy (Scandizzo 1993; Scandizzo et al. 2010). This methodology can be formalized as a problem of constrained maximization within the context of the generalized cross entropy (GCE) model proposed by Golan et al. (1996). In general terms, the estimation problem can be formulated as follows. Assume that a SAM is specified as a matrix of transactions between J sectors, factors and stakeholders. Consider each transaction (or, in normalized form, each coefficient) b_{ij} as the expected value of a random variable with support $[z_1, z_2, \dots z_M]$ and probabilities $[p_{1ij}, p_{2ij}, \dots p_{Mij}]$. The support values indicate the range of possible values for each coefficient. Since the SAM coefficients are shares of column totals, the interval of these values is comprised between 0 and 1. The corresponding range of the support values may be constituted, in the interval considered, by a discrete series of values or by a continuum. For simplicity, we assume that the first hypothesis holds and that it is possible to specify the same set of possible, but not equally probable, values M for each coefficient. Given a set of prior estimates q_{mij} of the probabilities associated to the possible values of each coefficient, posterior estimates can be obtained by solving the problem:

$$\max_{p_{mij} \geq 0} H = -\sum_m \sum_i \sum_j p_{mij} \log \frac{p_{mij}}{q_{mij}} \tag{1}$$

Under the constraints:

$$\sum_m p_{mij} = 1 \tag{2}$$

$$\sum_i \sum_m p_{mij} z_m = 1 \tag{3}$$

$$\sum_j \sum_m p_{mij} z_m v_{*j} = v_{i*} \tag{4}$$

where v_{*j} is the vector of the pre-defined column totals and v_{i*} the vector of the pre-defined row totals.

The objective function in (1) which is typically denoted as "cross entropy", in reality is not an entropy indicator, but the sum of the entropy measures, according with Shannon's definition (19448) for each column of the matrix and for each element of the probabilistic support $[z_1, z_2, \ldots z_m]$. More precisely, we can define as column entropy level for the m-th state of nature the function $H_{jm} = -\sum_i p_{mij} \log p_{mij}$. This function measures the quantity of information contained in the probability of each column for each state of nature as the logarithmic difference of the uniform distribution. When information is constituted only by the constraint that the probability sum must equal 1, the entropy is at a maximum, and the best estimate of the probabilities of the j-th column is that they are all equal to $1/M$. The entropy indicator thus measures the additional degree of information with respect to an informed prior distribution. If the analyst possesses a more informed prior, for example in the form of a prior probability q_{mij}, this can be incorporated in the logarithmic term of the entropy measure:

$$H_{jm} = -\sum_i p_{mij} \log \frac{p_{mij}}{q_{mij}} \tag{5}$$

Given a SAM, it will thus be possible to specify a different measure of entropy for each column (or each row) or even each value of the stochastic support z_m. The "cross entropy" is the sum of these row or column entropies and represents, not itself an entropy, but only one possible synthetic index of the entropy that can be associated to the SAM's rows and columns. Instead of a simple sum, in particular other weighting schemes can be used to reflect the different value that can be attributed to the information contained in a SAM according with the size or the variability of the flows, their statistical reliability and other special properties one may wish to consider.

Going back to the problem (1)–(4), the estimation of the coefficients b_{ij} is given by:

$$b_{ij} = \sum_m p_{mij} z_m$$

The corresponding Lagrangean is:

$$L = -\sum_m \sum_i \sum_j p_{mij} \log \frac{p_{mij}}{q_{mij}} \tag{6}$$

$$-\sum_i \sum_j \gamma_{ij} (\sum_m p_{ijm} - 1) - \sum_j \lambda_j (\sum_i \sum_m p_{mij} z_m - 1) - \sum_i \mu_i (\sum_j \sum_m p_{mij} z_m v_{*j} - v_{i*}) = 0$$

The Kuhn Tucker conditions for the solution of the problem (1)–(4), are given by the constraints (2), (4), assuming that they are binding and by the following expressions:

$$\frac{\partial L}{\partial p_{mij}} = \log \frac{p_{mij}}{q_{mij}} + 1 + \gamma_{ij} z_m + \lambda_{ij} z_m + \mu_i v_{*j} z_m \quad \text{or} \quad p_{mij} = 0 \qquad (7)$$
$$m = 1, 2 \ldots M;\ i = 1, 2 \ldots I;\ j = 1, 2 \ldots J$$

Solving for p_{mij}:

$$p_{mij} = q_{mij} \exp(-1 - \gamma_{ij} z_m - \lambda_j z_m - \mu_i z_m v_{*i}) \qquad (8)$$

Summing over m, we obtain:

$$\sum_m p_{mij} = 1 = \sum_m q_{mij} [\exp(-1 - \sum_m \gamma_{ij} z_m - \sum_m \lambda_j z_m - \sum_m \sum_j \mu_i z_m v_{*j})]$$
$$\sum_m p_{mij} = 1 = \sum_m q_{mij} [\exp(-1 - \gamma_{ij} z_m - \lambda_j z_m - \mu_i z_m v_{*i}] \qquad (9)$$

Implying:

$$\exp(1) = \sum_m q_{mij} [\exp(-\gamma_{ij} z_m - \lambda_j z_m - \mu_i z_m v_{*i}] \qquad (10)$$

And, substituting in (8):

$$p_{mij} = q_{mij} \exp[-z_m(\gamma_{ij} + \lambda_j + \mu_i v_{*i})] / \sum_m q_{mij} \exp[-z_m(\gamma_{ij} + \lambda_j + \mu_i v_{*i})] \qquad (11)$$

From (11) one can derive the estimate of a distribution of m matrices of $I \times J$ coefficients which are function of a prior value of the probabilities and the constraints' shadow prices:

$$b_{mij} = p_{mij} z_m = \frac{q_{mij} z_m [\exp(-(\gamma_{ij} + \lambda_j + \mu_i v_{*i}) z_m)]}{\sum_m q_{mij} [\exp(-(\gamma_{ij} + \lambda_j + \mu_i v_{*i}) z_m)]} \qquad (12)$$

With expected values:

$$b_{ij} = \sum_m p_{mij} z_m = \frac{\sum_m q_{mij} z_m [\exp(-(\gamma_{ij} + \lambda_j + \mu_i v_{*i}) z_m)]}{\sum_m q_{mij} [\exp(-(\gamma_{ij} + \lambda_j + \mu_i v_{*i}) z_m)]} \qquad (13)$$

In our experience a prior distribution q_{mij} may be typically characterized as a normal distribution with mean and variance equal to:

$$b_{ij}^0 = E b_{mij}^0 = \sum_m q_{mij} z_m, \quad i = 1, 2 \ldots I, \quad j = 1, 2 \ldots J \qquad (14)$$

$$Var(b) = E \left\| b_{mij}^0 - b_{ij}^0 \right\| \qquad (15)$$

This prior distribution is the distribution of non-balanced matrices derived from direct estimates of the totals from aggregating survey data, or using time series. The estimate proposed by Eq. (13), even though based on a constrained maximization, can be computed using a stochastic simulation and an iterative algorithm of the RAS type that re-proportions iteratively the columns and the rows of the matrix to estimate: The estimate can in fact be interpreted as an adaptation of an initial estimate proportional to a function of the expected value of the variable $x_{mij} = z_m \exp(-\gamma_{ij})$, to make this variable satisfy the constraints given by the sums of the rows and the columns.

References

Adelman, I., Taylor, J. E., & Vogel, S. (1988). Life in a Mexican village: A SAM perspective. *Journal of Development Studies, 25,* 5–24.

Bautista, V. H. (2008), *Análisis Estructural De La Economía Del Estado De Oaxaca, Un Enfoque De Insumo-Producto 1980, 1993, 1996, 2001*, Campus Montecillo.

Bell, C., Hazell, P., & Slade, R. (1982). *Project evaluation in regional perspective*. Baltimore: Johns Hopkins University Press.

Coleman, J. S. (1988). Social capital in the creation of human capital. *The American Journal of Sociology, 94.* (The University of Chicago Press).

Filipski, M., Manning, D., Taylor, J. E., Diao, X., & Pradesha, A. (2013). *Evaluating the local economywide impacts of irrigation projects: Feed the future in Tanzania*. IFPRI discussion papers 1247, International Food Policy Research Institute (IFPRI).

Golan, A., Judge, G. G., & Miller, D. (1996). *Maximum entropy econometrics: Robust estimation with limited data*. Wiley.

Leontief, W. W. (1966). *Input-Output economics*. Oxford: Oxford University Press.

Lewis, B. D., & Thorbecke, E. (1992). District-level economic linkages in Kenya: Evidence based on a small regional social accounting matrix. *World Development, 20*(6), 881–897.

Martinez Jimenez, M. A. (2012). *Analisis de los encadenamientos del sector forestal en el estado de Oaxaca*, Universidad de la Sierra Juarez.

Parikh, A., & Thorbecke, E. (1996). Impact of rural industrialization on village life and economy: A SAM approach. *Economic Development and Cultural Change, 44*(2), 351–377.

Ralston, K. (1992). *An economic analysis of factors affecting nutritional status of households in rural west Java, Indonesia.* University of California, Berkeley, California, USA.

Rodriguez, G., & Ramirez, A. S. (2009). *Matriz insumo-producto de Oaxaca y un análisis de su economía,* Revista Mexicana de Economia Agricola y de los recursos naturales.

Scandizzo, P. L., & Ferrarese, C. (2015). Social accounting matrix: A new estimation methodology. *Journal of Policy Modeling, 37*(1), 14–34.

Scandizzo, P. L., Ferrarese, C., & Vezzani, A. (2010). The social accounting matrix: a new estimation methodology. *Il Risparmio, 58*(3), 19–62.

Subramanian, A., & Qaim, M. (2009). Village-wide effects of agricultural biotechnology: the case of Bt Cotton in India. *World Development, 37*(1), 256–267.

Subramanian, S., & Sadoulet, E. (1990). The transmission of production fluctuations and technical change in a village economy: A social accounting matrix approach. *Economic Development and Cultural Change, 39*(1), 131–73 (University of Chicago Press).

Taylor, J. E., Adelman, I. (1996). *Village economies: The design, estimation, and use of villagewide economic models*. Cambridge University Press.

Taylor, J. E., Dyer, G. A., & Yunez-Naude, A. (2006). Disaggregated rural economy wide models for policy analysis. *World Development*, *33*(10), 671–1688 (Elsevier).
World Bank. (2015). *World development report 2015: Mind, society, and behavior*. Washington, DC: World Bank. https://doi.org/10.1596/978-1-4648-0342-0.

Part III
Static and Dynamic CGEs and Policy Applications

A CGE Model for Productivity and Investment in Kenya

Pasquale Lucio Scandizzo, Maria Rita Pierleoni and Daniele Cufari

Abstract This chapter develops an application of a CGE model to analyze some important economic features and policy problems for Kenya, one of the most dynamic African countries. The CGE model developed reflects the basic structure of the Kenya's economy and captures some of the key trade-offs affecting its policy choices, especially for what concerns aggregate growth, sustainability and inclusiveness. The results suggest that a policy strategy aimed to boost agricultural productivity and infrastructure investment would be the best choice for the long run development of the country.

Keywords General equilibrium · Total factor productivity · Agriculture Industry · Investment

JEL Codes C68 · D58 · D78

Parameters

ad_a	Production function efficiency parameter
aq_a	Shift parameter for composite supply (Armington) function
at_c	Shift parameter for output transformation (CET) function
$cwts_c$	Weight of commodity c in the CPI
ica_{ca}	Quantity of c as intermediate input per unit of activity a

P. L. Scandizzo
University of Rome Tor Vergata, Rome, Italy
e-mail: scandizzo@uniroma2.it

M. R. Pierleoni
Presidency of the Council of Ministers-Italian Government, Rome, Italy
e-mail: mr.pierleoni@governo.it

D. Cufari (✉)
OpenEconomics Srl, Rome, Italy
e-mail: daniele.cufari@openeconomics.eu

© Springer International Publishing AG, part of Springer Nature 2018

F. Perali and P. L. Scandizzo (eds.), *The New Generation of Computable General Equilibrium Models*, https://doi.org/10.1007/978-3-319-58533-8_6

$inta_a$	Quantity of aggregate intermediate input per activity unit
iva_a	Quantity of value-added per activity unit
mps_h	Share of disposable household income to savings
pwe_c	Export price (foreign currency)
pwm_c	Import price (foreign currency)
$qdtst_c$	Quantity of stock change
qbarinv(C)	Exogenous (unscaled) investment demand
$qinv_c$	Base-year quantity of private investment demand
$shry_{if}$	Share for domestic institution i in income of factor f
te_c	Export tax rate
tm_c	Import tariff rate
tq_c	Rate of sales tax
tr_{ii}	Transfer from institution i' to institution i
tva	Value added tax
ty_i	Rate of nongovernmental institution income tax
α_a^a	Efficiency parameter in CES function
α_a^{va}	Efficiency parameter in CES function for value added
β_{ch}	Share of commodity c in the consumption of household h
βtou_c	Share of commodity c in tourism consumption
δ_a^a	Share parameter in CES function
δ_{fa}^{va}	Share parameter for factor fin activity a, in value added CES function
δ_c^q	Share parameter for composite commodity supply (Armington) function
δ_c^t	Share parameter for output transformation (CET) function
θ_{ac}	Yield of commodity c per unit of activity a
ρ_a^a	CES function exponent
ρ_c^q	Armington function exponent
ρ_c^t	CET function exponent
ψ	Per capita consumption of tourist
σ_c^t	Elasticity of substitution for composite supply (Armington) function
σ_c^t	Elasticity of transformation for output transformation (CET) function

Variables

CPI	Consumer price index
$CDTOUR_c$	Tourists' consumption
EG	Government expenditures
EXR	Exchange rate
FSAV	Foreign savings
GSAV	Government savings
IADJ	Investment adjustment factor
PA*a*	Activity price
PD*c*	Domestic price of domestic output

PE*c*	Export price (domestic currency)
PM*c*	Import price (domestic currency)
PQ*c*	Composite commodity price
PVA*a*	Value-added price (factor income per unit of activity)
PX*c*	Aggregate producer price for commodity
QA*a*	Quantity (level) of activity
QD*c*	Quantity sold domestically of domestic output
QE*c*	Quantity of exports
QF*fa*	Quantity demanded of factor *f* from activity *a*
QFS*f*	Supply of factor f
QG_c	Government demand
QH*ch*	Quantity consumed of commodity c by household *h*
QINT*ca*	Quantity of commodity *c* as intermediate input to activity *a*
QINV*c*	Quantity of investment demand for commodity
QM*c*	Quantity of imports of commodity
QQ*c*	Quantity of goods supplied to domestic market (composite supply)
QVA	Quantity of value added
QX*c*	Aggregated marketed quantity of domestic output of commodity
–	Walras dummy variable
WF*f*	Average price of factor *f*
WFDIST*f*	Wage distortion factor for factor f in activity *a*
YF*if*	Transfer of income to institution I from factor *f*
YG	Government revenue
YI*i*	Income of domestic nongovernment institution
YT_{tt}	Tourists' income
OBJ	Object function (to maximize)

1 Introduction

Compared with structural econometric as well as simulation models used in economic policy analysis, computable general equilibrium (CGE) solve numerically abstract general equilibrium structure a la Arrow and Debreu with real data to find the equilibrium levels of supply, demand and price for specified markets. According to Wing (2004), CGE models are useful but they are nonetheless viewed with suspicion by some in the economics and policy analysis communities as "black boxes" (Panagariya and Duttagupta 2001), whose results cannot be meaningfully traced to any particular feature of their data base or input parameters, algebraic structure, or method of solution. Such criticism, mainly due to the lack of communication and information across broader economics and policy community, typically rests on the presumptions that CGE models contain a large number of variables and parameters and are structurally complex, both characteristics allowing questionable assumptions to be hidden within them that end up driving their results. Descriptions of models' underlying structure, calibration and solution methods abound, but tend to be spread across a broad cross section of materials, each subset of which focuses on a different aspect of the subject.

As explained in the introduction to this volume (Chap. 1), a recent revival of CGE models is based on several new facts and advancements of both theory and practice. CGEs have become the only point of encounter of macroeconomic policies with project evaluation, where they promise to perform a critical function to connect two frameworks that typically don't mingle and often risk to contradict each other. In a series of important research attempts, in large part conducted at the World Bank, several generations of computable general equilibrium models (CGE) since the late 70s were developed and gradually became important and useful tools for policy analysis. In these models, social accounting matrices (SAM) became the core of the representation of general equilibrium as a circular flow of production, consumption and incomes, with prices in all markets as the equilibrating variables. Solving algorithms started with fixed point (Scarf and Hansen 1973) and mathematical programming procedures (Norton and Scandizzo 1981; Walbroeck and Ginsburg 1981) and gradually developed into nonlinear equation systems and local or global search solution methods (Devarajan et al. 1997). At present, while the macro-econometric models prevailing in the 1970s have all but disappeared from the economic practice, CGEs are increasingly used around the world, both in their static and dynamic versions, as tools to analyze economic policy options.

In this Chapter we develop an application of a CGE to analyze some important economic features and policy problems for Kenya, one of the most dynamic African countries. We try to build a model that reflects the basic structure of the Kenya's economy and captures some of the key trade-offs affecting its policy choices, especially for what concerns aggregate growth, sustainability and inclusiveness. The plan of the Chapter is as follows: Sect. 2 provides a brief description of the Kenya economy and its recent trends and major development problems. Section 3 presents the mathematical structure of the CGE model and discusses its main assumptions and related characteristics. Section 4 reports and briefly discusses the estimates of Kenya's social accounting matrix (SAM). Section 5 presents the model simulations and discusses their implications from the point of view of the economic policy problems examined. Section 6 finally develops some conclusions and policy recommendations.

2 The Kenya's Economy

Kenya is a sizable country (580.400 km^2 with a population of 44 million) and an income per capita of about 1,400 US$ at the official exchange rate. According to the Kenya Economic Update (World Bank 2017), for the third consecutive year, economic activity gave rise to sustained economic growth. Kenya's economy expanded by 5.8% in 2016, 0.1% points higher than the previous year, in spite of a background of weaknesses in several emerging markets and Sub-Saharan economies where GDP growth decelerated. Unlike oil exporting countries, Kenya, being an oil importer, benefitted from the slump in oil prices, particularly in the first half of 2016.

Similarly, earlier good rains supported favorable harvests in 2016, particularly in the first half of the year. Further the tourism sector, which had slowed down since the 2013 terrorist attacks, rebounded in 2016. Finally a positive role was played by domestic developments such as the government's infrastructure drive aimed at easing supply side constraints and a stable macroeconomic environment, supported economic activity in 2016. These favorable compensated for the weakness in external demand and the sharp deceleration in credit growth to the private sector.

The service sector contributed 3.2% points to Kenya's GDP growth for the first three quarters of 2016; in other terms, some 54% of Kenya's growth in 2016 derived from the strength of the service sector. Performance among various service sub-sectors was, however, mixed. For examples thanks to the rebound in the tourism sector, accommodation and restaurant sub sector contributed to some 0.41% points to GDP growth; transport and storage also accelerated, as they benefitted from lower fuel prices. In contrast, Kenya's real estate sector presented a deceleration in 2016, that could be reflective of the slowing private sector credit growth. Similarly, in 2016, the financial sector contributed only 0.3% points to GDP growth compared to its contribution to GDP of some 0.6% points in 2015. The decline in the contribution of the financial services is consistent with tougher environment faced by Kenyan banks in 2016 as a tighter regulatory condition for the provisioning of bad debts and lower interest margins resulting from the Banking Amendment Act.

Agricultural output grew at 4.9% in first three quarters of 2016, the sector's contribution to growth increasing by 0.2% points from that of the 2015. For the first three quarters of 2016, Kenya's industrial sector expanded by 5.6% but the sector's contribution to GDP growth decelerated to 1.6% points from 1.8% points over the same period in 2015. Much of this deceleration in growth can be attributed to sluggish/below par growth in the manufacturing sector and lower dynamism in the construction sector. A key question for the Kenyan economy thus appears to be the role of total factor productivity and the consequences of its increase in productivity agriculture versus the other sectors. Given the weight of agriculture and demographic pressures for Africa, this question appears important to identify goals for technological innovation and diffusion, as well as to suggest alternative strategies of growth.

The macroeconomic environment was stable in 2016; in particular inflation was moderate in 2016. However unfavorable weather has led to a surge in food inflation in recent months. The fiscal deficit declined from 8.4% of GDP in Financial Year (FY) 14/15–7.5% in FY 15/16. Kenya's medium-term fiscal policy is anchored by its commitment to achieve convergence with the East African Community Monetary Union protocols. In recent years, the government has embarked on an ambitious infrastructure plan (roads, railways, ports and power projects) that drove the share of development spending to 8.8% of GDP in FY 14/15 from 6.3% a year earlier. However, in FY 15/16 development spending was moderate, thereby supporting the commencement of the fiscal consolidation. In contrast to development spending, recurrent spending increased to 15.6% of GDP in FY 15/16. Fiscal consolidation should help to: (i) anchor Kenya's macro stability, (ii) reduce crowding out pressures, (iii) contain the pace of debt accumulation and (iv) contribute towards a more favorable sovereign debt credit rating

However, in contrast to the consolidation that took place in FY15/16, the fiscal deficit is projected to rise to 8.9% of GDP in FY16/17. Given the projected increase in revenues (as a share of GDP), the increase in the deficit is being driven by an expansionary fiscal stance, with government expenditures increasing from 27.1% of GDP in FY15/16 to 30.0% in FY16/17. The significantly higher deficit, however, assumes that there will be a full execution of the development budget in FY16/17. Given the track record of 31% under-execution rate for development spending, deficit turnouts could be lower than current projections. There is a need to recreate fiscal space through reductions in the share of recurrent spending, and expansion of the revenue base in order to carry out the ambitious public investment drive without straining public finances.

From the point of view of development policy, the role of investment identifies a second policy question for the future of the Kenyan economy. This question concerns the trade-off between fiscal consolidation and growth faced by policy makers in the short run, even under the favorable assumption that the economy proceeds on a virtuous path of productivity improvement. While such an improvement may endogenously generate enough resources to fuel further growth, it is legitimate to ask whether an aggressive policy of investment in public goods, such as the one pursued by the government in recent years, might not be important to ensure stale support to a higher path of development for the country.

3 The Core CGE Model for Kenya

The CGE model is based on a social accounting matrix that provides a schematic portrayal of the circular flow of income in the economy: from activities and commodities, to factors of production, to institutions, and back to activities and commodities again. In particular, the equations of the core CGE model follows the same pattern of income generation of the SAM. These equations can be grouped in the following blocks: (1) equations which define the price system, (2) equations that describe production and value-added generation, (3) equations that describe the mapping of value added into institutional income, (4) equations which completed the circular flow, showing the balance between supply and demand for goods by the various actors, and (5) a number of "system constraints" that the model economy must satisfy; these include both market clearing conditions and the choice of macro "closure" for the model (Robinson et al. 1999).

The core of the model follows a standard structure (Robinson et al. 1999; Lofgren et al. 2002) based on CGE model specified in terms on non-linear algebraic equations and numerical solution techniques (Devris et al. 1982). While the model is designed as a neoclassical structure, different closure rules may be used to incorporate Keynesian hypotheses and mechanisms of income formation and to analyze differences in policy implementation (as explained later). In keeping with the private market orientation of the Kenyan economy, the core of the model is a process of maximization of profits by producers and utility by households. Labor is assumed to be mobile, markets are competitive within an open economy, and international trade and tourism.

Technology for producers and preferences for consumers are described by Cobb-Douglas functions and consumption demands are derived from the optimization process. Commodities are either sold in the domestic markets or exported to international markets. A constant elasticity of transformation function (CET) describe the relationship between the internal and external markets, with the determination of price ratios and elasticities of transformations to determine the level of output exported or sold domestically.

For imports, households and producers are assumed to utilize commodities based on Armington's composite commodity function, which describes the substitutions between imports and domestic commodities through a constant elasticity of substitutions (CES) function. The government's inflows are represented by taxes and transfers from other institutions and at the same time use the income to purchase commodities, make transfer to other institutions and savings. The commodities demanded by government are determined in fixed proportion and transfers from and to other institutions are also fixed in foreign currency. Enterprises are also included in the model as institutions, and receive inflows from factor of production and transfers from other institutions. As outflows, enterprises' incomes are used to pay taxes, savings and transfers but not to consume commodities. International tourists are also represented as institutions, that receive as inflows incomes from the rest of the world and consumes commodities and savings domestically.

The CGE model incorporates all the flows from the Social Accounting Matrix (production, consumption, distribution etc.) and simulate the product and factor markets role in setting equilibrium relative prices. Depending on a number of factors and the purpose of which the model simulation is used, model closure consists of choosing a particular set of exogenous variables in a way that allows a consistent and possibly unique set of solutions. Because a problem of over-determination (Sen 1963; Ratso 1982) may arise when the number of equations implied by the model exceeds the number of endogenous variables, we refer to four basic closure rules: the neoclassical, the Keynesian, the Joansen and the neo-Keynesian. In *the Neoclassical closure* investment is endogenous and savings driven (i.e. saving determine the level of the endogenous investment that adjust consequently). The *Keynesian closure* is characterized by unemployment in equilibrium, hence the level of employment is not fixed and variation in the level of output and employment will clear the market of saving and investments. In the *Joansen-closure*, the model is investment-driven, hence the level of savings adjusts, differently from the neoclassical closure as Johansen considers the government as an important source of savings. Government consumption or tax rates become endogenous and savings depend on tax rate and adjust to ensure the saving-investment gap. Finally, in the *neo-Keynesian* closure, the real wage is not equal to marginal product of labor and the functional distribution of income ensures the equality between savings and investments.

For an open economy with trade and international tourism the closure problem becomes more complex, with the introduction of a new equilibrium condition in the foreign exchange rate and new source of savings in the investment—savings balance

(Ratso 1982; Delpiazzo 2011; Robinson 2006). In the case of the Neoclassical closure, foreign savings are assumed to be fixed and the real exchange rate is fluctuating to ensure the equilibrium on the foreign market. Investment is saving-driven and the model behaves in a way similar to the closed economy (Robinson 2006). In the Keynesian closure usually foreign savings is assumed to be fixed exogenously, while the exchange rate adjusts to clear the foreign exchange market. As the exchange rate varies, real prices (including the wage rate) will adjust, generating employment, income, production and savings to match fixed investments. Since foreign savings are fixed, they have no role in the adjustment and the multiplier is operating similarly to the closed economy case (Robinson 2006; Taylor and Lisy 1979). The Johansen closure in the open economy case is characterized by saving driven investment as in the closed economy, but foreign savings are endogenous and adjust to ensure investments—savings balance (and not the domestic savings as the closed economy case). A change in the level of investments will adjust the real exchange rate that is the equilibrating variable and generate changes in foreign savings. These in turn adjust to investment levels. Furthermore, the model assumes that the wage rate is free to vary and ensure equilibrium in the labor market. In the neo Keynesian closure, a fixed wage is the numeraire, while the exchange rate is exogenous and foreign savings adjust. Changes in the real wage provoke adjustments in price level and exchange rate. If for example the price level increases, the real wage decreases and employment, income and savings all increase. On the foreign market the real exchange rate appreciates, with a consequent deterioration of the balance of trade and increase in foreign savings. The increment in both foreign and domestic savings ensure macro equilibrium, so that the investments level end the effect of the Keynesian multiplier is lower than that of standard Keynesian closure.

4 Social Accounting Matrix and Computable General Equilibrium for Kenya

Several social accounting matrices are available for Kenya. The 2003 Kenya Social Accounting Matrix (SAM), estimated by the Kenya Institute for Public Policy Research and Analysis (KIPPRA) and the International Food Policy Research Institute (IFPRI), is the main example of a family of models both at national and regional level, built in the past decade. This matrix is a consistent data framework that captures the information contained in the national income and product accounts and the input-output table, as well as the monetary flows between households, government and other institutions. The Kenya SAM also used surveys to estimate the production technology underlying different sectors of the economy. By combining this information with the country's household income and expenditure survey, the SAM provides a comprehensive picture of the structure of the Kenyan real economy built in 2003.

Using the 2003 KIPPRA-IFPRI SAM as a starting point, we updated the SAM for Kenya by applying the entropic methodology described in Scandizzo and Ferrarese (2015), to national and state account data and other statistics (Kiringai et al. 2006; KNBS 2014). These methods are based on the so called "maximum entropy econometrics" (Golan et al. 1996) and are able to handle the "ill-conditioned" estimation problems associated with the lack of the degrees of freedom typical of I-O matrices. The methods are very flexible in combining a variety of specific data with prior information and national accounts.

The matrix estimated contains detailed sector accounts for production, sales and purchase of goods and services. In total there are 35 production sectors, with primary activities including irrigated and non-irrigated agriculture as well as forestry and mining. Tourism is also detailed both from the point of view of various types of tourist demand and hotel and lodging supply. Three factors of production (labor, capital and land) are accounted for in connection with productive sectors and households. The latter are divided into four categories: namely rural poor, rural non-poor, urban poor and urban non-poor, using standard poverty level.

The SAM distinguishes between 'activities' (the entities that carry out production) and 'commodities' (representing goods and non-factor services exchanged on the market). SAM flows are valued at producers' prices in the activity accounts and at market prices (including indirect commodity taxes and transactions costs) in the commodity accounts. The government is disaggregated into a core government account and different tax collection accounts, one for each tax type. Taxes are disaggregated into commodity, direct and trade taxes, plus a core government account. Enterprises institutions are also considered, as well as the capital account (savings—investments) that comprehends all formal and informal transactions concerning the various forms of credit in the economy, including transactions from the formal banking system and all financial transactions that play a crucial role to supply an outlet to savings and a source of credit to consumer-producer households. The rest of the world account represents trade flows between the national economy and rest of the world, such as imports and exports of goods and services.

Figure 1 and Table 1 shows the Rasmussen indexes of backward and forward linkages computed from the matrix. These indexes describe the direct and indirect connections between the different actors of the economy, whose accounts are represented in the SAM. Introduced by Hirschman and defined by Rasmussen (1957), the indexes of backward linkages are based on the average multipliers (from the columns of the SAM inverse) and can be interpreted as the increase in output of the entire system of industries needed to cope with an increase in the final demand for the products of one industry by one unit (Rasmussen 1957, pp. 133–134). The indexes of forward linkages are instead based on the row multipliers and quantify the extent to which the system of industries draws upon a given industry. They are indexes of sensitivity of dispersion, as they measure the increase in the production of an industry driven by a unit increase in the final demand for all industries in the system. Both indexes are normalized by dividing the average multiplier for each sector by the total average multiplier for all sectors.

The magnitude of the multipliers depends on the number of the accounts considered exogenous and are lower the larger such a number, while the Rasmussen indexes, being normalized with the average multipliers, indicate only the relative

Fig. 1 Rasmussen backward and forward indicators. *Source* Our elaboration

importance of a sector linkage as compared to the mean. The multipliers used in the table correspond to the hypothesis that the capital formation account is exogenous. Under this hypothesis, as the table shows, the Rasmussen indexes of backward multipliers, which average 1 by construction, range from a minimum of 0.54 for metal and machines account to a maximum of more than 1.12 for sales taxes, indirect taxes and tariffs. This means that if the demand of one sector increases 100%, the average impact on the demand for the products of the other sectors is between a minimum of 54% and a maximum of 112% the average. The results show that the country enjoys a stronger than average backward connectivity for many sectors, like tourism, which are at the end of their value chain. The indexes of forward linkages, on the other hand, measure the degree of participation of each sector/institution to the overall economic activity, that is, on average, how much a sector demand increases in response to an equi-proportional increase in all sectors. They are much more diverse than the backward indexes, with especially large values for some sectors such as non-irrigated agriculture, transport and especially labor and capital. The lowest value of 0.06 is for

Table 1 Comparison of backward and forward linkages related to different elaborations the of Kenya SAM

Our elaboration of Kenya social accounting matrix			Elaboration of Kenya social accounting matrix (WB Report, 2017)			Elaboration of Kenya social accounting matrix (Wanjala and Were 2009)		
	B	F		B	F		B	F
Irrigated agriculture	1.05	0.25	Irrigated agriculture	1.26	0.36			
Non irrigated agriculture	1.05	2.83	Non irrigated agriculture	1.29	3.77	Agriculture	3.43	8.42
Forestry and fshing	1.06	0.20	Forestry and fishing	1.08	0.34	Fishing	3.15	1.14
						Forestry	3.4	1.23
Poaching	0.90	0.06	Poaching	1.08	0.12			
Mining	1.03	0.12	Mining	1.11	0.22	Mining	3.33	1.07
Food, beverage and tobacco	1.05	1.27	Food, beverage and tobacco	1.73	0.86	Beverage and Tobacco	2.64	2.77
Petroleum	0.92	0.21	Petroleum	0.86	0.31	Petroleum	1.96	5.21
Textile and clothing	1.06	0.20	Textile and clothing	0.85	0.31	Textile and footwear	2.8	2.73
Leather and footwear	0.99	0.11	Leather and footwear	0.98	0.16			
Wood and paper	0.93	0.25	Wood and paper	0.76	0.39	Wood and paper	3.2	1.15
Printing and publishing	0.83	0.56	Printing and publishing	0.50	0.71	Printing and publishing	2.59	2.34
Chemicals	0.56	0.48	Chemicals	0.54	0.70	Chemicals	1.84	3.09
Metals and machines	0.54	0.25	Metals and machines	0.47	0.32	Metals and machines	1.56	2.63
Non metallic products	1.04	0.13	Non metallic products	1.04	0.19	Non metallic products	2.97	1.61
Other manufactures	0.84	0.68	Other manufactures	0.76	0.84	Other manufactures	2.45	3.64
Distribution water	1.06	0.07	Distribution water	1.57	0.45	Electricity and water	3.11	2.12
Electricity	1.01	0.42	Electricity	0.93	0.57			
Construction	0.98	0.13	Construction	0.93	0.16	Bulding and construction	3.24	1.19
Trade	1.05	1.50	Trade	1.01	2.26	Trade	3.64	7.11
Hotel	1.10	0.13	Hotel	0.89	0.18	Hotel and restaurants	3.14	3.65
Lodge	1.09	0.08	Lodge	0.14	0.87			
Rent house	1.09	0.09	Rent house	0.15	0.87			

(continued)

Table 1 (continued)

Our elaboration of Kenya social accounting matrix			Elaboration of Kenya social accounting matrix (WB Report, 2017)			Elaboration of Kenya social accounting matrix (Wanjala and Were 2009)		
	B	F		B	F		B	F
Restaurant	1.10	0.13	Restaurant	0.87	0.18			
Park Tourism	1.08	0.12						
Beach Tourism	1.08	0.08						
Cultural Tourism	1.08	0.07						
Business Tourism	1.08	0.09						
Transport	0.91	1.52	Transport	0.93	2.02	Transport and communication	3.23	8.01
Information and communication	1.05	0.56	Information and communication	1.00	0.78			
Financial and insurance activities	1.03	0.95	Financial and insurance activities	0.94	1.35	Financial services	3.33	4.52
Real estate	1.03	0.93	Real estate	0.99	1.16			
Other services	1.05	0.56	Other services	0.97	0.77	Other services	3.30	3.92
Public administration and defence	1.02	0.82	Public administration and defence	0.27	0.90	Administration	3.29	1.19
Health and social work	1.05	0.35	Health and social work	1.02	0.32	Health	3.91	1.60
Education	1.06	0.92	Education	1.02	0.60	Education	3.62	1.67
LAB	1.01	4.31	LAB (skilled + semi-skilled + unskilled)	3.87	6.58			
CAP	0.99	5.08	CAP	0.89	6.58			
rp-hhd	0.95	1.60	rp-hhd	1.34	2.11			
rnp-hhd	0.95	2.46	rnp-hhd	1.33	3.22			
up-hhd	0.95	0.22	up-hhd	1.36	0.30			
unp-hhd	0.94	5.76	unp-hhd	1.16	6.37			
ent	0.93	5.49	ent	0.83	6.59			
gov	1.06	2.00						
stax	1.12	1.09	stax	0.12	1.73			
ytax	1.12	0.69	ytax (Direct taxes)	0.13	0.97			
tar	1.12	0.15						
h-tour	1.04	1.00	h-tour	1.19	0.39			
l-tour	0.99	1.00						

Source Our elaboration

poaching activity, while the highest value of 5.08 is for capital. Tourism activities have backward multipliers near to the average, but low forward multipliers. This suggests on the backward side that the local value chain, even though still weak, has already some depth, and, on the forward side, that the sector is not dependent on domestic economic activity and relies mostly on foreign demand.

5 Impact Analysis: Policy Simulations

Because of the importance of the agricultural sector in most developing economies, raising agricultural productivity appears a plausible and appealing choice for policy makers to promote economic growth. The literature provides abundant theoretical and empirical evidence that agricultural growth is essential to foster overall growth, especially in developing countries and identifies the diverse roles that agriculture plays in the process of growth and development as well. For example, for Johnston and Mellor (1961) agriculture contributes to economic development with food and raw materials, labor and capital, foreign exchange and markets for the outputs of other sectors. Agricultural productivity growth would generate increased demand not only for food but also for other industrial outputs and services via intermediate and final demand linkages (Adelman 1984; Mellor 1976). Bautista (1986) identifies increased agricultural production through productivity increase result in foreign exchange savings and reduction in food imports and increase the ability to export. Further, increased agricultural productivity may cause lower and more stable food prices making households better off (Adelman 1975; de Janvry and Sadoulet 2002).

It is important to underline the distinction between increases in output and productivity since these do not necessarily have similar impacts. In some cases, output and productivity increase together whereas in other cases they can vary inversely with differential consequences for poverty (Irz et al. 2001; Schneider and Gugerty 2011).

The effects of agricultural growth spread to the non-farm economy through different linkages; production, employment and incomes. Higher real incomes in the agricultural sector stimulate demands for the products of other sectors and labor within the sector, while higher agricultural outputs stimulate the creation of non-farm rural and urban employment opportunities through backward and forward linkages to manufacturing and services sector activities (Hanmer and Naschold 2000; Thirtle et al. 2001). Irz et al. (2001) summarize and review many possible arguments of effects of agricultural productivity growth on farm economy, rural economy as a whole and the national economy and the necessary conditions to achieve them. It is not clear that rural income will increase at all times with improvements in agricultural productivity, as a result of possible deterioration of agricultural terms of trade arising from price and income inelasticity of agricultural products (Bautista 1986). Arndt et al. (2000) suggest that price declines due to an increase in agricultural productivity would transmit most of the gains to urban households, to non-agricultural sectors and to non agricultural factors of production. Rural households who mostly engage in agricultural activities gain from greater availability of food. Further, Thirtle et al. (2001) argue that productivity gains may not trigger poverty reduction if the decline in output prices outweighs the gain from increased productivity. These complex relationships between direct and indirect general equilibrium effects emphasize the linkages between agricultural productivity, growth and poverty reduction. While many studies of productivity or technical change in developing country agriculture have been conducted (Mellor 1999; Self and Grabowski 2007; Thirtle et al. 2003), linkages of the agricultural sector with the rest of the economy have been the object of only a limited number of studies.

Studies investigating the multiplier effects of agricultural growth on the other sectors of the economy (Arndt et al. 2000; Bautista 1986; Coxhead and Warr 1991, 1993, 1995; Dorosh et al. 2003) include social accounting matrices and CGE models. Arndtet al. (2000), for example, use a CGE model to analyze improvements in agricultural productivity and reductions in marketing costs in Mozambique. Their results suggest that that increasing agricultural productivity may be a priority for Mozambique with large potential gains. However, increasing agricultural output with very high marketing costs leads to significant fall in prices transmitting most of the gains in factor income to non-agricultural sectors. Bautista (1986) developed a CGE model to investigate the effects of productivity increases in Philippine agriculture. The study simulates the impacts of productivity increases in three agricultural sectors; food crops, export crops and livestock and fishing sectors and the food manufacturing sector on sectoral prices and outputs, rural and urban income, trade balance and national income. The simulations imply a 10% increase in total productivity separately in the four sectors and increased productivity in all sectors simultaneously. The cause of the productivity increase is assumed to be the result of technological change and/or improved infrastructure. Increased productivity in the food crops sector results in a fall in food prices but promotes the food processing sector. Productivity improvements in the crop sector results in a decline in sector prices while improving sector production. Increased productivity in the food manufacturing sector stimulates growth in production and in the food crop sector as well. Simultaneous productivity increases in all four sectors show moderate positive impacts on household income while there are significant impacts on macroeconomic variables. Based on those results, the author argues that increasing agricultural productivity does not necessary result in reduced rural income but is more likely to benefit urban households.

Coxhead and Warr (1991) used a CGE model for Philippines to investigate the distributional effects of technical progress in Philippine agriculture. They show, in a small open economy, that technical improvements in farming are likely to benefit the poor, especially if the technical change is labor-using—land-saving. A technical change which substitutes capital for labor with no increase in output in irrigated agricultural sector triggers a reduction in real wage in the same sector. Households owning only labor lose while real incomes of households that do not depend on labor show a slight increase. Coxhead and Warr (1995) used the same model to trace the effects of differential rates of technical progress in the irrigated and non-irrigated agricultural sectors on income distribution of factor owning household groups, poverty and economic welfare within a small open economy with open agricultural trade and agricultural trade under restrictions. The results clearly showed that reduced poverty from technical progress is substantially greater when agricultural trade is unrestricted at constant world prices. Similar results are obtained by Coxhead and Warr (1993), who examine the distributional effects of technical change in Philippines' agriculture using a CGE model. They show that technical change in Philippine agriculture may lead to increased incomes, reduced poverty and improved income distribution.

Given this background of past studies and results, we propose a series of simulation with the CGE model that combine increases in productivity of agriculture and, alternatively, in the industrial sector. with investment increases. These simulations aim to measure the potential growth spill overs of technical change in agriculture and industry. They also aim to quantify the link between investment and productivity increase, since one of the most important reasons to invest in infrastructure is precisely to induce productivity growth and, on the other hand, any exogenous increase in productivity needs to be accommodate by further investment to spread to the rest of the economy. More specifically, we simulate the following scenarios:

(a) doubled agricultural total factor productivity (TFP) with a parallel increase in investment (50%) with a Keynesian closure;
(b) increased investment (50%) with a Keynesian closure.
(c) doubled agricultural TFP with a Neoclassical closure.
(d) doubled industrial TFP with a parallel increase in investment (50%) under a Keynesian closure;
(e) doubled industrial productivity with a Neoclassical closure.

The simulations are made with the GAMS (General Algebric Modelling System) software described in Brooke et al. (1996).

Tables 2 and 3 and Fig. 2 present a summary of the results obtained in the simulation scenarios mentioned above, in terms of real increases for the key economic variables of the model. The simulations representing a Keynesian scenario assume wage rigidity and exogenous investment, while in the neoclassical simulations wages are free to vary and investments are endogenized. In the former scenarios almost all variables are higher. The agricultural growth simulations also suggest that increases in agricultural TFP would be more beneficial, ceteris paribus, from the point of view of production, factor income and income redistribution. Furthermore, the impact of industrial TFP growth combined with the exogenous investment stimulus in the Keynesian scenario, would have rather modest effects and would display the greatest difference between relatively large benefits to factor remunerations and GDP increases, compared to rather low and uniform benefits to personal (disposable) incomes. These results are in line with empirical literature mentioned above where the authors found in a neoclassical economy a welfare gains from agricultural productivity increasing, but the differences of performance displayed between agriculture and industry TFP growth are a somewhat novel finding of our study. They suggest the intriguing hypothesis that a stage-wide pattern of growth may be mandated by the very structure of a developing economy.

In detail, simulations I, III, IV and V show the potential benefits of technical productivity change respectively in agriculture (scenarios I and III) and industry

Table 2 Agricultural TFP and investment increase simulations (Mln US$)

	Base case	Simulation I (Agricultural productivity doubled and investment increasing of 50%) Keynesian closure	Var. w.r.t. BC* (%)	Simulation II (investment increasing of 50%) Keynesian closure	Var. w.r.t. BC* (%)	Simulation III (Agricultural productivity doubled) Neoclassical closure	Var. w.r.t. BC* (%)
Absorption	56,246.90	74,333.90	32	68,625.10	22	70,893.80	26
Private consumption	42,208.70	56,325.20	33	50,616.40	20	54,679.20	30
Fixed investments	7,941.20	11,911.80	50	11,911.80	50	10,117.70	27
Government income	9,231.21	11,883.24	29	12,126.76	31	11,115.66	20
Exports	943.80	1,458.80	55	1,101.70	17	1,414.00	50
Imports	−7,471.50	−9,800.30	31	−9,795.90	31	−9,091.60	22
GDP (market price)	49,719.20	65,992.50	33	59,930.90	21	63,216.10	27
Indirect taxes	4,601.10	5,623.80	22	5,492.40	19	5,257.00	14
GDP (factor cost)	49,460.40	65,205.50	32	59,100.50	19	62,641.90	27
Agricultural production	17,121.20	29,994.87	75	23,213.87	36	30,390.18	78
Industrial production	29,159.44	35,176.71	21	35,004.30	20	31,802.87	9
Investment multiplier		3.47		2.61		6.06	
Households income							
Rural poor	8,762.35	10,810.66	23	11,184.14	28	10,241.66	17
Rural non poor	13,626.20	16,811.50	23	17,391.42	28	15,926.82	17
Urban poor	936.77	1,177.89	26	1,214.21	30	1,116.47	19
Urban non poor	32,083.47	40,392.68	26	41,634.34	30	38,280.81	19

Source Our elaboration based on CGE results
*% Var. w.r.t. BC = variation with respect to Base Case

Table 3 Industrial TFP and investment increase simulations (Mln US$)

	Simulation IV (industrial productivity doubled) Neoclassical closure	Var. w.r.t. BC* (%)	Simulation V (Industrial productivity doubled and investment increasing of 50%) Keynesian closure	Var. w.r.t. BC* (%)
Absorption	68,086.80	21	64,085.80	14
Private consumption	47,854.50	13	46,077.00	9
Fixed investments	14,135.40	78	11,911.80	50
Government income	10,909.20	18	9,344.99	1
Exports	1,266.80	34	1,358.00	44
Imports	−9,361.80	25	−8,159.60	9
GDP (market price)	59,991.70	21	57,284.20	15
Indirect taxes	5,908.60	28	5,549.10	21
GDP (factor cost)	59,434.60	20	57,360.30	16
Agricultural production	17,569.36	3	16,529.04	−3
Industrial production	44,102.40	51	41,809.14	43
Investment multiplier	1.61		1.99	
Households income				
Rural poor	9,854.81	12	9,180.97	5
Rural non poor	15,325.00	12	14,277.47	5
Urban poor	1,046.44	12	986.88	5
Urban non poor	35,844.00	12	33,813.65	5

Source Our elaboration based on CGE results
*% Var. w.r.t. BC = variation with respect to Base Case

(scenarios IV and V). They also aim to detect the potential gains due to a combination of investment and productivity increases in these two different sectors (scenarios I, II and V). The simulations permit also to compare the results obtained with different closures of CGE models—neoclassical and Keynesian—which correspond to different hypotheses on the functioning of the economic system. The increase in productivity, both agricultural and industrial, effectively reduce marginal production costs and increase the level of production and GDP. The results show that agricul-

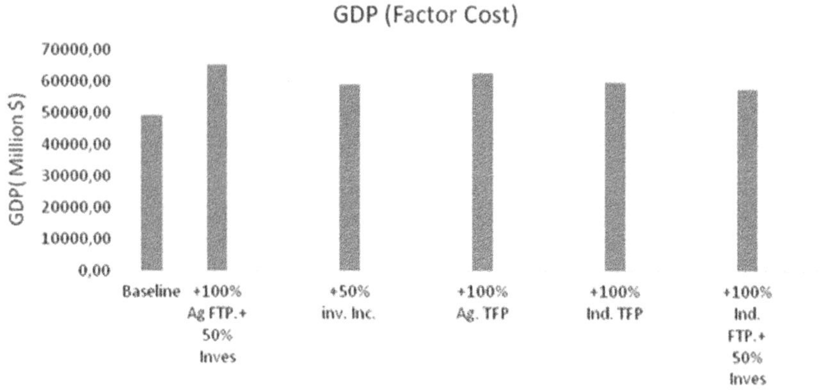

Fig. 2 Impact of simulation scenarios on GDP. *Source* Our elaboration based on CGE results

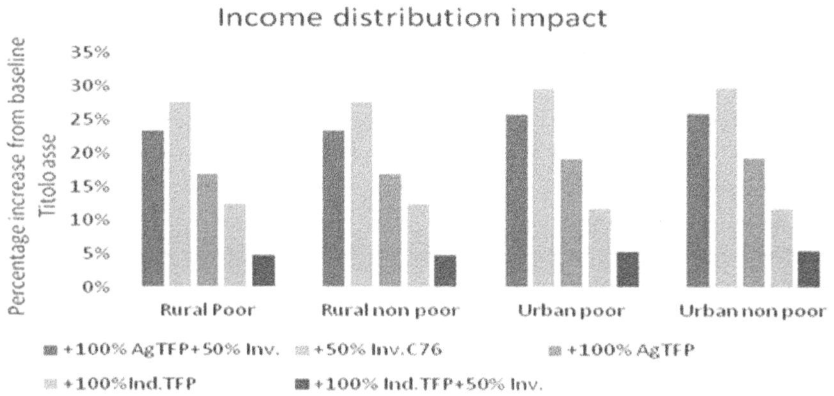

Fig. 3 Impact of simulation scenarios on household incomes. *Source* Our elaboration based on CGE results

tural production increases of 75 and 78% and industrial production increases of 51 and 43% with respect to the base case (respectively in the Keynesian and Neoclassical closure). The increase in real GDP drives up also Government income and both imports and exports in the foreign market.

Notably, in the first scenario the investment multiplier is higher than in the second one (3.47 vs. 2.61); however, the highest multiplier is in the third scenario (6.06), even though in this case it is the productivity increase that determines the surge of investment and some of the ensuing effects. In the first scenario almost all variables are higher with respect to the other simulations; for example the GDP increases around 33% respect to 21% in the second and 27% in the third (Fig. 3). This suggests that a mix of increasing productivity and investment is a better development vehicle as compared to just increasing investment demand. However in the third scenario the investment multiplier is greater, meaning that increasing agricultural productivity works well in a neoclassical economy when it is supported by a stimulus of endogenous demand.

Table 3 presents scenarios IV and V—concerning the industrial sector productivity increases—with the results obtained significantly lower than those in scenario I and III. For example the GDP increases by 21% in the fourth scenario and by 15% in the fifth one, with lower investment multipliers (respectively 1.61 and 1.99). This suggests that the combination in increasing industrial productivity and investment demand (scenario V) does not determine the best development path of the economy either from the production side or from the income one. The stimulus of domestic demand, in fact, is higher in simulation IV (neoclassical closure), where the increase in real GDP drives up the Government income by about 18% while real growth os much lower in the last scenario. Furthermore, private consumption is higher and in the IV scenario increases of 13% with respect to the base case, against an increment of only 9% in the last scenario.

6 Conclusions

In this chapter we have presented an application of a CGE model, estimated for Kenya on the basis of a recent estimate of a SAM matrix, to the analysis of the country economic structure and to some basic policy choices confronting the government. These choices concern the weight to give to agriculture in the development strategy of the country, and how to combine an infrastructure-centered policy of public investment with a drive for diversification and industrial growth.

The results of the analysis present us with a picture of Kenya as a country that, though still dominated by agriculture, may develop at a fast rate and diversify in multiple industries and services, including a very dynamic tourism sector. At the same time, the CGE simulation results suggest that agricultural growth may be the most important driver of the country economic development. This result appears to depend not only on the weight of agriculture on the economy (70% of Kenyans are still estimated to take their livelihood from agriculture), but also on its multiple connections, through backward and forward linkages, with the rest of the economy.

More intriguingly, the CGE simulations appear to indicate that a bias in favor of policies directed to industrial development may not only display an inferior performance, with respect to policies oriented towards increasing agricultural TFP, but may even be counterproductive. In particular, the combination of industrial TFP and investment increases may cause producer and consumer prices to diverge, and ultimately bring about a major gap between aggregate growth results and the improvement of living conditions by boosting disposable incomes. This is a phenomenon that has been experienced also by several advanced countries in the course of the past 30 years, and is at the root of the perceived inability of aggregate growth to create employment and lift people from poverty.

Appendix: Model Specification

Core Equations of the Model

Price Block

Import Price

$$PM_c = pwm_c\,(1 + tm_c) * EXR \tag{1}$$

Export price

$$PE_c = pwm_c\,(1 - te_c) * EXR \tag{2}$$

Absorption

$$PQ_c * QQ_c = PD_c * QD_c + PM_c * QM_c * (1 + tq_c) \tag{3}$$

Market output value

$$PX_c * QX_c = PD_c * QD_c + PE_c * QE_c \tag{4}$$

Activity price

$$PA_a = \sum_c PX_c * \theta_{ac} \tag{5}$$

Value-added price

$$PVA_a = PA_a - \sum_c PQ_c * ica_{ac} \tag{6}$$

In the price Block, PE and PM are the international prices for commodities traded with foreign economies. The prices are respectively export and import prices, which are reduced by governmental subsidies in the first case (te) and incremented by tariffs in the second case (tm). PQ is the price paid into the domestic market for the commodity demand and represents the composite price. PX is the producer's price, which is the combination of commodities sold domestically and exports. PA is the price received by each activity from commodities selling, that allows a multiple commodity production by each activity. PVA is the value-added price which reflect the price of activities net the price of output, thus reflecting the price of production factors.

Production Equation

Production function

$$QA_a = \alpha_a^a * (\delta_a^a * QVA_a^{-\rho_a^a} + (1 - \delta_a^a) * QINTA_a^{-\rho_a^a})^{\frac{-1}{\rho_a^a}} \tag{7}$$

Value added and factor demand

$$QVA_a = \alpha_a^{va} * \left(\sum_f \delta_{fa}^{va} * QF_{fa}^{-\rho_a^{va}} \right)^{\frac{-1}{\rho_a^{va}}} \tag{8}$$

Factor demand

$$WF_f * WFDIST_{fa} = PVA_a * (1 - tva_a) * QVA_a * \left(\sum_f \delta_{fa}^{va} * QF_{fa}^{-\rho_a^{va}} \right)^{-1} * \delta_{fa}^{va} * QF_{fa}^{-\rho_a^{va}-1} \tag{9}$$

Intermediate demand

$$QINT_{ca} = (ica_{ca} * QA_a) \tag{10}$$

Output function

$$QX_c = \Sigma_a (\theta_{ac} * QA_a) \tag{11}$$

Composite supply

$$QQ_c = \alpha_c^q * (\delta_c^q * QM_c^{-\rho_c^q} - (1 - \delta_c^q) * QD_c^{-\rho_c^q})^{\frac{-1}{\rho_c^q}} \tag{12}$$

Import-domestic demand ratio

$$\frac{QM_c}{QD_c} = \left(\frac{PD_c}{PM_c} * \frac{(1 - \delta_c^q)}{\delta_c^q} \right)^{\frac{1}{1+\rho_c^q}} \tag{13}$$

Output transformation function

$$QX_c = at_c * (\delta_c^t * QE_c^{\rho_c^t} + (1 - \delta_c^t) * QD_c^{\rho_c^t})^{\frac{1}{\rho_c^t}} \tag{14}$$

Export-domestic supply ratio

$$\frac{QE_c}{QD_c} = \left(\frac{PE_c}{PD_c} * \frac{(1 - \delta_c^t)}{\delta_c^t} \right)^{\frac{1}{\rho_c^t - 1}} \tag{15}$$

In the production block QA represents the value of activity production, modeled as a CES function but put in its simplest version as cob-douglass production function where alpha is the efficient parameter and rho is the share parameter (Eq. 7). Equation (8) states represents the quantity of value-added that is a CES function of disaggregated factor quantities. The optimality condition brings to factor demand (Eq. 9) that is in function of relative factor prices. The demand of intermediate inputs in Eq. (10) is fixed in proportion if intermediate input coefficients. Equation (12) reflects the Armington specification of the composite supply, where the supply is divided for domestic and international markets defined by share parameters of market supply. Equation (13) is the optimality condition for the Armington specification. The Output transformation function (Eq. 14) defines the substitution between output produces for domestic market and output produced for foreign market. Also in this case the optimality mix in in function of share parameters and production elasticities. Equation (15) represents the optimality condition.

Institution Block

Factor income

$$YF_{if} = shry_{if} * \Sigma_a(WF_f * WFDIST_{fa} * QF_{fa}) \tag{16}$$

Household demand function

$$PQ_c * QH_{ch} = \beta_{ch} * (1 - mps_h) + \left(1 - t_y\right) * YH_h \tag{17}$$

Investment demand function

$$QINV_c = \overline{\text{IADJ}} * \overline{qinv_c} \tag{18}$$

Government consumption function

$$QG_c = \overline{\text{GADJ}} * \overline{qg_c} \tag{19}$$

Government revenues

$$\begin{aligned} \text{YG} = {} & \Sigma_i ty_i * YI_i * EXR * trs_{gov,row} + \Sigma_c tq_c * PQ_c * QQ_c + shry_{gov,f} \\ & + r_{gov,ent} + \Sigma_c tm_c * QM_c * EXR + \Sigma_c te_c * pwe_c * QE_c * EXR \end{aligned} \tag{20}$$

Government expenditure

$$EG = \Sigma_c PQ_c * QG_c + \Sigma_i trs_{i,gov} \tag{21}$$

Tourist demand

$$CDTOUR_i = \frac{\Sigma_{tt}\,(\beta_{itt} * YT_{tt}*)}{PQ_c * EXR} \tag{22}$$

Objective function

$$OBJ = (warlas)\ \hat{}2 \tag{23}$$

The first equation of the institution block represents the factor income, which is the sum for each activity of the quantity demanded for the respective factor price. The household demand function, (Eq. 16), is a liner expenditure function of commodities demanded by households. In its simplest form (like in this case) the function can be stated as cobb-douglass demand function, where households demand commodities in function of their disposable income, after paying taxes, other transfers and savings. The demand for investments in Eq. (17) is fixed according to fixed investment coefficients and multiplied by an adjustment factor IADJ. The same specification is stated for Government demand of commodities according to fixed demand coefficients. The Government total revenue is the sum of taxes, tariffs and transfers from the rest of the World, while the spending are the sum of commodity consumption and transfers to other institutions. Tourist demand in Eq. (21) is stated as in function of an exogenous tourist income and enterprise revenue is the sum of capital shares and transfers from other institutions such as government and rest of the world.

System of constraint block

Factor market

$$\Sigma_a QF_{fa} = QFS_f \tag{24}$$

Composite commodity market

$$\mathrm{QQ}_c = \Sigma_a QINT_{ca} + \Sigma_h QH_{ch} + QG_c + QINV_c + qdst_c + CDTOUR_c \tag{25}$$

Current Account balance

$$\Sigma_c pwm_c * QM_c + \Sigma_i tr_{row,i} = \Sigma_a pwe_c + QE_c + \Sigma_i tr_{i,row} + \frac{\Sigma_c PQ_c}{EXR} * CDTOUR_c + FSAV \tag{26}$$

Savings-Investment balance

$$\Sigma_i MPS_i * (1 - ty_i) * YI_i + (\mathrm{YG} - \mathrm{EG}) + \mathrm{EXR*FSAV} = \Sigma_c PQ_c * QINV_c + \Sigma_c PQ_c * qdst_c + WARLAS \tag{27}$$

Price Index

$$\overline{CPI} = \Sigma_c PQ_c * cwts_c \tag{28}$$

The system of constraints is very important for the specification of the CGE model and represents the conditions for model equilibrium. The first Eq. (24) represents the equality between factor demand and supply. In the Neoclassical specification, factor supply is fixed, reflecting the fact of full employment, while in Keynesian specification this hypothesis can be relaxed. Equation (25) stated the equality between commodity supply and composite demand, where the supply is equal to demand from institutions (households, government), investment demands and demand for intermediate inputs.

Equation (26) states the current account constraint, where imports equal exports plus the foreign deficit and everything is expressed in foreign currency. Finally, Eq. (27) is the saving—investment balance, where savings from domestic institutions and the rest of the world, equals the total capital formation.

Changing the specification of the closure equations (i.e. allowing some variables to vary and fix others), changes the model closures, according to policies and hypothesis that we want to analyze with the study.

References

Adelman, I. (1975). Growth, income distribution and equity-oriented development strategies. *World Development, 3*(2–3), 67–76.

Adelman, I. (1984). Beyond export-led growth. *World Development, 12*(9), 937–949.

Arndt, C., Jensen, H. T., Robinson, S., & Tarp, F. (2000). Marketing margins and agricultural technology in Mozambique. *Journal of Development Studies, 37*(1), 121–137.

Bautista, R. M. (1986). Effects of increasing agricultural productivity in a multisectoral model for the Philippines. *Agricultural Economics, 1*(1), 67–85.

Brooke, A., Kendrick, D., Meeraus, A., & Raman, R. (1996). *GAMS, A user's guide, 1998*. Washington, DC: GAMS Development Corp.

Coxhead, I., & Warr, P. G. (1995). Does technical progress in agriculture alleviate poverty? A Philippine case study. *Australian journal of agricultural economics, 39*(1), 25–54.

Coxhead, I. A., & Warr, P. G. (1991). Technical change, land quality and income distribution: A general equilibrium analysis. *American Journal of Agricultural Economics, 73*(2), 360–435.

Coxhead, I. A., & Warr, P. G. (1993). The distributional impact of technical change in Philippine agriculture: A general equilibrium analysis. *Food research institute studies, 22*(3), 253–274.

de Janvry, A., & Sadoulet, E. (2002). World poverty and the role of agricultural trchnology: Direct and indirect effects. *Journal of Development Studies, 38*(4), 1–26.

Delpiazzo, E. (2011). *The mozambican participation in SADC. A liberalization process through different models and different closures*. Ph.D. Dissertation.

Dervis, K., de Melo, J., & Robinson, S. (1982). *Planning models and development policy*. London: Cambridge University Press.

Devarajan, S., Go, D. S., Lewis, J. D., Robinson, S., & Sinko, P. (1997). Simple general equilibrium modeling. In J. F. Francois & K. A. Reinert (Eds.), *Applied methods for trade policy analysis: A handbook*. Cambridge: Cambridge University Press.

Dorosh, P., El-Said, M., & Lofgren, H. (2003). *Technical change, market incentives and rural incomes: A CGE analysis of Uganda's agriculture*. Paper Presented at the 25th International Conference of Agricultural Economics. Durban, South Africa.

Golan, A., Judge, G., & Miller, D. (1996). *Maximum entropy econometrics: Robust estimation with limited data*. Chichester, England: Wiley.

Hanmer, L., & Naschold, F. (2000). Attaining the international development targets: Will growth be enough? *Development Policy Review, 18*, 11–36.

Irz, X., Lin, L., Thirtle, C., & Wiggins, S. (2001). Agricultural productivity growth and poverty alleviation. *Development Policy Review, 19*(4), 449–466.

Johnston, B. F., & Mellor, J. W. (1961). The role of agriculture in economic development. *The American Economic Review, 51*(4), 566–593.

Kiringai, J., Thurlow, J., & Wananjala, B. (2006). "A 2003 Social Accounting Matrix (SAM) for Kenya", Kenya Institute for Public Policy Research and Analysis, Nairobi, and International Food Policy Research Institute, Washington, D.C.

KNBS Kenya National Bureau of Statistic (2014), Statistical abstract.

Lofgren, H., Harris, R. L., & Robinson, S. (2002). A standard computable general equilibrium (CGE) model in GAMS (Vol. 5). Intl Food Policy Res Inst.

Mellor, J. W. (1976). *The new economics of growth: A strategy for India and the developing world*. London: Cornell University Press.

Mellor, J. W. (1999). Faster, more equitable agricultural—the relation between growth in agriculture and poverty reduction. *Agricultural policy development project research project research report no.4*. Cambridge.

Norton, R. D., & Scandizzo, P. L. (1981). Market equilibrium computations in activity analysis models. *Operations Research, 29*(2), 243–262.

Panagariya, A., & Duttagupta, R. (2001). The 'Gains' from preferential trade liberalization in the CGEs: Where do they come from? In S. Lahiri (Ed.), *Regionalism and globalization: Theory and practice* (pp. 39–60). London: Routledge.

Rasmussen, P. N. (1957). *Studies in inter-sectoral relations*. Amsterdam: North-Holland.

Rattso, J. (1982). Different macroclosures of the original Johansen Model and their importance in policy evaluation. *Journal of Policy Modelling, 4,* 85–98.

Robinson, S. (2006). Macro models and multipliers: Leontief, Stone, Keynes, and CGE models. In *Poverty, Inequality and Development* (pp. 205–232). Boston, MA: Springer

Robinson, S., Naude, A. Y., Ojeda, R. H., Lewis, J. D., & Devarajan, S. (1999). From stylised models: Building multisector CGE models for policy analysis. *The North American Journal of Economics and Finance, 10,* 5–38.

Scandizzo, P. L., & Ferrarese, C. (2015). Social accounting matrix: A new estimation methodology. *Journal of Policy Modeling, 37*(1), 14–34.

Scarf, H. E., & Hansen, T. (1973). *The Computation of economic equilibria*. New Haven, CT: Yale University Press.

Schneider, K., & Gugerty, M. K. (2011). Agricultural productivity and poverty reduction: Linkages and pathways. *The Evans School Review, 1*(1), 56–74.

Self, S., & Grabowski, R. (2007). Economic development and the role of agricultural technology. *Agricultural Economics, 36,* 395–404.

Sen, A. K. (1963). Neo-classical and neo-Keynesian theories of distribution. *Economic Record, 39,* 54–64.

Taylor, L., & Lisy, F. (1979). Vanishing income redistributions. Keynesian clues about model surprises in the short run. *Journal of Development Economics, 6,* 11–29.

Thirtle, C., Irz, X., Lin, L., McKenzie-Hill, V., & Wiggins, S. (2001). *Relationship between changes in agricultural productivity and the incidence of poverty in developing countires*.

Thirtle, C., Lin, L., & Piesse, J. (2003). The impact of research-led agricultural productivity growth on poverty reduction in Africa, Asia and latin America. *World Development, 31*(2), 1959–1975.

Waelbroeck, J. and Ginsburgh, V. (1981). *Activity analysis and general equilibrium modelling*.

Wanjala, B. M., & Were, M. (2009). Gender disparities and economic growth in Kenya: A social accounting matrix approach. *Feminist Economics, 15*(3), 227–251.

Wing, I. S. (2004). Computable general equilibrium models and their use in economy-wide policy analysis: Everything you ever wanted to know (But were afraid to ask), The MIT Joint Program on the Science and Policy of Global Change, Technical Note, no. 6.

World Bank Group. (2017). Housing—Unavailable and Unaffordable. *Kenya Economic Update, 15*. World Bank, Washington, DC (April 2017). ©World Bank. https://openknowledge.worldbank.org/handle/10986/26392, License: CC BY 3.0 IGO.

The Political Economy of the CAP Reform in Italy

Antonella Finizia, Riccardo Magnani and Federico Perali

Abstract This chapter analyses the ex-ante socio-economic impact of the CAP reform on Italian agriculture and the whole economy using a micro-funded general equilibrium model which differentiates the impact at the household level. The political economy analysis of the consequences of the reform has clearly revealed the positions of farmers and agro-food industries, consumers, and farming unions concerning the issue of a total or partial implementation of decoupling. The policy analysis permits both an understanding of the possible social conflicts arising from the implementation of the reform and a unique ranking of the policy alternatives.

Keywords CAP reform · Political economy · General equilibrium · Ex-ante policy analysis

JEL Classification F1 · D5 · Q1

1 Introduction

This chapter investigates the ex-ante impact of the Mid Term Review of the Common Agricultural Policy (CAP) on Italian agriculture and describes the political economy aspects associated with the execution of the reform. Although the reduction of domestic farm supports may lead to a net gain in national economic welfare, some sectors and households can be adversely affected. Tracking the aggregate impacts down at

A. Finizia
ISMEA, Rome, Italy

R. Magnani
Department of Economics, Université
Paris 13, and Sorbonne Paris Cité and CEPN, Paris, France

F. Perali (✉)
Department of Economics, University of Verona,
Via Cantarane 24, 37129 Verona, Italy
e-mail: federico.perali@univr.it

© Springer International Publishing AG, part of Springer Nature 2018
F. Perali and P. L. Scandizzo (eds.), *The New Generation of Computable General Equilibrium Models*, https://doi.org/10.1007/978-3-319-58533-8_7

the micro level is then crucial to understand the sources of political frictions that may hinder the process leading to the implementation of reforms. The *ex-ante* analysis of the possible causes of social conflicts, the identification of potential losers and the quantification of their losses may help designing accompanying policy actions making the reform politically feasible and enforcing the bargaining position of the institutions supporting the policy change. This motivation has markedly shaped the present research.

The effects of the CAP reform, in terms of producers, markets and levels of well-being of agricultural, rural and urban households, are first evaluated by using an applied general equilibrium model that permits to implement the CAP instruments by modeling the associated market failures, price rigidities and non-linearities. The general equilibrium results obtained with the MEG ISMEA model[1] are further elaborated in order to carry out the political economy analysis of the different scenarios.

The paper first illustrates the three policy alternative scenarios delineated by the Mid-Term Review of the CAP, one with full decoupling of aids and two with different options of partial decoupling. We then present the results of the simulations and their political economy interpretation aiming at ranking the policy scenarios accounting for the producers and consumers' point of view and society's changes in welfare.

2 The Mid Term Review and the Policy Scenarios

The present work analyses the impacts of the policy scenarios delineated in the Mid Term Review of the CAP as approved at the end of June 2003 in Luxembourg.[2] As it is well known, the aim of the reform is to substitute payments "coupled to specific farm activities"[3] with a lump-sum payment which has no distortive effects in the markets and farmers' allocation decisions. In essence, a price subsidy and/or an income subsidy coupled to a specific production is substituted with a decoupled income subsidy which in fact transfers support from the products to the producers. Farmers can thus optimize the activity portfolio according to the allocative information conveyed through the market, ensuring Pareto efficiency. The objective to decouple payments from specific farming activities is achieved while safeguarding

[1] In this paper, we only show the general equilibrium results that are useful to the political economy analysis of the implementation of the CAP reform in Italy. For a complete discussion of the general equilibrium results, see Finizia et al. (2005).

[2] EC Regulations 1782/2003 and subsequent ones.

[3] Since the Mac Sharry 1992 Reform the direct payments, for the majority of agricultural products, cannot be defined as coupled in strict sense, which is a term that more properly refers to a direct link of the support with the level of production. In fact, payments are computed on the basis of historical yields and are therefore independent of the current level of production. However, by being linked to the declared hectares devoted to a specific crop, they are in this sense coupled as compared to the payment introduced by the Luxembourg agreement, which is decoupled from specific crops. On the other hand, the premia given to the olive oil and tobacco sectors, which have not been reformed in the Luxembourg agreement, are in fact maintained coupled to production in the model.

agricultural incomes, by ensuring an income support as a single farm payment (SFP) representing a certain financial flow, which should help to keep farmers in business and to sustain the rural households' livelihoods (De Filippis 2004).

The main objective of favoring greater orientation towards more and better markets is accompanied by other important objectives such as: favoring greater sustainability of agriculture; assuring more attention towards issues of food security and animal welfare, by asking farmers to sign contracts of environmental cross-compliance in exchange for public support; rising equity in distributing the support with respect to coupled payments which are mainly benefited by large producers; realizing more integration and synergy with rural development; obtaining administrative simplification.

The reform can be summarized in three main pillars (European Commission 2003a, b):

1. modifications of the market policies through variations of the intervention prices and/or variations of the existing premia or introduction of new premia for some products;
2. decoupling of the premia: decoupling introduces a single payment per farm starting from year 2005, whose amount equals the mean of the total direct payments received by the farm during the years 2000–2002, for some productions (cereals, protein crops, oil seeds, rice, livestock, sheep and goats and, from 2008, milk as well). The payment corresponding to the set-aside area in possess during the reference period is attributed separately;
3. modulation of the premia: all direct payments given to farmers (the single decoupled payment and specific coupled payments for durum wheat, protein crops, rice, fruits in shell, olive oil, tobacco) will be reduced in the period 2005–2012 in the proportion of 3% in 2005, 4% in 2006, and 5% from 2007 to 2012. Premia below EUR 5,000 are exempted.

The objective of the modulation, which is mandatory, is to transfer an amount of aids from the first pillar (market support) to the second pillar of the CAP (rural development). The single farm payment (SFP) is the mean of the payments received by the farm during the reference period 2000–02 for cereals; protein crops; oilseeds; rice; dried fodder; bovine meat; sheep and goats and, from 2008, milk. Permanent crops are not eligible. Further, there is a specific payment for the set-aside area. The SFP does not account for: (a) the quality premium for durum wheat, (b) the special premium for protein crops, and (c) part of the rice premium (EUR 453/ha). The eligible land has to be kept in good agronomic and environmental condition and is constrained not to produce fruit, vegetables and potatoes. If the reform is implemented regionally by giving a uniform rate per hectare, then farmers are free to produce any good with the exception of permanent crops.

The SFP that will be received in the future by the farm is composed by the per hectare SFP multiplied for the number of eligible hectares. Because of the link with eligible land, if a farmer reduces in one year the number of hectares, either by selling or renting some land in or out, the SFP will be proportionally reduced. It is then impossible to exert the right to the Single Payment without being in possess of at least one hectare of land.[4,5] On the other extreme, it is possible to produce nothing on the eligible land, if the land is maintained in "good agronomic and environmental conditions".

The estimation of the transfers generated by the reform takes into consideration two impacts:

- the effects of the variations in the levels of intervention prices and premia in the involved Common Market Organizations (CMO), which modify the comparative advantage across agricultural activities and the absolute level of the premium;
- the non distortive effects of the decoupled lump-sum transfer which determines market-based reallocations of the activity portfolio of the farms.

The effects are microsimulated using farm level data to generate a base scenario depicting the situation of Agenda 2000 (scenario A).[6] Because for some commodities the changes due to the reform are introduced gradually, the impacts of the scenarios are simulated referring to an abstract situation where the reform is fully implemented at year 2008.

The implementation of the Luxembourg agreement requires adjustments to the common market organization mainly for certain arable crops (cereals, oilseeds, protein crops), and dairy products.[7] The adjustments to CMOs are summarized in Table 1. The adjustments in the CMOs, for the products which have been considered in the micro-simulation and in the MEG ISMEA model, are as follows:

1. **Cereals, oil seeds and set-aside**: The direct payment of 63 EUR/ton is the same as for the base scenario.
2. **Durum wheat**: The base premium remains the same as for the base scenario; a reduction has been decided of the supplementary premium for the traditional areas from EUR 344.5/ha to EUR 285/ha in 2006, along with the elimination of the premium for normal areas; a quality premium of EUR 40/ha has been introduced in traditional production zones to farmers who are using certified seed of selected varieties within the limits of current Maximum Guaranteed Areas. We make the hypothesis that all farmers in traditional areas are eligible and access the quality incentive.
3. **Protein crops**: The base premium remains the same as for the base scenario, but the current special payment of EUR 9.5/ton is converted into a crop specific area payment of EUR 55.57/ha; with respect to the base scenario, considering the average historical yields in Italy the premium results to be slightly higher.

---

[4]It is not necessary that the land be physically the same. It is in fact possible to sell or rent land that was available in the reference period and sell it or renting it elsewhere.

[5]Livestock production without farming land represents an exception with special right. To claim the right, at least 50% of the livestock which received a premium in the reference period should be maintained.

[6]The variation of the intervention price for a certain good is introduced through a change in both the intervention price and the import price since we assume that the CAP significantly affects the European market.

[7]Other products interested by the reform are dried fodder, seeds, energy crops and nuts; however, the policy regimes for these products are not incorporated in the simulation.

Table 1 A summary of the mid term review policy changes—base situation 2001–2002 and full implementation

	Base year	Full implementation
Durum wheat		
Supplementary aid in traditional areas (euro/ha)	344.5	285
Supplementary aid in other areas (euro/ha)	138.9	0
Quality payment	0	40
Rice		
Intervention price (euro/t)	298.35	150
Payment (euro/t)	52.65	177
Protein crops		
Special payment (euro/t)	9.5	0
Special payment (euro/ha)	0	55.57
Dry forage aid (euro/ha)	19	24
Nuts		
Payment (euro/ha)	0	120.75
Energy crops		
Payment (euro/ha)	0	45
Milk and Dairy products		
Intervention price for butter (euro/100 kg)	328.2	246.39
Intervention price for skimmed milk powder (euro/100 kg)	205.52	174.69
Payment per ton of quota (euro/t)	8.15	24.49
Effective payment per ton of quota in Italy, including national envelope (euro/t)	11.14	33.48
Quota (million tons)	118.891	120.545

4. **Rice**: The intervention price for rice is reduced by 50% to EUR 150/ton and 88% compensation is provided through higher payments. The final compensation increases existing direct payments from EUR 52.65/ton to EUR 177/ton. Hence, for an average productivity of 6.04 tons/ha in Italy the premium is about 1070 EUR/ha, which is about three times the premium of the base scenario.
5. **Milk and Butter**: Dairy quotas are extended until the 2014/15 season. The intervention price of butter is reduced by 25%. The skimmed milk price is cut by 15%. As a compensation, it is introduced a premium of EUR 24.49/ton. Keeping

Table 2 Level of micro-simulated direct payments and changes in intervention and import prices for the commodity disaggregation simulated in the general equilibrium model

Durum wheat	4.3% total reduction of premia
Proteic crops	1.9% total increase of premia
Rice	245% total increase of premia
	50% reduction of the intervention price
	0.6% reduction of the import price
Milk	Payment proportional to the quota owned by the farm
	6.1% reduction of the import price of dairy products
Butter	25% reduction of the intervention price
Skimmed milk powder	Not considered because Italy does not produce SMP

also into account a uniform distribution of the national envelope on a per quota basis, the Italian premium is EUR 34.87/ton.

The policy microsimulation was performed using the farm budgets of the ISMEA socio-economic survey (ISMEA 2005) in the following steps:

1. Determination of the level of premia received by each farm of the ISMEA micro-data considering the Agenda 2000 package and the 2001 situation in order to reconstruct the historical yields to compute the premia and the number of animals which effectively received a premium in the bovine meat CMO.
2. The changes in direct payments and variations in prices described above are then reported to the universe using the 2001 Census of Italian Agriculture and used to compute the SFPs on the basis of the estimated eligible land. The modulation is considered, that is the reduction by 5% of all premia when the reform reaches its full implementation regime.

The non-behavioral microsimulation exercise generates the sector-level impacts that are summarized in Table 2 according to the commodity disaggregation adopted in the MEG ISMEA model. The table also presents the changes in intervention and import prices introduced at the macro equilibrium level as a result of the adoption of the reform. For all other products, we do not consider changes in prices and premia.

Another important feature of the reform is the possibility given to the Member States (MSs) to partially adopt the decoupling regime. This possibility concerns only the application of the arable crops, cattle and sheep and goats regimes. In detail, for arable crops the MSs can choose to couple up to 25% of the base premium for or, alternatively, up to 40% of the supplementary premium for durum wheat.

For livestock, the MSs can choose to couple up to 50% of the actual premia for sheep and goats and up to 100% of the slaughter premium for calves and, further, one of the following alternatives: up to 100% of the suckler cows premium and up to 40% of the slaughter premium for adult bovines, or up to 100% of the slaughter premium for adult bovines, or up to 75% of the special male premium.

The Reform foresees also the constitution of national reserves by means of further percentage reductions of the premia, in addition to that coming from the modulation. The reserves are intended to permit the access to the activity to new farms, which are excluded by the SFP as they were not active in the reference period 2000–2. The decision about the premia cut for the constitution of the reserve is left to national governments. Other decisions for national governments concern the opportunity to cut part (up to 10%) of the crops, bovine meat and dairy premia and to use this amount of aids for special quality programs or to give incentive to specific productions in the same sectors. As no indication is available regarding the Italian decisions on these subjects at the moment of the simulations, we have not considered these options.

3 The General Equilibrium Model and Simulations' Design

The MEG ISMEA model, which is described in detail in Finizia et al. (2005), is a static multisectoral computable general equilibrium model of the Italian economy with two different trade areas, the European Union (EU) and the rest of the world (RoW). The aim of this distinction is to take into account that the Italian agricultural policy is a European policy (OECD 1988; Gohin et al. 1999, 2002; Gohin 2002). Table 3 reports a summary description of the main features of the MEG ISMEA model.

The MEG ISMEA represents a Walrasian economy where all markets are perfectly competitive, firms maximize their profits, households maximize their utility and the production factors are remunerated on the basis of their marginal productivity. In this "ideal" economic environment some rigidities are introduced, in the goods and in the factors markets, in order to reproduce the main features of the Common Agricultural Policy (CAP) (Weyerbrock 1998; Hertel 1999; De Muro and Salvatici 2001; FAPRI 2003; ISMEA 2004).

The MEG ISMEA model includes 41 sectors and places particular emphasis on the agricultural and agri-food sectors. As shown in Table 4, agriculture is disaggregated into 23 agricultural sectors, food industry in 9 sectors, other industries in 7 sectors, and services in 2 sectors. Each sector produces a single output, using intermediate goods and primary factors: self-employed farm labor, hired labor, land (distinguished in three types), agricultural capital, and animals (distinguished in four types). The other sectors use two production factors: non-agricultural capital and labor.

MEG ISMEA considers 11 household types: 7 farm-household types describing the agricultural sector, 1 rural household type, and 3 urban low-middle-high income classes. The classification of the 7 farm-household types has been derived from the cluster analysis of the ISMEA 1995 Survey about the Socio-Economic Conditions of Italian Agriculture (ISMEA 2005). The social accounting matrix is further articulated into a rural class, and three urban classes graduated in terms of income levels. The information is derived from the Bank of Italy Income Survey. This classification

Table 3 The structure of the MEG ISMEA general equilibrium model

– A single country, multi-sector CGE model of the Italian economy focused on agriculture and agri-food sector
– A static model calibrated on the 1995 ISMEA I/O table
– Perfect competition in all markets and neoclassical macroeconomic closure
– 41 sectors: 23 in the primary sector, 9 in the agro-food sector, 7 in the industrial sector, 2 in the service sector
– 2 trade areas: the rest of the European Union (EU) and the Rest of the World (RoW)
– 2 institutional sectors: the households (11 household categories) and the Italian government
– Two-stage constant-returns to scale production functions with imperfect substitution between inputs, including intermediate inputs using nested CES functions
– 11 types of primary production factors: labor (hired labor and farm self-employed labor); capital (capital and agricultural capital); land (three types of land); animals (four types of animals)
– Household preferences are described using a two-stage CES utility function. In the first stage, the utility depends on aggregate consumption and leisure. In the second step each class decides, on one hand, the optimal allocation of the aggregate consumption across the goods produced by the 41 sectors, and, on the other, the optimal allocation of labor supply between hired labor and self-employed farm labor
– International trade
On the export side, the relation between domestic sales and exports is described with a CET function.
On the import side, domestic and foreign goods are "Armington" imperfect substitutes. We have two cases:
(1) Large country hypothesis for some goods: imperfect substitution between production and import so that their prices are different and the market equilibrium price is endogenous
(2) Small country hypothesis with respect to the rest of the world for wheat, durum wheat, soy-bean assuming perfect substitution between production and import so that their prices are identical and the market equilibrium price is fixed at the world level
Modeling of the Common Agricultural Policy's main features such as the single farm payment, intervention price mechanism, import tariffs, production quotas, set-aside, decoupling
Political economy interpretation using collective choice rules

permits an accurate distributional and welfare analysis of the impact of agricultural policies upon policy relevant farm-household types (ISMEA 2005).

The MEG ISMEA model builds on a Social Accounting Matrix (SAM) describing the economic relations between the structure of production and the income distribution across household classes. The SAM is based on the 1995 input-output table of the agri-food sector (ISMEA 1997). The input-output table is based on the data gathered in 1996 through two ad hoc surveys, the Survey on the Socio-economic Conditions of the Italian Agriculture and the Survey on the Economic Conditions of the Italian Food Industry.

We use the MEG ISMEA model to simulate the general equilibrium effects of the following policy scenarios:

Table 4 Sector definitions

Agriculture		
1	Cereals	Soft wheat
2		Durum wheat
3		Rice
4		Corn and other cereals
5		Fodder (corn silage)
6		Non irrigated forage
7	Vegetables	Potatoes
8		Tomatoes
9		Other vegetables and legumes (beans, peas, garlic, cabbages, mushrooms…)
10	Industrial crops	Sugar beet
11		Soy-bean
12		Other industrial crops (hemp, linen, cotton, peanuts, sesame, other oil seeds)
13		Raw tobaccos
14	Viticulture	Grapes
15	Olive	Olives
16	Fruit	Citrus fruit, fresh and dry fruit
17	Floriculture	Floriculture and other products (flowers and seeds, spices, sugar, coffee…)
18	Milk	Bovine milk
19	Beef	Bovine meat livestock
20	Forestry	Forestry
21	Other livestock	Sheep and goats
22		Pigs, poultry, other animals
23	Fish	Fish and other sea products
Agro-food sector		
24	Meat	Fresh and preserved meat
25	Milk products	Milk and milk products
26	Bread, pasta, trasf. cereals	Cereal products, bread and pastry, pasta
27	Veg-fruit	Processed and preserved fruit and vegetables
28	Oil and fats	Olive oil, other vegetal oil, fats
29	Feed	Prepared animal feeds
30	Tobacco	Cigarettes
31	Other agro-food ind	Sugar and other products
32	Beverages	Wine, alcoholic beverages, beer, non alcoholic beverages, tea, coffee

(continued)

Table 4 (continued)

Agriculture		
Other industries sector		
33	Fuel and lubrif	Fuel and oils
34	Energy	Electric power
35	Water	Water
36	Fertilizers	Fertilizers
37	Pesticides	Pesticides
38	Other chemical and pharmaceutical prod	Other chemical and pharmaceutical products
39	Heavy industry	Maintenance, other industrial products, agricultural and industrial machinery, constructions and public works, other industrial productions (products of iron and steel, glass, motor vehicles, ships, aircrafts, spinning and webbing, footwear, furniture…)
Services sector		
40	TRCOMUNCRINS	Transports and communication, credit and insurance
41	Other services	Other services (business, hotels and public services, leisure—cultural services, Public Administration services, public and private health services…)

- Scenario A (the base scenario): our ex-ante situation refers to the premia established by Agenda 2000 in its full implementation (period 2001–2). This information has been constructed using a non-behavioral model that has been updated from the 1995 to the 2001 situation and incorporated in our Agenda 2000 situation, as implemented in Italy in the years 2001 or 2002, depending on the products. This is the benchmark against which we evaluate the effects due to the introduction of the reform as described in the following scenarios.
- Scenario D1 (total decoupling, with modulation): we consider both decoupling and modulation. Modulation has been implemented by assuming that all the direct premia, both those which are part of the SFP and those coupled, are cut by 5% with the exemption of the first EUR 5,000.
- Scenarios D2A and D2B (partial decoupling): these scenarios are based on the possible options for the countries to maintain a proportion of payments "coupled" to specific commodities.[8] In the scenarios names, the letters A and B refer respectively to:

[8]Originally, we considered ten scenarios describing the most important possible combinations of partial decoupling options. For the simulations with MEG ISMEA we selected only those four scenarios identified as the most relevant options for Italy.

Table 5 The partial decoupling scenarios: percentages of decoupled premia in detail

	D1	D2A	D2B
Soft wheat	100.0	75.0	100.0
Durum wheat	66.3	49.7	37.1
Rice	58.7	58.7	58.7
Corn	100.0	75.0	100.0
Forages	100.0	100.0	100.0
Potatoes			
Tomatoes			
Other vegetables			
Sugar beet			
Soy beans	100.0	75.0	100.0
Other industrial crops	100.0	75.0	100.0
Tobacco	0.0	0.0	0.0
Grapes			
Olives	0.0	0.0	0.0
Fruit			
Floriculture			
Milk	100.0	100.0	100.0
Bovine meat livestock	100.0	63.2	63.2
Forestry			
Sheep and goats	100.0	79.0	79.0
Other livestock			

Note 1 In the case of tobacco and olives the premium is coupled to production
2 Empty cells indicate that no premium has been proposed in the Review

i. the option of leaving 25% of the base payment of cereals coupled (scenario D2A);
ii. the option of leaving 40% of the supplementary payment per hectare of durum wheat coupled (scenario D2B).

The two scenarios should be considered "maximum" coupling options, where the minimum alternative is represented by the scenario D1 of complete application of the decoupling regime. In fact, each combination considers the maximum percentage of partial coupling admitted by the Regulation. Table 5 reports the percentages of decoupled premia generated by the reform for each product of the model, in the five scenarios. The political economy analysis of the different scenarios, presented in the next section, is carried out using the general equilibrium results obtained with the MEG ISMEA.

4 The Political Economy of the CAP Reform in Italy

The political economy question aims at reconciling the different views of the actors involved in the decision making process into a unique social outcome. With this objective in mind, the scenarios have been ranked according to (a) the agricultural producers' point of view as affected by the impact on value added and interested in production protection, (b) the general point of view of the agricultural and food industry, which includes other aspects besides valued added in agricultural production, and (c) the consumers and society's point of view based on the impact of the reform on the consumer price index of the basket of food goods and on social welfare level.

We use the *Borda* voting rule to aggregate the individual or sector-specific preferences. We also measure the impact of the Mid Term Review on the distribution of incomes among the socio-economic groups of interest and the related effects on society's welfare. The changes in welfare levels of each household class also influences the classes' preference orderings with respect to the policy alternatives, and the equity—efficiency trade-off implied by them, and the prospect that political coalitions are formed thus affecting the distribution of political power and the policy ranking. We examine these issues in sequence.

According to the *Borda* voting rule each person reports his preference relation. Suppose that there are N alternatives. The highest ranked alternative is assigned a fixed point k_i. The alternative in the second preference place is assigned a smaller fixed point $k_i - l_i$, $0 < l_i < k_i$ for $i = 1, \ldots, N$, a third place is assigned a yet smaller fixed point and so on to the last choice which is assigned l point. The sum of the weights gives the social preference ordering and the single best alternative. We assume that the *Borda* social decision function is incentive compatible, that is there are no incentives for strategic behavior by declaring false preferences, because in the present scheme there is only one voter.

The *Borda* aggregation method gives a rational collective preference but the outcome is not independent of irrelevant alternatives. As a consequence, the choice over the number of scenarios/candidates and the number of election outcomes to be aggregated, that is the control of the "agenda", is of critical importance for determining the final collective preference. The voting mechanism is designed for one voter in the vests of a benevolent social observer. It runs in two rounds. In the first round of the elections, the benevolent social observer is asked to vote for J elections by ranking the N alternatives forming the set of alternatives $A_l = \{D1, D2A, D3B\}$ as if each production sector were a separate industry in the economy in terms of the value added contribution of each sector.

The voting rule ranking the alternative reform scenarios assigns a higher vote to the highest positive percentage change and the lowest percentage change in value added. This is intended to reflect the producers' interest in maximizing profits from agriculture and protecting agricultural production in general. The least preferred gets 0 points, then the sequence increases by equal increments of 1 until N. The weights need not to be equally spaced. The *ex aequo* outcome is attributed when the differences are within the range [−0.3, 0.3] and receives the lowest vote. In other words, ties are not counted in the sum. In the second round, the social observer is asked to produce a social rule based on an objective weighting scheme reflecting the relative "importance" of the scenarios based on the value added share contributed by each sector. The weighting scheme changes the equal spacing rule of the votes cast in the first round. The intersectoral aggregation is the weighted sum of the *Borda* votes, which gives a unique voting outcome revealing the most preferred scenario from the producers' point of view.

The voting procedure can be summarized as follows:

The election
1 voter being a benevolent social observer
$i = 1, \ldots, N$ alternatives with $N = 3$
$j = 1, \ldots, J$ elections (one for each agricultural sector, $J = 23$)
The voting mechanism
Round 1—Vote for the best scenario per each sector
Round 2—Aggregate each vote using objective weights

4.1 The Producers' Interests

The voting outcome of both the first and second round of elections is presented in Table 6. The simulated changes determined with the MEG ISMEA in the production levels for each agricultural sector are presented in the first three columns of Table 6. The general equilibrium results show that the reform induces marked productive reallocations from cereal crops to fodder. The effect is particularly unfavorable for soft and durum wheat (respectively −27.64% and −36.11%), soy-bean (−80.67%) and other industrial crops (−20.68%). Soft wheat is also less competitive. Vice versa, livestock production is slightly encouraged from the cost reduction, given the higher availability of forage (and consequent cost reduction), with the exception of sheep and goats which are typically raised on extensive agricultural areas.

The outcome of the first round of the voting where all sectors have the same importance weight gives the total decoupling scheme D1 as the winner. The *Borda* score is 23 as compared to 13 for the partially decoupled scheme D2A and 8 for the D2B scheme. The results of the second round of elections can be read in the last row of the last three columns. The weights used to account for the different contribution of each sector to the agricultural value added are shown in the seventh column. The aggregation rule incorporating the weighting scheme preserves the same preference ordering.

Table 6 The agricultural producers point of view

Products	% simulated impacts on production			Borda vote			Value added weight	Weighted vote		
	D1	*D2A*	*D2B*	*D1*	*D2A*	*D2B*		*D1*	*D2A*	*D2B*
Soft wheat	−27.64	−18.84	−30.05	1	2	0	2.8	0.03	0.06	0.00
Durum wheat	−36.11	−25.67	−13.19	0	1	2	4.3	0.00	0.04	0.09
Rice	0.20	−0.44	−0.26	1	0	0	1.6	0.02	0.00	0.00
Corn and other cereals	−0.71	0.18	−1.10	1	2	0	9.7	0.10	0.19	0.00
Fodder (silage)	16.32	12.15	17.15	1	0	2	5.6	0.06	0.00	0.11
Non irrigated forage	30.36	22.30	17.45	2	1	0	3.7	0.07	0.04	0.00
Potatoes	1.80	0.27	0.27	1	0	0	1.7	0.02	0.00	0.00
Tomatoes	1.86	0.31	0.30	1	0	0	2.1	0.02	0.00	0.00
Other vegetables	−0.52	0.46	0.45	0	1	1	13.2	0.00	0.13	0.13
Sugar beet	2.48	1.90	2.10	2	0	1	2.3	0.05	0.00	0.02
Soy-bean	−80.67	−74.24	−80.80	0	1	0	0.9	0.00	0.01	0.00
Other industrial crops	−20.68	−13.58	−29.01	1	2	0	0.8	0.01	0.02	0.00

(continued)

Table 6 (continued)

Products	% simulated impacts on production			Borda vote			Value added weight	Weighted vote		
	D1	*D2A*	*D2B*	*D1*	*D2A*	*D2B*		*D1*	*D2A*	*D2B*
Raw tobaccos	2.19	1.52	1.74	2	0	1	2.6	0.05	0.00	0.03
Grapes	0.18	0.09	0.09	1	0	0	8.2	0.08	0.00	0.00
Olives	0.38	0.08	0.10	1	0	0	6.6	0.07	0.00	0.00
Citrus fruit	0.32	0.13	0.13	1	0	0	10.7	0.11	0.00	0.00
Floriculture	2.27	1.70	1.87	1	0	0	8.2	0.08	0.00	0.00
Milk	5.21	3.89	4.21	2	0	1	2.4	0.05	0.00	0.02
Bovine meat livestock	1.22	1.09	1.33	0	0	0	3.7	0.00	0.00	0.00
Forestry	2.19	1.41	1.51	1	0	0	1.6	0.02	0.00	0.00
Sheep and goats	−2.49	−1.94	−3.27	1	2	0	1.0	0.01	0.02	0.00
Other livestock	2.35	1.95	1.43	2	1	0	6.1	0.12	0.06	0.00
Total				23	13	8	100	0.95	0.57	0.40

4.2 The General Interest of the Agricultural and Food Industry

Table 7 proposes a more enlarged view, which includes the general interest of the agricultural and food industry and of other sectors related to agriculture such as the chemical sector. From this wider perspective, it is not just the performance of the single sectors that is important but other factors such as the size of the trade deficit of agri-food products, the impact on land prices, the changes in both farm and non-farm labor employment and the impact on income levels are of primary importance.

It is interesting to note that while the outcome of the voting procedure for the producers' point of view depends only upon the choice of the voting rule, the outcome of the more general interest at the industry level depends also on the choice of the weights, that are now subjective, and the "agenda setting" which selects the number and type of elections. While the agenda setting is less of a problem regarding the agricultural producers' view because the number of elections corresponds to the number of sectors included in the model, in the more enlarged view incorporating also the preferences of the agricultural and food industry, the selection of the number of elections is critical.

The subjective weights are assigned according to the following "conformity rule" based on the degree of proximity of a sector outcome to the objectives of the reform. We summarize the reform objectives as follows: (a) greater market orientation and efficiency, (b) income maintenance and employment, (c) low factor use where, in general, extensive choices are preferred to intensive choices, (d) low environmental impact, (e) sustainability of agriculture and incentives for rural development, (f) fairness in the distribution of the level of support. Based on the subjective evaluation of the social observer we assign a conformity score on the basis of a low (0 score), medium (0.5 score) and high (1 score) level of conformity.

The weighted outcome is presented in the right corner of Table 7, which reports the subjective scores assigned to each item of the agenda in the first column. The order of preference ranks total decoupling (D1) first and the partial decoupling scheme D2A as more preferred to the D2B scheme. Interestingly, the conformity weights change the preference ordering of the non-weighted count.

4.3 The Consumers' Interests

In general, consumers are worried about price instability and the impact of policy changes on the level of the consumption price index. Recently, agricultural and food products have been often blamed to be the main responsible for inflationary pressures. This situation justifies the growing public concern for the impact of the Mid Term Review on both the level and variability of the primary commodities composing the food basket and the associated impact of the consumer price index for food products on the overall level of the consumer price.

Table 7 The agricultural and food industry point of view

Products	% simulated impacts on production			Borda vote			Weights	Weighted vote		
	D1	*D2A*	*D2B*	*D1*	*D2A*	*D2B*		*D1*	*D2A*	*D2B*
Soft wheat	−27.64	−18.84	−30.05	1	2	0	0.5	0.5	1.0	0.0
Durum wheat	−36.11	−25.67	−13.19	0	1	2	0.5	0.0	0.5	1.0
Rice	0.20	−0.44	−0.26	1	0	0	0.5	0.5	0.0	0.0
Corn and other cereals	−0.71	0.18	−1.10	1	2	0	0.5	0.5	1.0	0.0
Fodder (silage)	16.32	12.15	17.15	1	0	2	0.5	1.0	0.0	2.0
Non irrigated forage	30.36	22.30	17.45	2	1	0	1	2.0	1.0	0.0
Other vegetables	−0.52	0.46	0.45	0	1	1	0.5	0.0	0.5	0.5
Soy beans	−80.67	−74.24	−80.80	0	1	0	1	0.0	0.5	0.0
Other industrial crops	−20.68	−13.58	−29.01	1	2	0	0.5	0.5	1.0	0.0
Milk	5.21	3.89	4.21	2	0	1	0.5	1.0	0.0	0.5
Bovine meat livestock	1.22	1.09	1.33	0	0	0	0.5	0.0	0.0	0.0
Sheep and goats	−2.49	−1.94	−3.27	1	2	0	1	1.0	2.0	0.0

(continued)

Table 7 (continued)

Products	% simulated impacts on production			Borda vote			Weights	Weighted vote		
	D1	*D2A*	*D2B*	*D1*	*D2A*	*D2B*		*D1*	*D2A*	*D2B*
Cereal products	−0.05	−0.03	−0.03	0	0	0	0.5	0.0	0.0	0.0
Fresh and preserved meat	0.44	0.38	0.37	0	0	0	0.5	0.0	0.0	0.0
Dairy products	0.47	0.27	0.30	0	0	0	0.5	0.0	0.0	0.0
Fats and oils	−0.60	−0.51	−0.73	0	0	0	0.5	0.0	0.0	0.0
Animal feeds	−2.41	−1.52	−2.22	1	0	1	0.5	1.0	0.0	1.0
Fertilizers	−7.13	−5.01	−4.27	2	1	0	0.5	2.0	1.0	0.0
Pesticides	−3.48	−2.51	−2.59	1	0	0	1	1.0	0.0	0.0
Agri-food trade deficit	7.19	5.62	5.57	0	1	1	1	0.0	0.5	0.5
Land price	18.27	14.23	15.45	0	2	1	1	0.0	1.0	0.5
Hired labor	−0.11	0.08	0.10	0	0	0	0.5	0.0	0.0	0.0
Farm labor	-0.76	−0.35	−0.29	0	1	1	0.5	0.0	0.5	0.5
Farm-hh income (non prof.)	−0.01	0.08	0.11	0	0	0	0.5	0.0	0.0	0.0
Farm-hh income (prof.)	0.55	0.50	0.54	0	0	0	0.5	0.0	0.0	0.0
Total				14	17	10		11.0	10.5	6.5

Table 8 reports the composition of the food basket as derived from the ISTAT Consumer Expenditure Survey for the base year 2001 for the household classes included in the general equilibrium model. The food budget shares are the weights used to compute the change in consumer price index and its variability. As it is apparent by inspecting the overall results, the reform has an impact that may have an economic interest only in the milk sector but the overall impact on both the levels of the consumer index and its variability is negligible. It follows that post-reform pressures on the consumption price index should not be imputed to the agricultural reform.

Also for consumers, the order of preference ranks total decoupling (D1) first and the partial decoupling scheme D2A as more preferred to the D2B scheme.

4.4 Social Welfare, Income Distribution and the Equity-Efficiency Trade-Off

The outcomes of the different policy scenarios affect the distribution of income among socio-economic groups and the level of social welfare. As expected due to the surgical nature of the reform that limits most of the changes to the agricultural sector and the related industries, changes in income are restricted mainly to farm-households (Table 9). In fact, the incidence of the effects varies among farm-household types. The groups experiencing the highest rise in real income are the professional medium-size, large and very large farm households.

The change in relative net output and input prices affects the distribution of value added between sectors and, within sectors, the distribution of value added between wages and rents. These changes, along with changes in the cost of living and lump-sum transfers in the form of SFPs associated with the reform, are responsible for the distribution of income among the household types. The magnitude of farm income changes depend on the size of the elasticity of substitution between labor, capital and land and the intensity of the factor uses due to the post-reform changes in output and factor prices and the size of the lump-sum transfer which is associated with the distribution of rights at the reference situation.

The impact of the different reform scenarios on the distribution of income of the overall society has been measured using Gini coefficients. As shown in Table 9, at the society level, where about 96% of the households are non-agricultural, the differential impact of the reform scenarios on the income distribution is indiscernible. The Gini index of 0.674 is not affected by the reform. The level of inequality for the agricultural society is much lower (0.371) as it is reasonable to expect for a relatively more homogeneous segment of society. As before, it does not vary across scenarios.

The effect on inequality is not the sole dimension of interest in ranking income distributions. It is in general of interest to combine the evaluation with considerations about efficiency as described, in the present context, by changes in society's average level of income. The social evaluation function that we choose to rank any pair of

Table 8 The consumers point of view

Products	Food budget share (%)	D1% change consumer price	D2A% change consumer price	D2B% change consumer price	Weighted D1	Weighted D2A	Weighted D2B
Fish	8.7	0.07	0.04	0.04	*0.006*	*0.003*	*0.003*
Meat	22.8	−0.13	−0.13	−0.12	*−0.030*	*−0.030*	*−0.027*
Milk and Dairy Products	13.8	−1.26	−1.16	−1.18	*−0.174*	*−0.160*	*−0.163*
Bread, Pasta, other cereal products	16.7	0.02	0.00	0.00	*0.003*	*−0.001*	*−0.001*
Vegetables and Fruits	17.6	−0.03	−0.05	−0.04	*−0.005*	*−0.009*	*−0.007*
Oils and Fats	3.8	0.01	0.00	0.06	*0.000*	*0.000*	*0.002*
Sugar, Coffee and Others	7.4	−0.31	−0.26	−0.28	*−0.023*	*−0.019*	*−0.021*
Beverages	9.2	−0.10	−0.09	−0.09	*−0.009*	*−0.008*	*−0.008*
Change in consumer price index for food products					***−0.231***	***−0.223***	***−0.221***
Variance of consumer price index for food products					0.0036	0.0030	0.0031

income distributions is in fact a function that aggregates both a concern for efficiency, as represented by the mean of the income distribution, and a concern for equity, as described by an index of inequality or dispersion of the income distribution:

$$W(x) = V(\mu, I) = \mu^{-G}$$

where $\mu(x) = \sum_{i=1}^{N} x_i / N$ with N being the number of household classes, $I = I(x)$ is an index of inequality of the distribution of income x such as the Gini coefficient,

Table 9 Social welfare rankings

Household class	Initial income level billions €	Population Share	Share of total income	D1% change	D2A % change	D2B % change
Limited-resources	0.353	0.003	0.001	−0.110	−0.008	0.020
Retirement	0.390	0.001	0.001	−0.130	−0.001	0.030
Residential/lifestyle	0.516	0.001	0.001	0.400	0.410	0.450
Farming occupation/lower-sales	0.387	0.005	0.001	0.010	0.100	0.120
Farming occupation/higher-sales	7.044	0.018	0.014	0.770	0.680	0.730
Large family farms	20.656	0.010	0.041	0.490	0.430	0.470
Very large family farms	19.662	0.002	0.039	0.210	0.210	0.240
Rural	61.401	0.130	0.123	0.008	−0.020	−0.030
High income	190.359	0.208	0.382	0.020	−0.010	−0.020
Mid income	163.974	0.415	0.329	0.010	−0.020	−0.020
Low income	33.577	0.208	0.067	0.020	−0.010	−0.020
Total/mean	498.320	1.000	1.000	0.053	0.022	0.020
	Initial	D1	D2A	D2B		
Abbreviated Social Welfare	55.858	55.856	55.867	55.876		
Gini index by scenario						
Society	0.67379	0.67398	0.67398	0.67399		
Agricultural society	0.37110	0.37055	0.37062	0.37060		

and V is a function increasing in its first argument but decreasing in the second argument. Lambert (1989) terms this social evaluation function as the abbreviated social welfare function. As the last row of Table 9 shows, the welfare level of the Italian society, incorporating both a concern for equity and efficiency, is not affected by the reform.

4.5 *Social Conflicts and the Distribution of Political Power*

Different agricultural reform schemes have a significant impact on the distribution of welfare levels especially, as it is rational to expect, within the farming sector. This affects the distribution of political power among the interest groups representing the different farm-household types. Are there conflicts among society? How does political power affect decisions? How will the political bargaining weight of the groups of gainers and losers affect the final policy outcome?

To investigate these questions, we use the Pareto criterion to rank the policy scenarios according to the preferences of each household class, and then inquire whether there are common interests across household classes that can be grouped. This process may identify the existence of possible class conflicts among coalitions representing the interests of the groups.

Pareto optimality ranks possible outcomes (economic states) by constructing a preference ordering among the elements of the choice set using the binary relation xRy stating that "welfare at state x is at least as high as welfare at state y". The ordinal preference relation R is complete and transitive and says nothing about the intensity, or cardinality, of the preferences. A strict preference is indicated as xPy; an indifference situation is indicated as xIy. According to the Pareto principle, the economic state x is Pareto superior to state y if xR_jy for all agent j and xP_jy for at least one agent j. In the context of the present social experiment, every household class is at least satisfied with the outcome of policy scenario y and x and at least one household class is strictly better off with x.

Inspection of Table 10, reporting the changes in welfare levels with respect to the base scenario per each policy alternative under consideration, reveals that it is not possible to establish a unique ranking across scenarios because there is at least one class that is worse off with respect to one of the binary comparisons of interest. However, some classes of households show a consistent preference ranking across scenarios. The limited resources, retirement, residential lifestyle, small farms rank *D2 B.P.D2A.P.D1* as shown in the no-shadow area in Table 10. The medium size, large and very large farm-households consistently rank *D1.P.D2 B.P.D2A*. The urban and rural households, the dark shadow area in Table 10, do not consider the agricultural reform as a political issue of interest as a consequence of their revealed indifference *D2B.I.D2A.I.D1* to the different policy scenarios.

We then assume that the less professional classes of farm-households (limited resources, retirement, residential lifestyle, small farms) form a coalition kept together by the common interests of preferring the D2B partial decoupling scenario to the D2A scheme and total decoupling D1 that we term the "small farm coalition." On the other hand, the professional agriculture (the medium size, large and very large farm-households) coalesce to form the "large farm coalition."

Under a political economy perspective, it is interesting to inquire whether the different interest groups have same bargaining power, as reflected by different political weights, to the point that one of the coalitions dominates the policy arena. Does the choice of political weights affect the Pareto ranking?

We assume that these interest groups know the level of gain or losses that is going to occur, that it has perfect knowledge of Table 10 gains and losses with respect to the base scenario, and that the coalitions are self-interested groups. Further, the intensity with which any group cares about a given policy change is proportional to the relative difference in welfare levels between the three alternative policy scenarios. We also define two weights describing the likely political importance of the two coalitions.

Table 10 Pareto rankings

Pareto ranking—% welfare changes	D1 % change	D2A % change	D2B % change
Limited-resources	−0.07	−0.005	0.004
Retirement	−0.08	0.02	0.04
Residential/lifestyle	0.43	0.45	0.49
Small family farms	0.03	0.12	0.14
Medium family farms	0.78	0.68	0.72
Large family farms	0.54	0.43	0.45
Very large family farms	0.28	0.21	0.21
Rural	−0.02	−0.02	−0.03
High income	−0.01	−0.01	−0.02
Mid income	−0.02	−0.02	−0.02
Low income	−0.008	−0.01	−0.02

Note The different shadow areas identify a unique Pareto ranking

The population share weight is based on the number of the farm-households entering each coalition in line with the one person, one vote paradigm; the value added weight is defined in terms of the value added contribution of each farm-household class.

Table 11 shows that the preference rankings are not affected by the different bargaining power of the two coalitions as captured by the population and value added weight. Comparing the differences in weighted welfare levels at the coalition level, it is reasonable to expect a more intense political action capable to dominate the policy arena from the coalition of the professional farmers who would enjoy a much larger gain in welfare by pursuing the total decoupling scheme (0.3 or 0.5 % change depending on the political weight) as compared to the less professional farmers who are expected to have a weaker motivation to pursue their own interest due to the small expected welfare gains from the reform.

5 Conclusions

The analysis of the impact of the CAP reform on Italian agriculture and the whole economy has been carried out within a micro-funded general equilibrium model capable of differentiating the impact by household type of policy concern. The political economy analysis of the ex ante impact of the reform on the interests of the society's groups has revealed the following positions about the issue of a total or partial implementation of decoupling:

- the producers and agro-food industry's interests: both producers, which give each agricultural sector a different importance based on the value added, and the

Table 11 Ranking and political power

	Value added weight	Welfare level (weighted by value added)			Population share weight	Welfare level (weighted by population			Ranking
		D1	D2A	D2B		D1	D2A	D2B	
Small farmers									
Limited-resources	0.075	−0.005	0.000	0.000	0.007	0.000	0.000	0.000	
Retirement	0.025	−0.002	0.001	0.001	0.024	−0.002	0.000	0.001	
Residential/lifestyle	0.025	0.011	0.011	0.012	0.034	0.015	0.015	0.017	
Small family farms	0.125	0.004	0.015	0.018	0.005	0.000	0.001	0.001	
Mean		**0.007**	**0.026**	**0.031**		**0.012**	**0.016**	**0.018**	*D2 B.P.D2A.P.D1*
Large farmers									
Medium family farms	0.450	0.351	0.306	0.324	0.024	0.019	0.016	0.017	
Large family farms	0.250	0.135	0.108	0.113	0.129	0.070	0.055	0.058	
Very large fam. farms	0.050	0.014	0.011	0.011	0.778	0.218	0.163	0.163	
Mean		**0.500**	**0.424**	**0.447**		**0.306**	**0.235**	**0.239**	*D1.P.D2 B.P.D2A*
Rural and urban households									*D2B.I.D2A.I.D1*

agro-food industry, which weights the industry activities on the basis of the conformity of the impact with the goals of the reform, rank the total decoupling scheme as the most preferred;

- the consumers' interests: the overall impact of the reform on the consumer price index for food products is negligible. As a consequence, potential post-reform pressures on the consumption price index should not be imputed to the agricultural reform;
- the society's interests—the level of inequality and social welfare of the Italian society, incorporating both a concern for equity and efficiency, is not significantly affected by the reform;
- the farming unions' interests: based on the impact of the reform on the welfare levels of the Italian farm-household types, the small less professional farms prefer partial to total decoupling. Professional farm-household types invert the ranking. The urban and rural households are indifferent with respect to the marginal impact of the reform on their levels of well-being. In general, it is reasonable to expect a more intense political action from the coalition representing the interests of the professional farmers who would enjoy a much larger gain in welfare by pursuing the total decoupling scheme.

The implementation of a totally decoupled reform gives back to the market both the allocative and the redistributive function thus favoring greater efficiency in the use of resources in activities and areas of greater comparative advantage. Income levels of farming households are maintained by granting a non distortive lump-sum corresponding to the amount of premia received in the reference situation of year 2001–2002. In general, a totally decoupled scheme would mitigate the problem of distributive justice associated with coupled payments which, by design, benefit mainly the large producers.

The adjustment process induced by the reform may encourage farmers to adopt least cost practices and activities with the objective of minimizing the use of labor and other inputs in agriculture. The increase in pasture production at the expenses of durum wheat in the Italian south is an example of such a change. This modification of the activity portfolio does not lead to an exit from the agricultural industry, but induces the rational adoption of cost-efficient activities and the abandonment of activities, such as durum wheat, that, without the coupled premium, do not cover operating costs in the less efficient farms. This type of change, that we term "disactivation," releases resources which can be employed more efficiently in other sectors of the economy. Agricultural surplus labor may give rise to unemployment, especially in the south, where employment opportunities lack. Lower demand for agricultural inputs coupled with higher costs of the chemical industry has a positive impact on the environment. Higher land prices are expected to curb transactions of land properties but may activate the rental market for land. The land market may also suffer from legal conflicts due to the unclear definition of property and rental rights in the reference situation leading to higher transaction costs.

The adoption of a partially decoupled scheme would reveal greater society's aversion to inequality in recognition of the fact that most of the benefits would accrue to non professional farmers. The evaluation of the pros and cons associated with the adoption of a partially decoupled scheme shows that the benefits would not be sufficient to mitigate the marked structural adjustments associated with the totally decoupled choice, especially in the cereal and sheep and goat production and as a consequence of the "disactivation" process, and would cause a loss of efficiency for the entire sector. Further, a partially decoupled solution has no significant effects in the livestock industry whose productivity is sensitive to market conditions and to the opportunity to gain from the reduced costs of feeding as it can be expected as a consequence of the greater availability of fodder.

In general, an obstacle to reforms is represented by the real or presumed costs of the adjustment imposed on farmers. The New Zealand experience, where in the past decade a market oriented reform without income compensations has been undertaken, teaches that farmers' incomes and the agricultural industry in general, recovered promptly from the initial shocks of *de-regulation* also thanks to other reforms in connected markets and outside of agriculture and the related general equilibrium effects (Rae et al. 2003).

The reform forces a change in the professional farmers' mentality who, despite the larger financial possibilities generated by the single farm payment, have to make production decisions without counting on the previously guaranteed returns stemming from each single activity. As a consequence, the post-reform marketing strategies have to take into account the changed competitive environment, the characteristics of the demand for their products, their competitive advantages and the special strengths of each farm organization.

The reform also imposes a "cultural" change in the quality of the Italian agricultural policy product towards greater market orientation which would foster a restructuring process in favor of better products, more efficient and competitive industries and a more effective integration between agricultural and rural policy. The push towards greater exclusion of the farms already at the margins of agriculture, especially in the South, is not so strong thanks to the single farm payment. It should be remarked, however, that these "less professional" farms are not the main object of interest of agricultural policies, but, more properly, of rural policies, which, curiously not enough, can be financed by the modulation of agricultural policy. What is relevant is then the "coupling" of agricultural with rural policies.

The reform will then be an opportunity rather than a problem, if State and Regions will be using in a modern way market policies that activate (a) the land market in order to favor the consolidation of those farms going out of market, (b) the insurance market and (c) the financial market. This action concerns mainly agricultural policies. If central and regional governments will also intervene by targeting non-professional farms, which do not fully benefit from the reform, by adopting effective rural policies, then, equity *cum* efficiency is a concrete objective.

Acknowledgements The authors would like to thank Jonathan Brooks, Fabio Del Bravo, Joe Dewbre, Jim Johnson, Stefania Lovo, Eleonora Matteazzi, Eugenio Peluso, Maria Sassi and Cristina Salvioni for helpful comments and suggestions. The present research has been carried out with the financial contribution of the Italian Ministry of Agricultural and Forestry Policies.

Appendix

See Tables 1, 2, 3, 4, 5, 6, 7, 8, 9, 10 and 11.

References

De Filippis, F. (2004). Verso la nuova PAC: la riforma del giugno 2003 e la sua applicazione in Italia. Forum Internazionale dell'Agricoltura e dell'Alimentazione, Quaderno n. 4.

De Muro, P., & Salvatici, L. (2001). La PAC nei Modelli Multisettoriali. In G. Anania (Ed.), *Valutare gli effetti della PAC*. Roma: INEA.

European Commission—Directorate-General for Agriculture. (2003a). Reform of the common agricultural policy: A long term perspective for sustainable agriculture, impact analysis.

European Commission—Directorate-General for Agriculture. (2003b). Reform of the common agricultural policy: Medium term prospects for agricultural markets and income in the European Union, 2003–2010.

FAPRI-Ireland partnership. (2003). The Luxembourg CAP reform agreement: Analysis of the impact on EU and Irish agriculture. TEAGASC Rural Economy Research Centre.

Finizia, A., Magnani, R., Perali, F., Polinori, P., & Salvioni, C. (2005). "Il modello Meg-Ismea per l'analisi delle politiche agricole", La Questione Agraria, 4.

Gohin, A., Guyomard, H., & Le Mouël, C. (1999). L'analyse des politiques agricoles: l'apport de l'équilibre général calculable. *Economie Rurale, 25,* 4–10.

Gohin, A., Guyomard, H., & Le Mouël, C. (2002). Tariff protection elimination and Common Agricultural Policy reform: implications of changes in methods of import demand modelling. Working paper 02-01. Rennes: INRA.

Gohin, A. (2002). Le Proposition de Révision à Mi-Parcours de la PAC: Evaluation des Impacts sur l'Agriculture Française à Partir du Modèle MEGAAF. Research Report 02-01. Rennes: INRA.

Hertel, T. W. (1999). Applied general equilibrium analysis of agricultural and resources policies. Staff Paper 99-2, Dept. of Agricultural Economics, Purdue University.

ISMEA. (1997). La tavola delle interdipendenze settoriali del sistema agroalimentare italiano, 1995.

ISMEA. (2004). L'impatto della Riforma PAC sulle imprese agricole e sull'economia italiana. Franco Angeli editore.

ISMEA. (2005). Agricoltura e ruralità: l'indagine socioeconomica dell'agricoltura italiana. Franco Angeli editore.

Lambert, P. (1989). *The distribution and redistribution of income*. Cambridge: Basil Blackwell.

OECD. (1988). Quantifying the Economy Wide Effects of Agricultural Policies. A general equilibrium model. Working Paper, n. 55.

Rae, A., Nixon, C., Lattimore, R. (2003). Adjustment to agriculture policy reform—Issues and lessons from the New Zealand Experience. Paper presented at policy reform and adjustment workshop.

Weyerbrock, S. (1998). Reform of the European Union's common agricultural policy: How to reach GATT-compatibility? *European Economic Review, 42,* 375–411.

A CGE Model for Mauritius Ocean Economy

Pasquale Lucio Scandizzo, Raffaello Cervigni and Cataldo Ferrarese

Abstract The study presented in this chapter applies a dynamic CGE model to the analysis of the "ocean economy", a somewhat new economic construct that reflects both a renewed attention to the potential contribution of ocean resources to economic growth as well as the perception of the ocean as an endangered ecosystem. The empirical application of the study concernes the economy of Mauritius, a dynamic country that has a recently chosen to base its strategy for long term growth on the valorization of its ample and promising ocean resources. A dynamic CGE model especially designed to address some of the key issues of an ocean economy investment strategy was developed as part of a World Bank project, and is based on a large data base and a joint effort with the Statistics Mauritius and a local technical team.

Keywords Ocean · Dynamics · Investment · General equilibrium

JEL Codes C68 · D58 · D78 · E17

The CGE model presented in this chapter has been developed with the support of the World Bank (see Cervigni and Scandizzo 2017 for further references).

P. L. Scandizzo (✉)
University of Rome "Tor Vergata", Rome, Italy
e-mail: scandizzo@uniroma2.it

R. Cervigni
World Bank, Washington, DC, USA
e-mail: rcervigni@worldbank.org

C. Ferrarese
OpenEconomics, Rome, Italy
e-mail: ferrarese@openeconomics.eu

© Springer International Publishing AG, part of Springer Nature 2018
F. Perali and P. L. Scandizzo (eds.), *The New Generation of Computable General Equilibrium Models*, https://doi.org/10.1007/978-3-319-58533-8_8

1 Introduction

The concept of the "ocean economy" is at the same time, a new perspective on sources of economic growth and a recent contribution to the debate on the desirable characteristics of global development policies. While the basic definition is limited to identify the set of activities whose inputs depend on the ocean, the interpretation from the international institutions has hinged on the opportunities offered by the concept to identify and pursue a strategy of green inclusive growth. Thus, for example, according to UNCTAD (2014) "... The concept of the oceans economy, also referred to as the blue economy, is one that simultaneously promotes economic growth, environmental sustainability, social inclusion and the strengthening of oceans ecosystems".

An ocean economy, from the point of view of its definition and concerns, therefore, appears to have a distinctive dual nature, as it reflects both a new attention to the potential contribution of ocean resources to economic growth as well as the perception of the ocean as an endangered ecosystem. For example, technological advances and greater capacity and scope for deep water marine research have led to increased awareness of greater opportunities for undertaking commercial extraction of deep water living and non-living resources. At the same time, however, the same advances have shed a new light on fragility of deep water systems (Norse 1993). The very potential for commercial use of the deep seas, in turn, has both furthered concern about ecosystemic vulnerability and advanced the development of marine research technologies (Broad 1997).

Mauritius launched its first ocean economy roadmap in 2013, which seeks to take advantage of the economic potential of oceans. This potential appears to be especially large for Mauritius, which commands an ocean extension much broader and unknown than its land based economy. Mauritius roadmap places emphasis on the need to make use of the untapped value locked up in the EEZ by ensuring sustainable and coordinated utilization of living and non-living resources. Business opportunities are expected to develop, organized into five clusters: marine services (marine tourism and marine pharmaceuticals); petroleum, minerals and ocean energies; fisheries and aquaculture; seaport related activities. Deep ocean water applications (DOWA) include a plan to pump cold water of approximately 5 C from depths of around 1,000 m to the surface to cool buildings.

Consistent with these expectations, as shown in Table 1, the Ocean Economy (OE) includes a large number of promising sectors where current activities can expand and new activities can be created anew, depending on ongoing and future innovation in technologies and modes of production. Industrial organization, competition and trade are also likely to evolve in response to the expansion of the ocean sectors and to the innovation engendered by many emerging technologies and investment opportunities. As a comparison between Table 1 (all OE sectors) and Table 2 (current OE sectors) readily shows, at present the traditional sectors account for the totality of Mauritius ocean related activities, which offer a limited direct contribution (about 10%) to GDP with tourism and marine leisure accounting together for more than 70% of it. While our preliminary calculations with the 2007 SAM matrix show that these

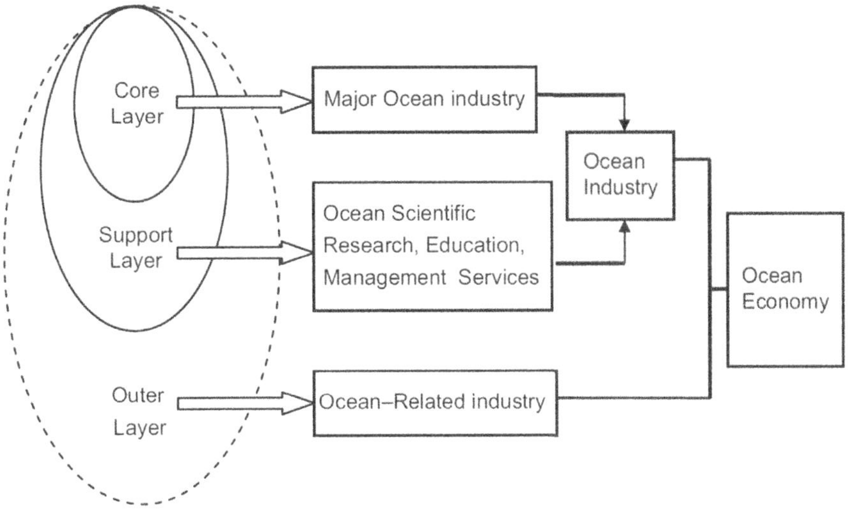

Fig. 1 A graphic concept of the ocean economy

percentage would about double if indirect effects were taken into account, it seems clear that the bulk of any substantive OE expansion should be in the new sectors or, in some new and revitalized form, in some of the traditional sectors hampered by insufficient productive and technological capacity.

Any OE expansion would also entail the need to cope with the environmental threats, coming from a combination of anthropogenic influences and the effects of climate change. Damages to the local biotopes are already significant from climate change and water pollution for coral reefs, lagoons habitats and biodiversity. Coastal erosion and higher vulnerability to weather extreme events has also been increasing.

Given the importance of traditional OE sectors in Mauritius, the above considerations suggest that the economic analysis should focus on the evaluation of the potential of the present OE model of resource exploitation, production and industrial organization. At the same time, new technologies, sectors and investment patterns, broadly included in the government OE programs should also be carefully evaluated. For the traditional sectors the data base available appears adequate for a quantitative analysis based on SAM-CGE modelling. For the new sectors, however, and several technological developments that may affect both traditional and modern sectors, data of current operations are not available in Mauritius. They may have to be either obtained from experimental/simulation data or from benchmarking countries that have already investing in the new technologies (Table 3).

For the "new sectors", past experience and theoretical knowledge can both provide basic narratives on the evolution and the prospects of related innovations, in addition to specific data, as shown for the exemplary cases of marine engineering and marine biotechnologies in Tables 4 and 5 (Fig. 1).

The classification proposed is from Zhao et al. (2014).

Table 1 Classification and definition of major ocean industries and related data needs for Mauritius study

Ocean sectors	Definition	Data available	Data needs
Marine fishery	Includes mariculture, marine fishing, marine fishery service industry and marine aquatic processing, etc	Present technologies and modes of exploitation	New modes of exploitation. New technologies. Input output data
Offshore oil and gas industry	Refers to production activities of exploring, exploiting, transporting and processing raw oil and natural gas in the ocean	None	Current and new technologies Input output data
Ocean mining industry	Includes the activities of extracting and dressing beach placers, beach soil chloride and sand, submarine geothermal energy, and coal mining and deep-sea mining, etc	None	Current and new technologies Input output data
Marine salt industry	Refers to the activity of producing salt products with the sodium chloride as the main component by utilizing seawater, including salt extracting and processing	Present technologies and modes of exploitation	New technologies Input output data
Shipbuilding industry	Refers to the activity of building ocean vessels, offshore fixed and floating equipment with metals or non-metals as main materials as well as repairing and dismantling ocean vessels	Present technologies	Investment typologies and technologies. Possible Benchmarking with other countries for Input output data
Marine chemical industry	Includes the production activities of chemical products of sea salt, seawater, sea algal and marine petroleum chemical industries	No data available, but this is one sector singled out in the government road map	Investment typologies and technologies Benchmarking Input output data

(continued)

Table 1 (continued)

Ocean sectors	Definition	Data available	Data needs
Marine biomedicine	Refers to the production, processing and manufacturing activities of marine based medicines and marine health care products by using organisms as raw materials (or by extracting these organism's useful components)	No data available, but this is one sector, as the chemical one, included in the government road map	Investment typologies and technologies Benchmarking
Marine biotechnology	Refers to the production, processing and manufacturing activities of marine based biotechnological products	No data available, but this is one sector, as the chemical and the biomedical ones, which appears to be important for the government road map	Investment typologies and technologies Benchmarking
Marine engineering and building industry	Refers to the architectural projects construction and its preparations in the sea, at the sea bottom and seacoast for such uses as marine production, transportation, recreation, protection, etc., including constructions of seaports, coastal power stations, coastal dykes, marine tunnels and bridges, land terminals of offshore oil and gas fields as well as building of processing facilities, and installation of submarine pipelines and equipment, but not the projects of house building and renovation	Some data on ongoing activities (see Table 2) and from the port master plan on investment programs	Investment typologies and technologies Input output data Benchmarking
Marine electric power	Refers to the activities of generating electric power in the coastal region by making use of ocean energies and ocean wind energy. It does not include the thermal and nuclear power generation in the coastal area	No data from ongoing activities and research	Investment typologies and technologies Input output data Benchmarking

(continued)

Table 1 (continued)

Ocean sectors	Definition	Data available	Data needs
Seawater utilization	Refers to the activities of the direct use of sea water and the seawater desalination, including those of carrying out the production of desalination and applying the seawater as water for industrial cooling, urban domestic water, water for fire fighting etc., but not the activity of the multipurpose use of seawater chemical resources	No data from ongoing activities	Investment typologies and technologies Input output data Benchmarking
Marine communications and transportation Marine ICT	Refers to the activities of carrying out and serving the sea transportations with vessels as main vehicles, including ocean-going passenger transportation, auxiliary activities of water transportation, pipeline transportation, loading, unloading and transport as well as other transportation and service activities. The sector includes also a whole new wave of ICT innovations, including radar, sensor and satellite based on ship and deep see applications	Some data from ongoing activities (see Table 2)	Investment typologies and technologies Input output data Benchmarking
Coastal tourism	Refers to the tourist related activities that take place in the coastal zone, on sea islands as well as recreational activities that use the ocean; including water based sports, marine mammal and bird watching, etc	Data available	New investment typologies and technologies

Source Our elaboration

Table 2 Mauritius ocean economy activities

				Contribution to GDP (%)		
	2012	2013	2014	2012	2013	2014
GDP at basic prices	**302.617**	**322.937**	**342.179**			
Activities/organisation						
Salt Production	**12**	**11**	**10**	0.00	0.00	0.00
Seafood fishing and processing	**4.314**	**4.833**	**4.687**	1.43	1.50	1.37
of which:						
Aquaculture	***28***	***30***	***37***	*0.01*	*0.01*	*0.01*
Fishing other than aquaculture	***281***	***398***	***754***	*0.09*	*0.12*	*0.22*
Fish processing	***4.005***	***4.405***	***3.896***	*1.32*	*1.36*	*1.14*
Ship building and maintenance	**292**	**433**	**471**	0.10	0.13	0.14
Storage	**689**	**698**	**753**	0.23	0.22	0.22
Sea transport	**141**	**152**	**132**	0.05	0.05	0.04
Services allied to transport	**4.776**	**5.007**	**5.473**	1.58	1.55	1.60
Hotels and restaurants	**16.181**	**14.773**	**16.134**	5.35	4.57	4.72
Leisure boat activities	**3.442**	**3.801**	**4.174**	1.14	1.18	1.22
Ship store and bunkering	**920**	**842**	**717**	0.30	0.26	0.21

(continued)

Table 2 (continued)

				Contribution to GDP (%)		
	2012	2013	2014	2012	2013	2014
Freeport activities	**1.596**	**1.735**	**1.884**	0.53	0.54	0.55
Shipping division (of Ministry of land transport and shipping)	**16**	**19**	**20**	0.01	0.01	0.01
Mauritius oceanography institute	**17.4**	**31.6**	**20.7**	0.01	0.01	0.01
External communications division	**6.1**	**7.7**	**8.5**	0.00	0.00	0.00
Ministry of fisheries	**120.1**	**140.8**	**149.2**	0.04	0.04	0.04
Tourism authority	**23.6**	**42.9**	**45.1**	0.01	0.01	0.01
Beach authority	**14.6**	**18.5**	**19.3**	0.00	0.01	0.01
Fisherman welfare fund	**2.2**	**3.9**	**4.2**	0.00	0.00	0.00
National coast guard	**323.6**	**389.2**	**419.1**	0.11	0.12	0.12
Total	**32.887**	**32.939**	**35.122**	**10.9**	**10.2**	**10.3**

Source Our elaboration on data from the Mauritius office of statistics

Table 3 Climate changes: narratives and related threats, opportunities and options

Narrative main plots	Threats	Opportunities	Options
The curse of modern economic activities: • Loss of biodiversity; • Inflexible practices; • Overfishing and loss of stock • Pollution • Inappropriate water management	• Irreversible loss of biodiversity; • Mismatch between fishing techniques and changes caused by CC; • Higher frequency and intensity of storms and inundations	• Protection of natural habitats • Re-generation of local species and varieties; • Re-introduction of local species and varieties; • Experimentation in ecological fishery and aquaculture	• Boost research on appropriate biotechnologies; • Rehabilitate fishermen towards more flexible practices; • Improve land planning/zoning • Introduce appropriate water management techniques/improve the existing ones
The impact developments in the economy: • Loss of traditional skills; • Proliferation of commercial fishery and loss of biodiversity; • Pollution; • Inadequate treatment of waste water	• Shift of population and migration toward the coast and the urban areas; • Education systems neglecting environmental management; • Sea level rising; • Rising temperatures and dwindling water supplies; • Increasing pollution	• Diversification of economic activities (industry, ecotourism and selective fishery); • Relocation of population and modernization of infrastructure	• Higher levels of training; • Educational system focused on ecological applied knowledge; • Cooperation among the fishermen to purchase implements and to manage the commons; • Re-building and maintenance of the mangrove ecosystems; • Urban agriculture
Loss of biodiversity: • Degradation of mangrove forests and lagoons systems and territory maintenance; • Low prices for fishery products due to marketing (transport and distribution) monopolization by few traders	• Degradation of mangrove ecosystem; • Increase in frequency and intensity of storms and destructive floods; • Rise in sea level; • Lowering of the water table; progressive salinity of underground water; • CC as an aggravating factor, mainly for drought prone areas; • Dune breaching and coastal erosion along the coastal zone associated with wave, storm surge and river discharges peak and duration during tropical storm events	• Liberalization of marketing activities; • Modernization of infrastructure and production processes; • Rehabilitation of mangrove forests; • Young people training and education on sustainable fishery and other environment sensitive activities; • Infrastructure maintenance techniques and CC risks; • If new well planned infrastructure is develop, new jobs and more sustainable activities could be generated	• Improvement of the governance structure; • Training and human capital formation; • Ensuring remunerative prices for fishermen; • "Blue" energy obtained from the sea (currents and waves); • Coastal resilience; solution of erosion problems; • Improvement of water quality in coastal areas (lagoons and sea); • Innovative coastal solutions for ports and tourist beaches

(continued)

Table 3 (continued)

Narrative main plots	Threats	Opportunities	Options
Property rights and market power: • Economic growth stifled by over-fishing, land tenure problems, unemployment, lack of investment	• Population growth; • Threats to the aquifers • CC aggravating the conditions of poor fishermen	• Investment in post-harvesting and marketing activities • Relocation • Recovery of the aquifers • New infrastructure and more rational settlements	• Cooperation for water management; • Post-harvesting and marketing activities; • Diversification of fishery and aquaculture • Marine and forest ecotourism; • Vesting of property rights and governance improvements; • Community programs: insurance and credit
Modernizing fishery: • Diffusion of part-time fishing; • Diffusion of aquaculture; • Increasing scarcity of catch; • Overfishing; • CC accelerating this movement	• Increasing erosion of the land and the water base; • Deepening poverty; • CC as a source of structural degradation of the environment • CC as a major threat to coastal and ocean assets	• Develop marine eco- tourism; • Improve the governance of the marine commons	• Training for new types of jobs (e.g. fishermen as tourist operators); • Marketing and processing activities; • Cooperation among fishermen to exploit scale economies in processing and marketing • Community level programs of cooperation, microcredit, marketing, processing and development
Women in the local economy: responsibilities without power: • Crucial role of women in the local economic activities; • Women have little power in socio-economic decision-making processes; • Women lack of adequate education, skills and tools to efficiently manage and perform economic activities	• No advances in the women socio-economic conditions; • Failure of the empowerment process; • Increasing health risks; • CC aggravates women's plight in fishery and commerce; • CC causes further impoverishment of women's conditions given their lack of technical knowledge	• Improve women capabilities and education level; • Develop women capacity to efficiently cope with future CC uncertainty; • Raise men awareness about the potential role of women in resource management, tourism, fishery and commerce; • Improve future development projects' design and management at the institutional level	• Constant training for a long-term sustainable empowerment of women on water and other resource management techniques, health and environmental risks; • Involvement of men aimed at their full understanding of women potential capabilities and contributions to community; • Rethinking of the design and management of development projects at the institutional level

(continued)

Table 3 (continued)

Narrative main plots	Threats	Opportunities	Options
Coastal erosion and the threat of sea rise • Coastal erosion is mainly the result of lack of planning and unregulated construction of barriers along the coast • Sea level rise increases the gravity of erosion, but inundation is probably go-ing to be limited • The basic strategy of adaptation is through low cost measures aimed to re-constitute the original habitat and hydraulic conditions	• Progressive erosion of the coastal area. • Destruction of the natural habitat (mangroves and lagoons) • Inundation of cities and inhabited areas • Greater vulnerability to extreme meteorological events	• Re-establish the original habitat • Re-equilibrate the ecosystems. • Recover the biological equilibrium	• Re-constitute the hydraulic conditions that allow mangroves to grow and prosper • Parcel coast land to ensure continuous care of beaches and water-sand circulation • Move part of the population of the areas exposed to inundation from sea level rise • Enact measures to care for the ecosystem (e.g. vulnerable species) • Environmental education • Marine planning and Integrated coastal zone management

Source Our elaboration

Table 4 Marine engineering and building industry: data needs from benchmarking

Main technologies	Technology and process components	Data needs
• Business-process technologies	• The principal "up front" management processes and other management activities, notably technologies for preliminary design, bidding, estimating, and sourcing, that are linked to the marketing capabilities of shipbuilders;	Description of processes and technologies. Input output data
• System technologies	• Engineering systems, such as process engineering and computer-aided design and manufacturing, that support shipyard operations;	Input output and cost data, labor requirement by skill
• Shipyard production processes technology	• Fabricating, assembling, erecting, and outfitting vessels;	Input output, Investment costs, labor requirements, employment
• New materials and new product technologies	• Innovations, including new designs and new components, that meet particular market needs	Input output, effects on productivity and the value chain. Effects on competitiveness

2 Building a SAM for the Ocean Economy

Constructing an economic-environmental SAM for the Mauritius Ocean Economy (OE) requires the integration of data from different sources. Building a SAM at the national level with the same aggregate detail of national accounts is straightforward. However, the construction of a SAM disaggregated around specific OE subsectors is more challenging because it requires the integration of micro and macro data and investigating in greater depth monetary and non-monetary transactions within some of the major sectors and stakeholders' communities, that are represented in the national accounts. As Table 6 shows, monetary transactions can be distinguished in a conventional SAM by sector of origin and destination, thus conveniently partitioning the economy in two broad sectors concerning, respectively, the Ocean and the non-Ocean based activities. Non-monetary transactions are particularly important for natural resources and environmental goods, which constitute a localized stock of wealth. This can be fully taken into account in the Social Environmental Accounting Matrix (SEAM), as shown in Table 7. Furthermore, because such a localized wealth is subject to the challenge of various forms of dynamic uncertainty, among which climate change looms increasingly larger, it can be considered *contingent wealth*, in the manner that can identified and measured through *real option theory* (Scandizzo 2010).

Table 5 Marine biotechnology: data needs from benchmarking

Pursuing human health and well-being Sustainable supply of high quality food Developing sustainable sources of energy alternatives to crude oil and gas New industrial products and processes with lower GHG emission Protection and management of the already stressed marine environment … and as a much needed new source of innovation and economic growth in many countries	The global market for marine biotechnology is projected to reach US $4.6 billion (2017) with growth rates of 4.0% per year. Possible prospects and/or targets for Mauritius	Investment prospects in different areas. Ongoing R&D in Mauritius and elsewhere
Main activities in EEZ and deep sea—open ocean	Commercial activity driven by existing companies or biotechnology companies (e.g. Aquaculture, drug companies)	Prospects for marine biotechnology SMEs Investment costs Investment and i-o data on production or extraction-based aquaculture
Investment challenges	Allocation, analysis, advocacy, accountability different drivers	Financial needs of established companies and SMEs;
Need to support marine specific capabilities and needs • (e.g. Vessels, ocean monitoring, collection, model organisms)	• Take advantage of other biotechnology RI • (e.g. genomics platforms, big data analytics, synthetic biology, nanotechnology) Standardization, integration, harmonization	Input output data from R&D
Existing frameworks for 'access and benefits' for EEZ and beyond	Coordination of marine resource 'activities' within EEZs (balancing productivity with sustainability)—'regional' Development of tools to monitor impact on ocean Monitoring frameworks at regional and International level	Ongoing and planned initiatives in Mauritius and other countries
Global challenges	Technology transfer/capacity building to allow for development of local solutions in • Aquaculture/food production • Integrated biorefineries	Ongoing and planned initiatives in Mauritius and other countries. Data on R&D, experimental projects

Source Our elaboration

Table 6 Structure of a social accounting matrix for the ocean economy

		Expenditures										
		1a. Ocean based activities	1b. Non ocean based activities	2. Products	3. Factor income	4. Natural resources	Households	5. Enterprises	6. Government	7. Capital accounts	8. Rest of World	9. Total incomes
Incomes	1a. Ocean based activities			Production supply by industries								Total domestic production
	1b. Non ocean based activities			Production supply by industries								
	2. Products	Intermediate consumption by industries	Intermediate consumption by industries				Household consumption		Government consumption	Investment	Exports	Total use of products
	3. Factor income	Ocean based value added	Non ocean based value added								FI from Abroad	Total factor income
	4. Natural resources	Blue value added	Green value added									
	5. Households				Compensation of employees		Inter-household transfers	Distributed profits	Transfers to households		Remittances from Abroad	Household income
	6. Enterprises				Net operating surplus				Transfers to enterprises		Enterprise transfers from Abroad	Enterprise income

(continued)

Table 6 (continued)

	7. Government	Taxes and subsidies on production		Taxes and subsidies on products			Taxes on household income	Corporate taxes			Other transfer from Abroad	Government revenue
	8. Capital account	Consumption of fixed capital				Maintenance and reconsti-tution	Household saving	Enterprise saving	Government saving		Capital account BOP	Total savings
	8. Rest of world			Imports	Factor payments to abroad		Remittances to abroad	Transfers to abroad	Government payments to ROW			ForEx outlays
	9. Total expenditures	Total production cost		Total product supply	Factor outlays, NDP at factor cost		Household expenditure	Enterprise expenditure	Government expenditure	Investment	ForEx earnings	

Source Our elaboration

Table 7 Structure of a social environmental accounting matrix (SEAM)

	Production sectors	Households	Capital formation	Government	Rest of the world	Primary factors Income	Natural resource supply	Natural resources as primary factors	Natural capital formation	Residual and natural resource recovery activities
Production sectors	A_{XX}	C_{XH}	I_X	C_{XG}	E_X		A_{XQ}			A_{XR}
Households						Ω_{HZ}		Ω_{HW}		
Capital formation		S_H		S_G	S_r					
Government		T				Ω_{GZ}		Ω_{GW}		
Rest of the world	M_X	M_H				Ω_{rZ}		Ω_{rW}		
Primary factors demand	F_{ZX}						F_{ZQ}			F_{ZR}
Natural resource supply	A_{QX}		I_Q				A_{QQ}			A_{QR}
Demand of natural resources as primary factors	F_{WX}						F_{WQ}			F_{WR}
Natural capital		S_{HK}		S_{GK}			Ω_{KZ}	Ω_{KW}		
Residual recovery activities	A_{RX}		I_R				A_{RQ}		K_R	A_{RR}

The accounts in the tables above and related data needs and availability can be briefly described as follows:

(a) **Production sectors**

1. **Structure**

Production sectors accounts describe production, sales and purchase of goods and services produced. In the case of the Ocean Economy, sea food related activities will be analyzed in some detail, both in terms of disaggregated production of commodities, post harvesting activities, and **investment**. Production sectors can be disaggregated into different accounts such as: **pelagic and demersal fisheries, aquaculture,** agriculture; Industry, including metal work, **food preparation** and chemistry, textiles, construction and carpentry; Services, including stores, communications, hotels and other accommodations, transportation, port services, public and private school, other services.

(b) **Natural resource sectors**

This category includes all activities that transform natural resources as primary inputs into intermediate goods for industrial processing. For the Ocean Economy they may include **the fish canning industry, energy production and distribution, water distribution as well as the new projects envisaged for marine biotechnology, and other technological innovations involving the use of ocean resources.**

(c) **Residuals**

All activities to transform, recycle and/or re-process residuals of production and consumption activities, including emissions, all forms of pollution, water treatment, safety processing of industrial waste, waste disposal etc. Also included are all activities that reconstitute or make easier to reconstitute, renew or replace, wholly or partially natural capital.

(d) **Capital formation**

The Capital formation account includes all formal and informal transactions concerning the formation of savings as well as the various typologies of credit in the economy, including transactions from the formal banking system, microcredit and all financial transactions that play a crucial role to supply an outlet to savings and a source of credit to consumer-producer households.

(e) **Rest of the world**

Transactions between the local economy and rest of the world are recorded in the rest of world accounts. The Rest of the World account can be disaggregated into three different components including **Rest of the Area**, **Rest of the Country** and **Rest of the World** to describe domestic and international trade.

(f) **Primary production factors**

Accounts for primary factors include various types of material and immaterial capital. Material capital comprehends the result of various forms of physical investment, such as cultivated land, orchards, irrigation equipment, tractors, vehicles, local roads, post harvest facilities, factories, machinery etc. Non material capital may include, inter alia, human capital (e.g. education), technical knowledge, and various forms of social capital.

(g) **Natural resources as primary factors**

This category includes **fresh and marine water (including lagoons), forests, sources of energy (including water falls, wind, fossil combustibles and other underground resources)**, land, and commons and public goods that are typically the object of intense governance activities from communities as well as governments.

(h) **Natural capital formation**

Natural capital includes all formal and informal transactions concerning the growth, the renewal as well as the depletion of the existing stock of natural resources.

(i) **Stakeholders**

Stakeholders' accounts include various types of institutions and groupings of individuals, such as households and household members, enterprises, local and nonlocal government, cooperatives, associations, traders, rest of the sub-regional economy, rest of the world. **Workers** and **Households** are disaggregated by socio-economic criteria such as education and income. Enterprises can also be disaggregated, for example by distinguishing small, medium and large firms, and formal and informal business by size. Government collects taxes, and distributes them through transfer payments. Other important stakeholders are the various forms of association and community level institutions that provide the governance of the commons as well as a variety of social functions.

Formally, the accounts in the SAM-ESAM described by Tables 6 and 7 can be represented as a system of equations, based on the following definitions:

X Vector of production activities for goods and services
Q Vector of production/distribution activities for natural resources
R Vector of activities of treatment/recycling/disposal of Residuals
C Matrix of Consumption Goods by Households
E Vector of net exports of final goods
M Vector of net imports of intermediate goods
Z Vector of employment levels of primary factors of production by industrial activity,
V Vector of incomes of primary factors
P_Z Vector of prices of primary factors

P_W Vector of prices of natural resources (as primary factors or a form of natural capital)
P_X Vector of prices of activities providing goods and services
P_Q Vector of prices of activities providing natural resources
P_R Vector of prices of activities of treatment/recycling/disposal of residuals
Y_H Vector of households' incomes from primary factors
Y_H Vector of Households' incomes from natural resources as primary factors

According to the balance of each column and row, the key equations of the economic model that can be constructed on the SAM in Table 1 can be developed as follows:

$$X = A_{XX}X + A_{XQ}Q + A_{XR}R + C_{XH}Y_H + C_{XG}Y_G + I_X + E_X \quad \text{Material Balance for Goods and Services} \tag{1}$$

$$Q = A_{QX}X + A_{QQ}Q + A_{QR}R + E_Q \quad \text{Material Balance for Natural Resources} \tag{2}$$

$$R = A_{RX}X + A_{RQ}Q + A_{RR}R + K_R \quad \text{Material Balance for Residuals} \tag{3}$$

$$Z = F_{ZX}X + F_{ZQ}Q + F_{ZR}R \quad \text{Primary Factor Employment} \tag{4}$$

$$V_Z = P_z^{'}Z \quad \text{Primary Factor Incomes} \tag{5}$$

$$W = F_{WX}X + F_{WQ}Q + F_{WR}R \quad \text{Natural Resource Flow} \tag{6}$$

$$V_W = P_W^{'}W \quad \text{Incomes from Natural Resources} \tag{7}$$

$$Y_H = \Omega_{HZ}V_Z + \Omega_{HW}V_W \quad \text{Household Income Formation} \tag{8}$$

$$Y_G = \Omega_{GZ}V_Z + \Omega_{GW}V_W \quad \text{Government Income Formation} \tag{9}$$

$$Y_K = \Omega_{KZ}V_Z + \Omega_{KW}V_W \quad \text{Natural Capital Income Formation} \tag{10}$$

$$EDP = P_X'X + P_Q'Q + P_R'R \quad \text{Environmentally Adj. Net Domestic Product} \tag{11}$$

$$GVA = V_Z + V_W \quad \text{Green Value Added} \tag{12}$$

Note also that the budget constraint will imply that the following equality will have to hold for natural capital:

$$K_R = \Omega_{KZ}V_Z + \Omega_{KW}V_W + S_{HK} + S_{GK} \tag{13}$$

This equality means that the value of investment in natural capital recovery must equal the payments to natural capital provided by natural resources and by the capital formation sector (savings). If these payments are sufficient to recover (in form of maintenance, decontamination, recovery etc.) the capital consumed by economic activities, the ecosystem will be in equilibrium. If they are not sufficient, some level of depletion will occur.

A dynamic interpretation of (13) can also be obtained by considering the logistic model for natural capital:

$$d\Psi = K = g\Psi(1 - \frac{\Psi}{\Psi*}) + K_R - (\Omega_{KZ}V_Z + \Omega_{KW}V_W + S_{HK} + S_{GK}) \tag{14}$$

where Ψ is the stock of natural capital and $\Psi*$ its maximum sustainable level. Equation (14) implies that in the steady state:

$$K_R + g\Psi(1 - \frac{\Psi}{\Psi *}) = \Omega_{KZ} V_Z + \Omega_{KW} V_W + S_{HK} + S_{GK} \tag{15}$$

In other words, in equilibrium, natural capital is naturally recovered through the natural dynamics of the resource and may also be recovered through industrial processes. The total amount of recovery (i.e. total expenditure) must equal total revenue, i.e. the income provided by the primary factors of production (including natural resources) plus institutions' savings. The two equilibrium conditions in (13) and (14) will coincide If the stock of natural capital is at its maximum sustainable level ($\Psi = \Psi *$).

3 The Base CGE Model

We developed estimates for a Mauritius Social Accounting Matrix (SAM) for 2015 based on the definitions in Tables 6 and 7 above, using a combination of macro and micro-economic statistics as well as the National SAM estimated by Statistics Mauritius (under the aegis of the Ministry of Finance and Economic Development) and published for year 2007. The estimates were obtained by applying a maximum entropy algorithm using the methodology outlined in Scandizzo and Ferrarese (2015). Altogether, the SAM estimated comprises 18 products (goods and services), 7 factors of production (6 classes of labor and one class of capital), and seven institutional accounts (households by income group, corporations, government, capital formation and rest of the world). The model in base version is estimated at 2015 values, using national account historical series made by Statistics Mauritius, and disaggregated on the basis of the 2007 Input-Output classification for 30 economic sectors.

In order to provide a full representation of the ocean economy in the model, we identified both current and potential developing sectors based on the direct and indirect use of the ocean, drawing on the classification of the Mauritius Office of Statistics. To complete the model with the environmental sectors, we used FAO statistics on water. The environmental sectors, included as factors of production are: ocean, green water, blue water, and wetland, and, as recipients of rents from environmental capital, water resources, natural capital, and emissions. The contribution of these variables to value added (as a form of non-remunerated environmental costs) and in the distribution of rents are estimated through a combination of the maximum entropy and the Wolsky disaggregation algorithm (Scandizzo and Ferrarese 2015).

4 Key Features of the Dynamic CGE Model

The dynamic CGE model aims to capture some of the relevant features of the Mauritius economy today and their potential evolution over time. The model's core is the SAM matrix described above for the entire economy of 118 sectors, commodities, factors, and institutions, and it projects the economy over time based on a moving equilibrium algorithm and a basic Solow-like structure of capital accumulation and growth. As a result, the time path followed by the model converges to a steady state where growth is solely determined by technological progress (productivity increases) and population growth. Considering both these factors as exogenous (in other words, the model converges to a zero growth steady state), the model can be used to investigate differential capital accumulation and resource allocation strategies and their trajectories over time.

The dynamic model presents several characteristics that should be noted carefully before interpreting its results. First, the model is based on the idea that the SAM coefficients represent budget shares. This implies that the SAM sector columns can be interpreted as value shares corresponding to Cobb Douglas production functions, and the consumption columns as linear expenditure functions from Cobb Douglas utility functions. Since the shares change with the model solutions under the impact of capital accumulation and technical changes, technology and preferences across several periods are represented by Cobb Douglas splines. As a consequence, the model exhibits decreasing returns to scale due to convex technologies and general equilibrium effects (price changes) in each simulation period, but its behavior over time depends on the combination of convexity with the transition from the older to the new technology/preferences. This means that investing in one sector will tend to increase less than proportionally its output within a single simulation period. However, the effects across periods will be path dependent—that is, they may be positive or negative and larger or smaller, depending on whether the share of the sector has increased or decreased in the preceding periods. Second, large changes may find equilibria that are unstable, in the sense that subsequent small shocks may tend to produce large effects. Third, the impact of a single project or program will be different based on whether it is considered by itself or as part of a more complex strategy. This is also due to effects of scale, but in this case, because of possible complementarities between different projects, economies rather than diseconomies of scale may ensue. Fourth, while all simulations should be compared to a counterfactual, unlike the static CGE case, the counterfactual cannot be unique and is necessarily specific, as its first best alternative, of each project or program that is simulated. Fifth, the pattern of productivity increases depends on specific hypotheses on the parameter changes of each simulation. These effects are important to understand the potential of alternative policy strategies in the long run, since productivity changes may significantly modify, and even reverse, the merit order of policy alternatives that are ranked on the basis of current production and consumption parameters.

5 Calibrating and Testing the Dynamic Model

To assess the capacity of the model to replicate the performance of the economy of Mauritius in the past, we simulated the effect of the investments of the past 9 years.

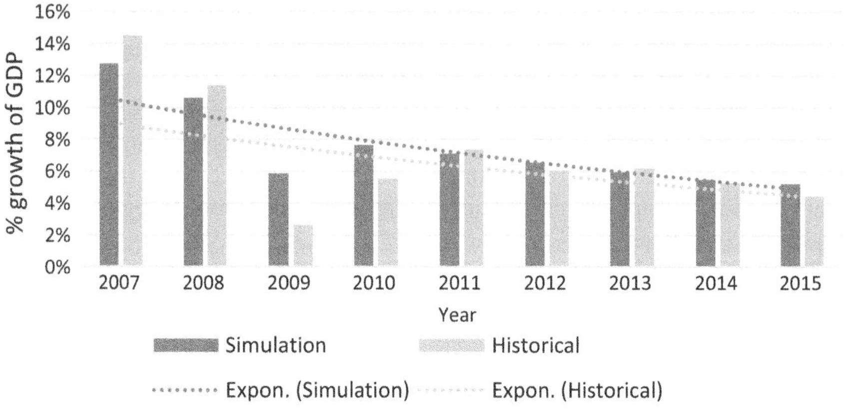

Fig. 2 Simulated versus historical growth in GDP. *Source* Authors' computations and statistics Mauritius

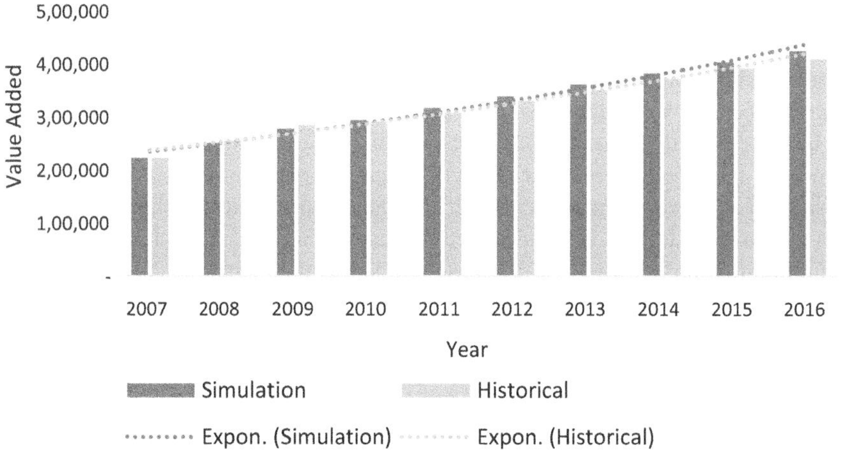

Fig. 3 Simulated versus historical value added. *Source* Authors' computations and statistics Mauritius

This allows us to evaluate if the CGE model captures the same trends observable from the historical data.

The first simulation is based on the rate of economic growth and the levels of value added generated by the model. The validation data are from Statistics Mauritius and show the gross fixed capital formation in years 2006–2014 (Tables 8 and 9).

Figures 2 and 3 show the results of model simulations, conditioned to the investment figure in Table 8, to backtrack historical figures of value added growth (Fig. 6) and value added level (Fig. 7).

Table 8 Gross fixed capital formation at current prices by type and use, 2006–2016, Million MR

	2006[1]	2007[1]	2008[1]	2009[1]	2010[1]	2011[1]	2012[1]	2013[1]	2014[1]
A. Building and construction work	27,501	35,987	45,278	48,809	52,166	53,165	54,405	50,111	47,016
Residential buildings	9,768	11,663	15,281	16,531	18,769	22,298	22,043	23,286	21,532
Non-residential buildings	10,666	17,794	22,162	22,016	21,530	17,698	18,837	15,925	12,877
Other construction work	7,067	6,530	7,835	10,262	11,867	13,169	13,525	10,900	12,607
B. Machinery and equipment	24,194	25,253	22,251	25,621	22,230	24,402	24,779	27,507	26,973
Aircraft	5,675	2,515	0	3,400	0	0	0	0	0
Marine vessel	0	0	600	0	0	0	0	2,630	2,013
Passenger car	2,497	3,406	3,635	2,864	3,459	3,548	3,953	3,714	3,630
Other transport equipment	1,945	2,433	2,288	2,228	2,395	2,678	2,976	2,618	2,645
Other machinery and equipment	14,077	16,899	15,728	17,129	16,376	18,176	17,850	18,545	18,685
Gross fixed capital formation	51,695	61,240	67,529	74,430	74,396	77,567	79,185	77,618	73,989

Source Statistics Mauritius

Table 9 Characteristics of OE2 and CF best investment trajectories (Year 1–Year 10)

	OE2	CF
Total OE supply year 10/Total OE supply year 1	1.88	1.47
Average contribution to growth rate (%)	3.17	2.93
Additional OE2 investment (000MR)	78,691,571	
Additional OE2 investment (million US$)	2,248	
GDP growth (000MR)	105,182,087	86,012,113
NPV (5%) GDP increase 000MR)	210,619,461	164,983,628
NPV (5%) Additional OE2 investment (000MR)	191,594,091	191,594,091
NPV (5%) Additional OE2 investment (million US$)	4,046	4,046
Total investment/GDP at the 10th year	0.29	0.28
NPV (5%) GDP increase/ADD INV	1.49	1.23
NPV (5%) export-import increase (000MR)	1,582,736	998,055

6 Investing in the Ocean Economy

In this set of experiments, we investigate the possible deployment of an investment strategy aimed at developing Mauritius' ocean economy (OE) by simulating trajectories for different autonomous investment increases with respect to the baseline. The model is thus solved by imparting an exogenous shock of investment expenditure, concentrated in the ocean economy sector, with the general aim of doubling the size of the ocean economy within a time horizon of 10–15 years. The investments are assumed to be concentrated on four main OE sectors—(i) fishery and sea food processing, (ii) sea transport ports and related services, (iii) ICT, and (iv) sewage and water treatment. The trajectories simulated are evaluated on the basis of a welfare function of the Stone-Geary variety, with consumption weights being a negative function of households' wealth. The best trajectory found corresponds to an investment path characterized by a gradual increase in OE investment from $25 million in year 1 to ten times as much in year 10. Tables 10 and 11 report other characteristics of the OE2, and show the trajectories of two of the main macro-effects of the OE2 simulated scenario. The ocean economy would more than double in 10 years, even though the investment stimulus only increases gradually to ten times its first year size, from $25 million, or 2.37% of total investment, to $2.5 billion, or about 30% of

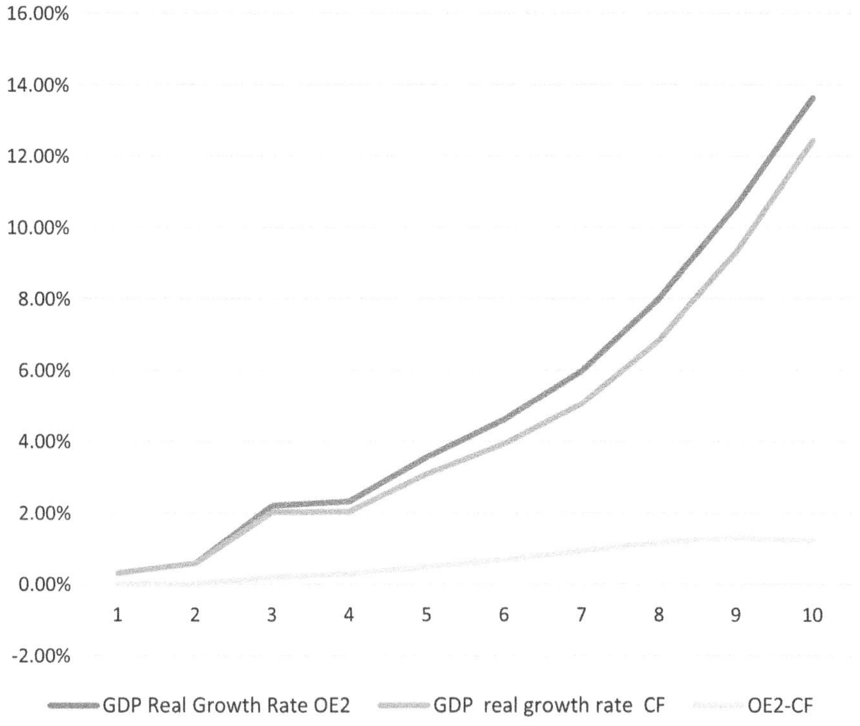

Fig. 4 Growth paths under OE2 and CF. *Source* Authors' computations

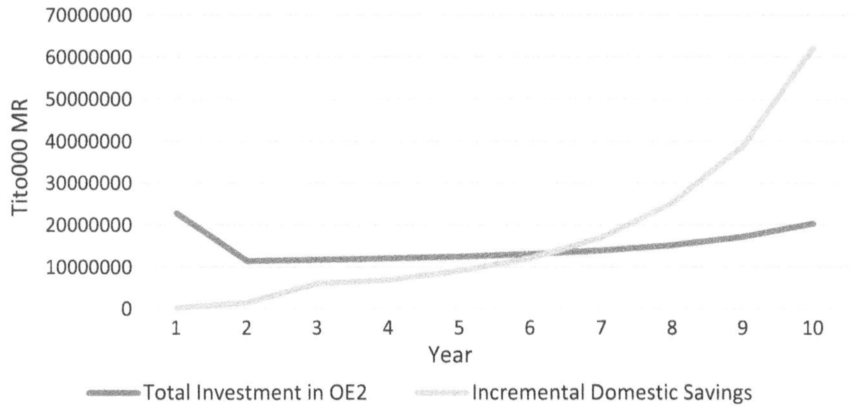

Fig. 5 Effect on domestic savings. *Source* Authors' computations

the total. Even though the OE more than doubles its size with respect to the first year of the development scenario, because of the expansion of the rest of the economy, its share of GDP (including the indirect effects) only increases from 12.6 to 20%.

Table 10 Macroeconomic impact of the best OE2 trajectory

	Year 1	Average[a] Year 1–Year 10	
		OE2	CF
Government deficit/GDP (%)	1.63	1.36	1.43
Government debt as % of GDP (%)	61.63	61.40	61.45
Government debt increase (000 MR)	6,200,006	6,209,550	6,376,374
GDP growth (%)	0.58	3.27	2.33
GDP real growth (%)	0.30	3.17	2.84
Implicit GDP deflator (%)	0.28	0.11	−0.51

[a]Geometric averages for % rates

Table 11 Sector impact of the best OE2 trajectory

	Year 1 000MR	Year 1 (%)	Year 10 000MR	Year 10 (%)
Fishery and sea food processing	17,823,352	18	40,230,512	22
Sea transport and related services	10,084,921	10	24,464,899	13
Marine ICT	6,845,453	7	15,613,962	8
Tourism	63,734,701	64	101,964,001	55
Sewage and water treatment	1,872,311	2	4,284,710	2
Total	100,360,738	100	186,558,084	100

Thousand MR

As Fig. 9 shows, the impact on GDP growth of the investment simulated is high and increasing over time, with very little inflation.

The performance of the OE2 strategy can also be seen in Tables 9 and 10 in comparison to a counterfactual (CF) scenario based on an equal amount of demand and investment increases as the OE2 scenario, but with investment allocated according to historical shares. The OE2 strategy appears to over-perform the CF scenario in all the macro indicators considered, and, as suggested by Fig. 9, differences tend to increase over time. However, the OE2 scenario appears to generate a general increase in capital income and its environmental costs (and the implicit investment costs to neutralize them) are much higher for OE2 than for the CF (Table 12).

Figures 4, 5, 6, 7, 8 and 9 indicate that the performance of the OE2 strategy, even though at the beginning is very demanding in terms of domestic and foreign savings, eventually results in a high and growing degree of domestic capital formation. Figure 7 shows that the (incremental) surplus of domestic savings is matched in the later years by an incremental trade surplus. Table 11 shows the evolution of

Table 12 Value added impact of the best OE2 and CF trajectory

	Value added increases (000 MR)			
	OE2 scenario		CF scenario	
	Year 1–2	Year 9–10	Year 1–2	Year 9–10
Primary education	148,161	318,818	148,161	315,861
Secondary education < SC	108,248	2,312,890	108,248	1,659,207
Secondary education SC and above	79,278	685,068	79,278	1,139,311
Tertiary education	146,855	1,562,083	146,855	2,664,605
Own account	216,976	2,303,242	216,976	4,474,613
Employer	53,841	99,904	53,841	691,514
Operating surplus	402,925	76,013,288	402,925	60,197,920
Ocean	8,639	181,560	8,639	133,116
Green water	2,574	253,598	2,574	219,919
Blue water	1,532	146,904	1,532	128,748
Wetland	48	156,432	48	137,407
Total VA	1,156,283	83,295,294	1,156,283	71,143,031
Natural resource cost	12,793	738,494	12,793	619,189

the OE sectors over time, pointing to the less than proportional expansion of some of the traditional sectors (such as the coastal hotels and restaurants). Other traditional activities, however, (for example, fishery and sea food processing, and the services allied to marine transport) and relatively new activities display more than proportional growth. Tables 12, 13 and 14 show that the OE2 scenario also has a more equitable impact on the formation of value added, job creation, and the income of the poor.

7 Conclusions

Developing a dynamic CGE developed for Mauritius ocean economy presents a number of methodological and empirical challenges, including the multiplying effects of demand expansion, capital accumulation and impact on natural resources. The model presented mimics the mechanism of demand driven growth through the multiplier effects of investment expansion, but also incorporates feedbacks on the supply side through capital accumulation and productivity changes. As a result, the CGE constructed presents characteristics that are common to Keynesian and neoclassical

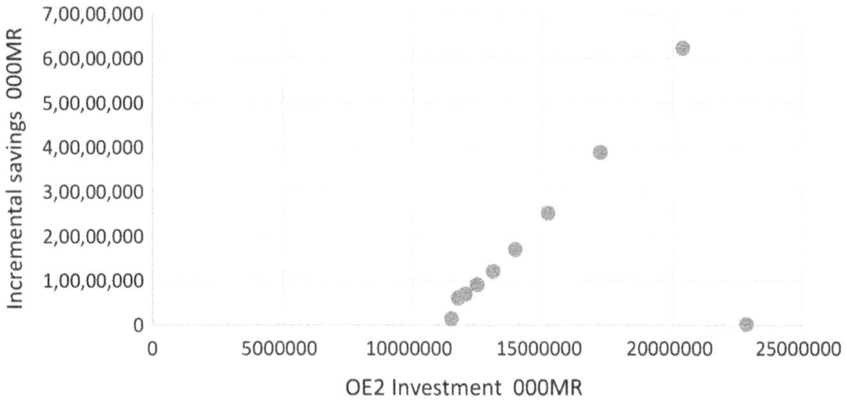

Fig. 6 Incremental domestic savings effect of OE2 investment. *Source* Authors' computations

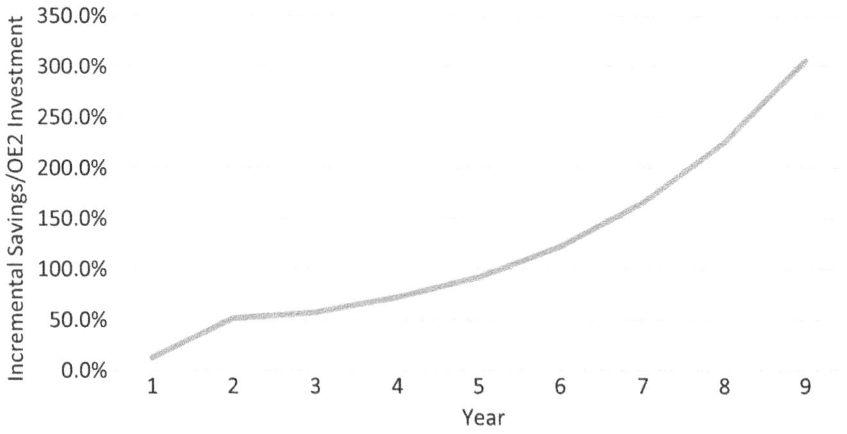

Fig. 7 Incremental savings/OE2 investment. *Source* Authors' computations

models, with growth critically depending on the forward and backward linkages of the sectors where investment expenditure is concentrated, and on the productivity and the potential for productivity increases of the sectors where capital accumulation occurs. These effects are compounded by the assumption of flexible technology, with input output coefficients fixed in any single period, but changing according to Cobb-Douglas splines across periods.

Given these characteristics, the simulation results suggest that concentrating investment in the ocean economy sectors may be a winning strategy, if compared with one of merely continuing the historical pattern of resource allocation. Such a strategy would yield sustained growth over the ten year period considered, with several desirable characteristics concerning diversification, high value added shares and inclusiveness, as compared with a plausible counterfactual. At the same time this strategy

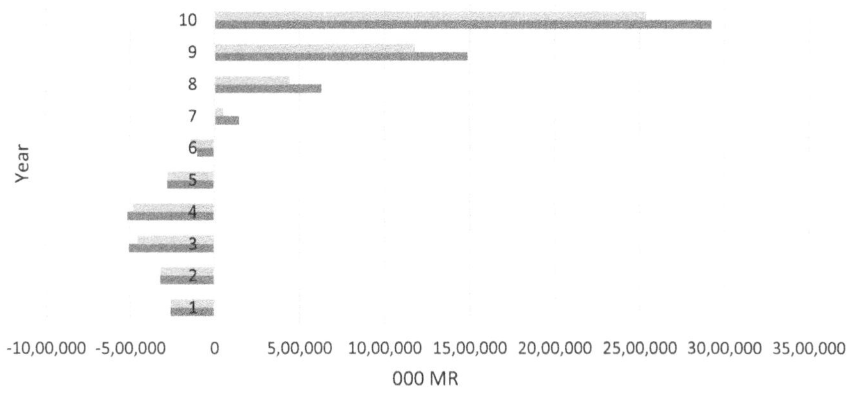

Fig. 8 Exports—Imports. *Source* Authors' computations

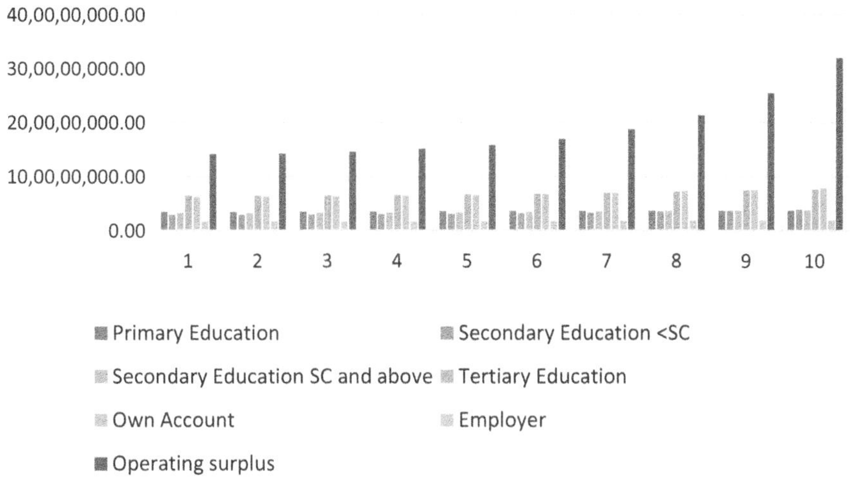

Fig. 9 Value added evolution in OE2. *Source* Authors' computations

appears vulnerable under the two critical aspects of natural and human resource management. It would put further pressure on natural resources, especially in and off shore water, and on the social and health infrastructure. On the human resource side, potential growth could be thwarted by the growing gap between demand and supply of labor due to skill mismatch and educational and training pitfalls. Large complementary investment in these areas would thus be necessary to make the ocean economy strategy fully successful and this would compound the need for financial resources.

These findings point at the usefulness of the CGE structure in producing an integrated framework to analyze alternative scenarios, but also to some of the weaknesses due to its representation of the natural and human resources accumulation mecha-

Table 13 Labor Real income increases and job creation in comparison with the counterfactual (CF)

	Real value added cumulative increases (thousand MR)				Job creation (estimate of 20 year tenure job vacancies)	
	OE2	OE2	CF	CF	OE2	CF
Labor qualification	Year 1–2	Year 1–10	Year 1–2	Year 1–10	Year 1–10	Year 1–10
Primary education	148,161	6,581,529	148,161	8,354,108	2388	3032
Secondary education < SC	107,844	8,089,315	107,844	6,059,886	2936	1649
Secondary education SC and above	79,082	3,348,899	79,082	4,504,748	1215	981
Tertiary education	146,523	7,424,015	146,523	10,432,028	2694	2271
Own account	216,229	9,622,295	216,229	15,960,195	3492	1390
Total	697,839	35,066,053	697,839	45,310,965	12,726	9,324

Table 14 Increases in incomes by recipient groups

	NPV (at 5% discount rate)		
	OE2	CF	OE2/CF
Government and NPISH	9,414,565	7,917,621	1.19
Poor	1,046,639	1,019,671	1.03
Lower middle	10,316,235	9,638,649	1.07
Higher middle	21,063,871	20,127,055	1.05
Wealth	12,577,645	12,546,702	1.00
Firms	145,785,223	114,647,792	1.27
Total	200,204,179	165,897,491	1.21

nisms. While the model accounting system has been enlarged to take incorporate the income flows emanating from both sets of resources, stock flow relation modeling is still rather crude and unsatisfactory. It also represents an area of major challenges for further research and applications.

References

Broad, W. J. (1997). *The universe below: Discovering the secrets of the deep seas*. New York: Simon & Schuster.

Cervigni, R., & Scandizzo P. L. (2017). *Mauritius ocean economy: Making it happen, making it last*. The World Bank.

Norse, E. A. (1993). *Global marine biological diversity: a strategy for building conservation into decision making*. Washington, DC: Island Press; Scandizzo P. L., & Ferrarese C. (2015). Social accounting matrix: A new estimation methodology. *Journal of Policy Modeling*, *37*(1), 14–34, January–February 2015.

Scandizzo, P. L., & Ferrarese C. (2015). CGE model an application for Tanzania. The World Bank Technical Report.

Scandizzo, P. L. (2010). *Adapting to climate change: How to evaluate strategies and projects*. USA and UK: VDM Verlag.

Scandizzo, P. L. (2011). *Climate change adaptation and real option evaluation*. World Bank and SSRN.

Scandizzo, P. L., & Ferrarese, C. (2015). Social accounting matrix: A new estimation methodology. *Journal of Policy Modeling, 37*(1), 14–34.

Tol, R. S. J. (2002). Estimates of the damage costs of climate change. Part II: Dynamic estimates. *Environmental and Resource Economics, 21*, 135–160.

Tol, R. S. J. (2007). The double trade-off between adaptation and mitigation for sea level rise: An application of FUND. *Mitigation and Adaptation Strategies for Global Change*, *12*, 741–753.

Zhao, R., Hynes S., & Shun He G. (2014). Defining and quantifying China's ocean economy. *Marine Policy*, 164–173.

UNCTAD. (2014). The oceans economy, opportunities and challenges for small islands developing states.

Sources of the Data and the Methodology

FAO, FAOSTAT, Aquastat database for Mauritius.

Ministry of Finance and Economic Development Statistics Mauritius National Accounts—Economic and Social Indicators.

Ministry of Finance and Economic Development Statistics Mauritius National Accounts—Social Accounting Matrix 2007.

Ministry of Finance and Economic Development Statistics Mauritius National Accounts—Input-Output table 2007.

Ministry of Finance and Economic Development Statistics Mauritius National Accounts—National Accounts of Mauritius.

Ministry of Finance and Economic Development Statistics Mauritius National Accounts—Historical Series National Accounts (2006–2016).

Ministry of Finance and Economic Development Statistics Mauritius National Accounts—Institutional Sector Accounts.

WTTC World Travel and Tourism Council. (2015). Travel and Tourism Economic Impact 2014.

United Nations Food and Agriculture Organization—Integrated Environmental and Economic Accounting for Fisheries, 2004.

Ministry of Finance and Economic Development Statistics Mauritius National Accounts—Value added and contribution to GDP for activities identified as ocean based.

Ministry of Finance and Economic Development Statistics Mauritius National Accounts—Ocean Input-Output table.

A Micro-Macro Simulation Model Applied to the French Economy: The Case of a Euro's Real Depreciation

Riccardo Magnani, Luca Piccoli, Martine Carré and Amedeo Spadaro

Abstract In this chapter, we use a Micro-Macro simulation model to evaluate the distributional effects of a real depreciation of the Euro on the French economy. Our Micro-Macro model consists of a microsimulation model and a CGE model which are integrated using an iterative approach. We find that a 10% real depreciation of the Euro stimulates the aggregate demand by increasing exports and reducing imports, which increases real GDP by 0.7% and reduces the unemployment rate in the economy by 2 percentage points. At the individual level, we find that the macroeconomic shock reduces poverty and, to a lesser extent, income inequality. In particular, the decrease in the equilibrium real wage slightly reduces disposable income for the employed, while the reduction of unemployment substantially increases disposable income of people who find a job, often bringing them out of poverty.

Keywords CGE models · Microsimulation models, inequality
Currency devaluation

JEL Classification D31 · C63 · C68 · F40

R. Magnani (✉)
CEPN - Université Paris13 and Sorbonne Paris Cité, Paris, France
e-mail: riccardo.magnani@univ-paris13.fr

L. Piccoli · A. Spadaro
Universitat de les Illes Balears, Palma, Spain
e-mail: luca.piccoli@uib.es

A. Spadaro
e-mail: amedeo.spadaro@uib.es

M. Carré
LEDa - Universite Paris-Dauphine, Paris, France
e-mail: martine.carre-tallon@dauphine.fr

© Springer International Publishing AG, part of Springer Nature 2018
F. Perali and P. L. Scandizzo (eds.), *The New Generation of Computable General Equilibrium Models*, https://doi.org/10.1007/978-3-319-58533-8_9

1 Introduction

CGE models and microsimulation models are widely used in policy analysis, respectively from a macro and micro economic perspectives. In particular, microsimulation models are used to evaluate the effects of a shock or of a policy reform at the household level, i.e. on the individual choices about consumption and labor supply and, then, on income distribution, inequality, poverty, etc. In contrast, general equilibrium models are used to determine the effects at the macro level, at the sectoral level, on the equilibrium prices, etc.

A drawback of microsimulation models is that they are implicitly carried out using a partial equilibrium framework, which implies that the effects on individual behavior are computed without taking into account for the general equilibrium changes determined at the macro level. Suppose, for instance, that the effect of a policy at the micro level is an increase in the individuals' labor supply and, thus, in their disposable incomes. In addition, suppose that at the macro level wages are rigid, which implies that labor demand is constant. Thus, the actual effect of the policy is an increase in the level of unemployment since more people want to work, but they cannot find a job.

On the other hand, an important drawback of general equilibrium models is that they are built on the basis of the representative agent paradigm. In other words, the representative agent is assumed to perfectly aggregate the individual decisions about consumption and labor supply. It is well-known that the aggregation of individual preferences is possible only under very restrictive and unrealistic assumptions.

For this reason, the integration between microsimulation models and general equilibrium models appears very appealing in policy evaluation. The integration between these two types of models allows to avoid the shortcomings of both types of models since, in integrated Micro-Macro models, the individual effects are computed by taking into account for the general equilibrium effects and the macro effects are computed by taking into account for the individual heterogeneity at a very detailed level instead of considering a representative agent.

Different procedures exist in the literature to integrate general equilibrium and microsimulation models[1]: The fully-integrated approach consists in introducing in the CGE model all the individuals of the micro dataset (see, for instance, Cockburn 2004). The top-down approach (see, for instance Bourguignon et al. 2008) consists in transmitting the variations of macro variables computed in the CGE model into the microsimulation model. The problem of this approach is that there is no feedback from the microsimulation model back to the macro CGE model. The sequential approach, used also in this study, consists in transmitting the variations of macro variables computed in the CGE model into the microsimulation model and in transmitting the variations concerning the individual behavior computed in the microsimulation model into the CGE model, until a fixed point is reached (see, for instance, Savard 2003). Another approach, developed by Magnani and Mercenier (2009), consists in

[1]For a review concerning the integration of microsimulation and CGE models, see Vaqar and O' Donoghue (2007).

using in the macro model several representative agents that aggregate the preferences of individuals who make discrete choices. This approach permits to avoid the iterations between the microsimulation and the CGE models and to avoid the presence of an excessive number of individuals in the CGE model.

In this study, we present an illustration of a Micro-Macro simulation model, where the integration is done using the sequential approach. In particular our Micro-Macro model is focused on the French economy and consists of a Microsimulation model that includes an arithmetical model for the French fiscal system and two behavioral models used to simulate the individual consumption behavior and the individual labor supply discrete choices, and of a multisectoral and static CGE model.

We use our Micro-Macro model to analyze and quantify, both at the macro and micro levels, the economic consequences for the French economy[2] of a real depreciation of the Euro against other currencies. Our model allows to assess the long-term macroeconomic and distributional effects of a structural change in the real exchange rate. In particular, it permits to evaluate the impact of a real depreciation of the Euro (i) on macroeconomic variables such as GDP, current account, employment and real wages, the relative competitiveness of domestic firms, and the purchasing power of households, (ii) on sectoral production and on the allocation of production factors across tradable and non-tradable sectors, (iii) on individual choices concerning labor supply and consumption, and (iv) on income distribution, inequality and poverty. Even though our model does not allow to analyze the dynamic path towards the long-run equilibrium, our approach permits to properly quantify the redistributive impact of a macroeconomic structural change such as a real currency devaluation.

In particular, the real depreciation of the Euro is considered in this study as an exogenous and common shock for the Eurozone. Thus, determining what kind of monetary policy would be needed to obtain the level of the real depreciation simulated is not an objective of this study. In addition, the approach used is not normative, with no aims of determining the optimal level of the real exchange rate. In this context, the real depreciation can be interpreted as an exogenous and structural shock in the rest of the world, namely a productivity shock, that is common for all countries of the Eurozone.

Focusing on redistributive aspects among sectors and among households is particularly relevant because real currency devaluation is by definition an asymmetric shock which affects the relative prices between monetary zones but also between sectors. Concerning the economic consequences at the sectoral level, Gourinchas (1999) states that a variation in the real exchange rate induces a reallocation of production factors across tradable and non-tradable sectors. Campa and Goldberg (2001) analyze the effects of exchange rate movements on employment and wages for manufacturing industries in the US. They find that for lower markup industries, the effect of a variation of the exchange rate on wages and employment is larger than for higher markup industries. International evidence on the effects of exchange rates on labor markets is provided among others by Burgess and Knetter (1998) who focus on the

[2]The French case is particularly interesting since starting from 2005 the current account displayed important deficits.

G7 countries. They confirm differences among industries in employment elasticities with respect to exchange rates, but also across countries. However, to our knowledge, the effect of real depreciation on inequality has not been analyzed in the literature.

It is important to highlight that the effects of a real currency depreciation, and more generally of any shock, strongly depend on the closure rule used in the macro model.[3] Real devaluation improves the external financial position by increasing exports and reducing imports. In a neoclassical framework in which investments are savings-driven and the elasticity of labor supply is small, the effect of a real depreciation on real GDP is negligible[4] since the level of output at the macro level depends on the quantities of labor and capital available in the economy which are assumed to be fully employed. Consequently, the increase in one of the components of the aggregate demand is compensated by a strong reduction in investments and consumption Hall (2009). In contrast, using the Keynesian closure where investments are fixed at a pre-determined level, real devaluation, by stimulating net exports and aggregate demand, reduces the level of unemployment and increases real GDP.[5] Álvarez-Martínez and Polo (2012) compare the neoclassical and the Keynesian macro closures and find that, in the case of a negative external shock for the Spanish economy, the neo-classical closure predicts an implausible investment boom. However, as shown by our sensitivity analysis, the use of the Keynesian closure rule implies an unrealistic positive reaction of the unemployment rate in the case of a real currency devaluation. This is why in the present study we propose an intermediate macro closure where investments and unemployment are both allowed to react when a shock occurs (Magnani 2015). In particular, an investment function estimated on French data is used to take into account the crowding-out effect on investments produced by a change in the components of the aggregate demand.

Our results show that a real depreciation of the Euro stimulates the aggregate demand by improving the trade balance, reduces unemployment and poverty. In particular, a 10% Euro's devaluation stimulates real GDP (+0.7%), reduces the unemployment rate (−2.0 p.p.) and induces significant effects at the sectoral level, with a noticeable heterogeneity of reactions in terms of employment and production. At the individual level, given the reduction in the unemployment rate determined at the macro level, some of the involuntary unemployed find a job and, given the change in real wages and consumption prices determined at the macro level, individuals modify their labor supply and consumption choices. The model predicts a significant reduction of poverty and a slight reduction of income inequality. In particular, the decrease in the equilibrium wage determined in the macro model moderately reduces the available income for people who already have a job, while the more conspicuous

---

[3]For a review of the macro closure rules see Löfgren et al. (2001), Rattso (1982) and Taylor and Lysy (1979).

[4]The only effect is due to the reallocation of factors across sectors.

[5]Rosensweig and Taylor (1990) used a CGE model with a Keynesian closure to simulate the effect of currency devaluation in Thailand. They find that a 10% devaluation could increase real GDP by 3.3%.

income increase of the involuntary unemployed who find a job often brings them out of poverty.

This article is organized as follows. In the next section we describe the main characteristics of our Micro-Macro model. Section 3 presents the results of our simulation, while Sect. 4 presents a sensitivity analysis. The last section concludes.

2 The Micro-Macro Model

2.1 Introduction

In this section, we briefly present our Micro-Macro model for the French economy which is composed by a microsimulation model, named SYSIFF 2006 (*SYStème d'Imposition Fiscale Français de 2006*), and a macro CGE model. For a full description of the model, see Magnani et al. (2017).

2.2 The CGE Model

The CGE model, which represents the macro component of our Micro-Macro model, is a multisectoral and static model with two foreign zones: the Eurozone and the rest of the world. The model is built by using the 2006 French input-output dataset provided by Insee. The input-output table, which includes 118 sectors, is aggregated into 19 sectors, 11 of which correspond to the sectors used in the microsimulation model concerning the consumption decisions (Table 1). The construction of the SAM (*Social Accounting Matrix*), necessary to calibrate our CGE model, is completed by using national accounts concerning the government account and the balance of payments.

The CGE model is solved by considering several variables as exogenous: the total quantity of labor supplied; the total demand for some type of goods and services; the total amount of employees' and employers' contributions; the total amount of income taxes; the total amount of transfers from the government. However, it is important to highlight that this does not mean that these variables are exogenous in our Micro-Macro model. In fact, the value of these variables is fixed at the initial level only in the first iteration of our Micro-Macro model. Starting from the second iteration, the value of these variables is fixed at the level computed in the microsimulation model, i.e. by taking into account individual heterogeneity.

In what follows we describe the main characteristics of our CGE model. The detailed description of the model could be found in Magnani et al. (2017).

Table 1 List of the sectors in the CGE model

			Microsimulation	CGE
1	Food	Agriculture, hunting, forestry, fishing. Food	X	
2	Beverage	Beverages	X	
3	Tobacco	Tobacco	X	
4	Energy	Mining and quarrying. Coke, refined petroleum products and nuclear fuel. Production, collection and distribution of electricity. Manufacture of gas; distribution of gaseous fuels through mains. Steam and hot water supply. Collection, purification and distribution of water		X
5	Mineral products	Chemicals excluding pharmaceuticals. Rubber and plastics products. Other non-metallic mineral products		X
6	Textile	Textiles, textile products, leather and footwear	X	
7	Housing	Wood and products of wood and cork	X	
8	Mechanic industry	Machinery and equipment, nec		
9	Electric industry	Office, accounting and computing machinery. Electrical machinery and apparatus. Medical, precision and optical instruments		X
10	Metallurgy	Iron and steel. Non-ferrous metals. Fabricated metal products, except machinery and equipment		X
11	Health	Health and social work. Pharmaceuticals. Education	X	
12	Construction	Construction		X
13	Transports	Motor vehicles, trailers and semi-trailers. Building and repairing of ships and boats. Aircraft and spacecraft. Railroad equipment and transport equip nec. Manufacturing nec; recycling. Land transport; transport via pipelines. Water transport. Air transport. Supporting and auxiliary transport activities. Activities of travel agencies	X	
14	Hotels and restaurants	Hotels and restaurants	X	

(continued)

Table 1 (continued)

			Microsimulation	CGE
15	Leisure	Pulp, paper, paper products, printing and publishing. Radio, television and communication equipment. Other community, social and personal services. Private households with employed persons and extra-territorial organizations and bodies	X	
16	Communications	Post and telecommunications	X	
17	Public administration	Public admin. and defense; compulsory social security		X
18	Non-financial services and R&D	Real estate activities. Renting of machinery and equipment. Computer and related activities, Research and development. Other business activities		X
19	Financial services	Finance and insurance	X	

2.2.1 The Production Side

For each sector, we use a multi-stage CES production function. In the first stage, the demand of total intermediate goods Z_i, labor L_i and capital K_i is optimally chosen by each sector i in order to maximize its profit given a technological constraint represented by the following production function:

$$Y_i = \left[(\alpha_{Z,i})^{\frac{1}{\sigma_i}} \cdot Z_i^{\rho_i} + (\alpha_{L,i})^{\frac{1}{\sigma_i}} \cdot L_i^{\rho_i} + (\alpha_{K,i})^{\frac{1}{\sigma_i}} \cdot K_i^{\rho_i}\right]^{\frac{1}{\rho_i}}$$

In the second stage, each sector i chooses the optimal repartition of the total intermediate good into different intermediate goods sold by sector j, Z_{ji}. The choice is made in order to minimize the total cost and to respect the following technological constraint:

$$Z_i = \left[\sum_j (\alpha_{Z_{ji}})^{\frac{1}{\sigma Z_i}} \cdot Z_{ji}^{\rho Z_i}\right]^{\frac{1}{\rho Z_i}}$$

In the third stage, each sector i chooses the optimal repartition of the intermediate goods sold by sector j between the quantity that comes from the domestic market Z_{ji}^h and from abroad Z_{ji}^f. The repartition is chosen in order to minimize the total cost and to respect the following technological constraint:

$$Z_{ji} = \left[(\alpha_{ji}^h)^{\frac{1}{\sigma Z_{ji}}} \cdot (Z_{ji}^h)^{\rho Z_{ji}} + (\alpha_{ji}^f)^{\frac{1}{\sigma Z_{ji}}} \cdot (Z_{ji}^f)^{\rho Z_{ji}}\right]^{\frac{1}{\rho Z_{ji}}}$$

In the last stage, each sector i chooses the optimal repartition of the intermediate goods sold by sector j that come from abroad between the quantity that comes from the Eurozone Z_{ji}^{Ez} and from the rest of the world Z_{ji}^{Row}. The repartition is chosen in order to minimize the total cost and to respect the following technological constraint:

$$Z_{ji}^{f} = \left[\left(\alpha_{ji}^{Ez}\right)^{\frac{1}{\sigma Z_{ji}^{f}}} \cdot \left(Z_{ji}^{Ez}\right)^{\rho Z_{ji}^{f}} + \left(\alpha_{ji}^{Row}\right)^{\frac{1}{\sigma Z_{ji}^{f}}} \cdot \left(Z_{ji}^{Row}\right)^{\rho Z_{ji}^{f}} \right]^{\frac{1}{\rho_i}}$$

The optimal repartition depends on the relative price, i.e. the ratio between the price in the Eurozone P_j^{Ez} and the world price expressed in Euros $P_j^{Row} \cdot \varepsilon$. In particular, (i) the exchange rate ε is assumed to be exogenous (while financial flows are endogenously determined in order to equilibrate the balance of payments) given that it is used to simulate the macroeconomic shock in our model; (ii) the world price of good j expressed in foreign currency P_j^{Row} is exogenous; (iii) the price in the Eurozone P_j^{Ez} is treated as endogenous since it is reasonable to assume that Euro's depreciation would affect prices in the whole Eurozone. In particular, for each sector j, the price in the Eurozone P_j^{Ez} is computed as a weighted average between the domestic price in the Eurozone (which is assumed to vary in the same proportion as the domestic price in France) and the world price expressed in Euros. This implies that we consider in our model a symmetric equilibrium in the sense the Euro's devaluation does not affect competitiveness within the Eurozone.

A fraction of the production is sold in the domestic market and the complementary fraction is exported. Goods that are exported are supposed to be identical to those sold in the domestic market, implying that the selling price is the same. Exports, towards the Eurozone and the rest of the world, are defined by a demand function that is decreasing in the relative price, i.e. the ratio between the domestic price and the foreign price expressed in domestic currency:

$$E_i^{Ez} = \alpha_i^{Ez} \cdot \left(\frac{P_i^{Ez}}{P_i^h}\right)^{\sigma E_i}$$

$$E_i^{Row} = \alpha_i^{Row} \cdot \left(\frac{P_i^{Row} \cdot \varepsilon}{P_i^h}\right)^{\sigma E_i}$$

Considering that the real devaluation of the Euro represents a shock affecting the whole Eurozone, it is reasonable to presume that also real GDP in the Eurozone is affected by the shock. For this reason, the terms α_i^{Ez}, which represent a measure of the purchasing power in the Eurozone, are assumed to be endogenous and to vary in the same proportion as the French real GDP.

2.2.2 The Demand Side

(a) **Consumption**

Concerning households, we consider one representative agent who maximizes his well-being by choosing the consumption level of different goods and services. As indicated in Table 1, we consider 8 "CGE sectors", the consumption level of which is endogenously determined in the CGE model, and 11 "microsimulation sectors", the consumption level of which is exogenous. In particular, the level is fixed at the initial level in the first iteration of our Micro-Macro model and, starting from the second iteration, the level is fixed at the level computed in the microsimulation model.
The preferences of the representative agent are modeled using a multi-stage utility function. In the first stage, the level of total consumption for the "CGE goods" C^{cge} is determined as a fraction of the total disposable income. In the second stage, the representative agent chooses, for each "CGE good" i, the optimal consumption C_i^{cge}. In the third stage, the representative agent chooses the optimal repartition of the consumption demand of good i between domestic goods C_i^h and foreign goods C_i^f. In the last stage, the consumption demand of the foreign good i is divided into foreign goods coming from the Eurozone C_i^{Ez} and from the rest of the world C_i^{Row}.

(b) **Investments**

The second component of the aggregate demand is given by the investment. As for consumption, we use a multi-stage structure. In the first stage, the aggregate investment I, that is determined as described in Sect. 2.2.6, is allocated into different sectors I_i. Then, we determine the repartition of the investment of sector i between investment coming from the domestic market I_i^h and the foreign market I_i^f. In the last stage, the foreign investment of sector i is divided into foreign investment coming from the Eurozone I_i^{Ez} and from the rest of the world I_i^{Row}.

(c) **Government expenditure**

The third component of the aggregate demand is given by the government expenditure. Here, we also use a multi-stage structure. In the first stage, the total government expenditure G, that is determined by assuming that the ratio with respect to real GDP remains constant, is allocated into different sectors G_i. Then, we determine the repartition of the government expenditure of good i between goods coming from domestic and foreign markets (respectively G_i^h and G_i^f). In the last stage, the government expenditure of the foreign good i is divided into foreign goods coming from the Eurozone G_i^{Ez} and from the rest of the world G_i^{Row}.

(d) **Total demand**

For each sector i, the total quantity demanded depends on the demand of the domestic good (that is given by the difference between the domestic production and exports) and on the demand of the foreign good. In particular, for each sector i, the total domestic demand of the domestic good X_i^h is given by the sum of domestic intermediate goods, private and public consumption and investments. For each sector i, the total imports respectively from the Eurozone M_i^{Ez} and from the rest of the world M_i^{Row} are given by the sum of intermediate goods, private and public consumption and investments imported respectively from the Eurozone and the rest of the world:

$$X_i^h = \sum_j Z_{ij}^h + C_i^h + I_i^h + G_i^h$$

$$M_i^{Ez} = \sum_j Z_{ij}^{Ez} + C_i^{Ez} + I_i^{Ez} + G_i^{Ez}$$

$$M_i^{Row} = \sum_j Z_{ij}^{Row} + C_i^{Row} + I_i^{Row} + G_i^{Row}$$

2.2.3 Budget Constraints

(a) **Household budget constraint**

The gross income earned by the representative agent is given by the labor and capital incomes earned in France and abroad, and the transfers from the government:

$$\begin{aligned} Y_{gross} = {} & w \cdot (1 - cot_{empl}) \cdot L_{Fr-Fr} \cdot (1 - u) + w^{Ez} \cdot L_{Fr-Ez} \\ & + r \cdot PI \cdot A_{Fr-Fr} + r^{Row} \cdot \varepsilon \cdot A_{Fr-Row} + \Gamma_{ms} + \Gamma \end{aligned}$$

In particular, the labor incomes earned in France depend on the endogenous domestic wage w, on the contribution rate paid by the employees cot_{empl} and the quantity of labor supplied by French people who work in France $L_{Fr-Fr} \cdot (1 - u)$. The latter variable depends on the quantity of labor that people decide to supply L_{Fr-Fr} that is exogenously fixed at the level determined in the microsimulation model, and on the unemployment rate u which can be exogenous or endogenous according to the macro closure that is chosen in the CGE model. By assuming that French people who work abroad work in the Eurozone, labor incomes earned abroad depend on the exogenous foreign wage rate w^{Ez} and the exogenous quantity of labor supplied by French people who work abroad L_{Fr-Ez}. The capital incomes earned in France depend on the endogenous domestic interest rate r and the value of assets owned by French people in France A_{Fr-Fr}, while the capital incomes earned abroad depend on the exogenous world interest rate r^{Row}, the exchange rate ε and the value of assets owned by French people in the rest of the world A_{Fr-Row}. We consider two types of transfers from the government: transfers Γ_{ms} that affect the labor incomes (and thus the labor market choices), the value of which is fixed at the level determined in the microsimulation model, and transfers Γ that do not affect individual labor choices that are treated as exogenous.

The disposable income is computed as the difference between the gross income and taxes on labor and capital incomes:

$$Y_{disp} = Y_{gross} - Tax_{lab} - \tau_{cap} \cdot r \cdot PI \cdot A_{Fr-Fr}$$

In particular, the value of the taxes on labor incomes Tax_{lab} is fixed at the level determined in the microsimulation model, while taxes on capital incomes are supposed to be proportional to the capital incomes earned, where τ_{cap} is the tax rate on capital incomes.

The budget constraint implies that the difference between the disposable income and the consumption of goods and services represents private savings S_H:

$$S_H = Y_{disp} - \sum_i PC_i \cdot C_i$$

(b) **Government budget constraint**

Government revenues come from direct taxes on labor and capital incomes, indirect taxes on production and on the value added, and social contributions on employers and employees, while government expenditures are represented by the total public expenditure G, interests on the public debt B and transfers to households Γ_{ms} and Γ. The difference between government revenues and expenditures determines public savings S_G:

$$\begin{aligned} S_G = & \sum_i \tau_{y_i} \cdot P_i^h \cdot Y_i^h \\ & + \sum_i \tau_{VAT,i} \cdot \left[P_i^h \cdot (C_i^h + I_i^h + G_i^h) + P_i^{Ez} \cdot (C_i^{Ez} + I_i^{Ez} + G_i^{Ez}) \right. \\ & \qquad \left. + P_i^{Row}.\varepsilon.\left(C_i^{Row} + I_i^{Row} + G_i^{Row}\right)\right] \\ & + Tax_{lab} + \tau_{cap} \cdot r \cdot PI \cdot A_{Fr-Fr} + \sum_i w \cdot \left(cot_{patr} + cot_{empl}\right) \cdot L_i \\ & - (P_G \cdot G + r \cdot B + \Gamma_{ms} + \Gamma) \end{aligned}$$

(c) **Balance of payments**

The balance of payments states that the current account surplus plus the capital account surplus must be equal to zero. In particular, the current account surplus is given by the net exports plus the net factor incomes from the rest of the world, while the capital account surplus is given by the net capital inflows, i.e. the difference between the flow of foreign assets to France ΔA_{Row-Fr} and the flow of domestic assets to the rest of the world ΔA_{Fr-Row}:

$$\begin{aligned} & \left[\sum_i P_i^h \cdot (E_i^{Ez} + E_i^{Row})\right] - \left[\sum_i \left(\sum_j PZ_i^f \cdot Z_{ij}^f\right)\right. \\ & \quad \left. + PC_i^f \cdot C_i^f + PI_i^f \cdot I_i^f + PG_i^f \cdot G_i^f \right] \\ & \quad + \left[w^{Ez} \cdot L_{Fr-Ez} + r^{Row} \cdot \varepsilon \cdot A_{Fr-Row}\right] \\ & - \left[w \cdot \left(1 - cot_{empl}\right) \cdot L_{Row-Fr} + r \cdot PI \cdot A_{RoW-Fr}\right] \end{aligned}$$

$$+ PI \cdot (\Delta A_{Row-Fr} - \Delta A_{Fr-Row}) = 0$$

Given that the real exchange rate is assumed to be exogenous and given that the flow of domestic assets to the rest of the world ΔA_{Fr-Row} is determined by the optimal asset allocation (see Sect. 2.2.4) the balance of payments determines the flow of foreign assets to France ΔA_{Row-Fr}.

2.2.4 Optimal Asset Allocation

We assume that the representative agent has to choose, at the beginning of the period, how to allocate his (exogenous) initial wealth A_{Fr} between investments in France A_{Fr-Fr} and abroad A_{Fr-Row}. We suppose that the two alternatives are not perfect substitutes and that the optimal allocation depends on the ratio between the rates of return on the two assets. In particular, the rate of return on assets invested in France is the (net of depreciation) marginal productivity of capital r, while the rate of return on assets invested abroad is given by the sum between the foreign interest rate r^{Row} and the percentage variation of the exchange rate, i.e. $\frac{\varepsilon-\varepsilon(-1)}{\varepsilon(-1)}$.

The total wealth owned by the representative agent at the beginning of the next period $A_{Fr}(+1)$, that is given by the initial total wealth A_{Fr} plus private savings S_H, must be also allocated between assets invested in France $A_{Fr-Fr}(+1)$ and abroad $A_{Fr-Row}(+1)$, on the basis of the expected ratio between the rates of return. Here, we consider extrapolative expectations implying that the expected rate of return on assets invested in France is fixed at the (net of depreciation) marginal productivity of capital of the first period; the expected foreign interest rate is fixed at the level of the first period; and the expected percentage variation of the exchange rate is fixed to zero.

The allocation of the total wealth in the two periods allows us to determine the flow of domestic assets to the rest of the world ΔA_{Fr-Row} which affects the balance of payments.

2.2.5 Equilibrium Conditions

For each sector i, domestic prices P_i^h adjust in order to guarantee the equilibrium between the quantity produced Y_i and the domestic and foreign demands:

$$Y_i = X_i^h + E_i^{Ez} + E_i^{Row}$$

In the labor market, the total labor demanded by all sectors $\sum_i L_i$ must be equal to the sum between the quantity of labor supplied by French people (that depends on the quantity of labor, determined in the microsimulation model, that French people want to supply L_{Fr-Fr}, and on the unemployment rate u) and the (exogenous) quantity of labor supplied by foreign people L_{Row-Fr}:

$$\sum_i L_i = L_{Fr-Fr} \cdot (1-u) + L_{Row-Fr}$$

In the capital market, the total capital demanded by all sectors $\sum_i K_i$ and by the government B must be equal to sum between the capital supplied by French people A_{Fr-Fr} (that depends on the optimal asset allocation choice) and the (exogenous) capital supplied by foreign people A_{Row-Fr}:

$$\sum_i K_i + B = A_{Fr-Fr} + A_{Row-Fr}$$

This equation determines the equilibrium domestic rate of remuneration of capital r.

Finally, the *numéraire* chosen is the domestic price index. Thus, the exchange rate ε represents the real exchange rate implying that the macroeconomic shock simulated in this paper is a depreciation of the real exchange rate.

2.2.6 Macro Closure

The macroeconomic equilibrium condition states that aggregate investments must be equal to aggregate savings (i.e. the savings of the representative agent, of the government and with respect to the rest of the world):

$$PI \cdot I = S_H + S_G + PI \cdot (\Delta A_{Row-Fr} - \Delta A_{Fr-Row})$$

The neoclassical closure, that is the most frequently used in general equilibrium models, implies that investments are savings-driven, i.e. the macroeconomic equilibrium condition determines the level of aggregate investment. The use of the neoclassical closure implies that a shock which increases the value of a component of the aggregate demand (for example, an increase in the current account induced by currency devaluation) produces a strong and unreasonable negative effect on investments, while the effect on the GDP is quite negligible since GDP is determined by the supply of productive factors that are supposed to be fully employed in the economy. Thus, currency devaluation can stimulate real GDP only by removing the hypothesis of full-employment of production factors and by assuming that the involuntary unemployment is provoked, according to the Keynesian view, by the weakness in aggregate demand.

With respect to the neoclassical closure, the Keynesian closure consists to fix the level of investments at a predetermined level (see Álvarez-Martínez and Polo 2012) and to endogenize the unemployment rate. The unemployment rate is then determined in order to satisfy the macroeconomic equilibrium condition between investments and aggregate savings, implying that aggregate production is demand-driven. In particular, and in contrast to neoclassical models, the macroeconomic equilibrium

Table 2 Estimation results of the investment function on French data from 1949 to 2012

	Coefficient	Std. Error	t-Statistic	Prob.
Constant	−116.769	10.220	−11.425	0.000
GDP	0.581	0.042	13.914	0.000
Detrended CONS	−0.428	0.185	−2.321	0.024
Detrended G	−0.315	0.157	−2.009	0.049
DetrendedCA	−0.910	0.187	−4.860	0.000
R-squared	0.998			
Adjusted R-squared	0.997			

may be an under-unemployment equilibrium, implying that unemployment appears in the case in which the level of the aggregate demand is insufficient.

However, even the Keynesian closure presents a major shortcoming since the reduction in the unemployment rate produced by the currency devaluation simulated in our paper would be excessively high. This is why we chosen to use a closure rule which is between the neoclassical and the Keynesian ones (Magnani 2015). The idea is the following: with a neoclassical closure, in which investments are savings-driven, an increase in the current account produces a crowding-out effect on investments which compensates the positive effect on the current account; in contrast, with a Keynesian closure, the same shock produces no effects on investments (since they are fixed at a given value) and a positive effect on real GDP due to the reduction in unemployment. The closure rule used in our model consists in introducing an investment function which takes into account the (partial) crowding-out effect on investments produced by a change in the components of the aggregate demand. In particular, using yearly French data from 1946 to 2012 provided by Insee, we estimate the following investment function:

$$I = \alpha_0 + \alpha_1 \cdot GDP_{real} + \alpha_2 \cdot \Delta C + \alpha_3 \cdot \Delta G + \alpha_4 \cdot \Delta CA$$

The previous equation implies that investments depend on the real GDP, the change in aggregate consumption ΔC, the change in public expenditures ΔG, and the change in the current account ΔCA. The detrended variables ΔC, ΔG and ΔCA are constructed using a HP filter with a smoothing parameter equal to 100. The estimation results, reported in Table 2, show that an increase in each component of the aggregate demand produces a crowding-out effect on investments. However, the crowding-out effect is only partial, i.e. it is lower than that obtained using a neoclassical closure. The introduction of this investment function allows us to build a CGE model with a macro closure that is between the neoclassical and the Keynesian ones. This implies that real devaluation, which permits to improve the current account, provokes a partial crowding-out effect on investments. Thus, the aggregate demand increases and the unemployment rate, determined in order to guarantee the equilibrium between aggregate savings and aggregate investments, decreases.

Table 3 Fiscal instruments simulated in SYSIFF 2006

Social contributions and VAT	Income tax, local taxes and public benefits
Employer social contributions	*Deductions*
Assurance Maladie/Solidarité	Retraites complémentaire volontaire
Assurance Vieillesse Plafonnée	Frais réels
Assurance Vieillesse Déplafonnée	Abattement général
Allocation Familiales	Déductions Enfant
FNAL	Déductions Ascendants
Allocation Chômage	Pertes en capital
Retraites Complémentaire	
AGFF	*Income tax*
Régime de Prévoyance Cadres	Foyer fiscal
Versement Transport	Impôt sur le Revenu
Réduction Fillon	
	Tax credits
Employee social contributions	Contribution non profit
Assurance Maladie/Solidarité	Assurance décès-sante
Assurance Vieillesse Plafonnée	Prestation compensatoire
Assurance Vieillesse Déplafonnée	Personnes âgées dépendantes
Allocation Chômage	Salarié à domicile
Retraites Complémentaires	Garde d'enfants
AGFF	Frais de scolarisation
Assurance Santé extra/complémentaire	Comp. Taxe Carbone
CAPS (Capital)	Réduction d'impôts DOM
Prélèvement Social (Capital)	Prime pour l'emploi
Allocations Familiales (self empl.)	
Formation professionnelle (self empl.)	*Local taxes*
Assurance Maladie (self empl.)	Taxe Habitation
Assurance Invalidité décès (self empl.)	Taxe Foncière sur le non-bâti
Assurance Vieillesse (self empl.)	Taxe Foncière sur le bâti
Régime d'Indemnités Journalières (self empl.)	
	Public benefits
Special contributions	AF—Allocations Familiales
CSG	PAJE—Prestation d'Accueil du Jeune Enfant
CRDS	(a) Child born before 01-01-2004
	APE (Allocation Parental d'Education)
VAT	APJE (Allocation Pour Jeune Enfants)
Food	AAM (Aide Assistant Maternelle)

(continued)

Table 3 (continued)

Social contributions and VAT	Income tax, local taxes and public benefits
Beverages	(b) Child born after 01-01-2004
Clothing	Prime à la Naissance
Energy	AB (Allocation de Base)
Furniture	CLCA (Complément de Libre Choix d'Activité)
Household appliance	Paje Emploi
Housing (products)	CFAM—Complément Familial
Health	API—Allocation Parent Isolée
Transports	ARS—Allocation Rentrée Scolaire
Communications	Prime de Déménagement
Amusements	Minimum Vieillesse
Books and newspapers	Aide au Logement
Cinema	RMI (Revenu Minimum d'Insertion)
Museums	RSA (Revenu de Solidarité Active)
Leisure	
Teaching	
Meals	
Beauty	
Other goods	

2.3 The SYSIFF 2006 Microsimulation Model

SYSIFF 2006 is a microsimulation model for the French fiscal system which integrates the arithmetical simulation model with two behavioral models concerning consumption and labor supply decisions. SYSIFF 2006 is a microsimulation model in the sense that it is based on a microeconomic dataset concerning a sample of families representative of the French population on which simulations can be performed. The micro dataset used in our paper is the *Budget de Familles 2006* (from now on BDF2006) by Insee. It has been chosen in virtue of the fact that it is the only dataset available in France with sufficient information to perform all computation required by a complex fiscal system as the French one. Moreover, it includes data on family expenditure on goods and services and the labor supply of households' members. This information is fundamental for the estimation of the demand system and labor supply functions necessary to integrate micro-level behavioral responses within the Micro-Macro model.

Table 4 Heckman estimation for salaries

		Single		Married	
		Female	Male	Female	Male
Log of hourly wage					
	Age	0.025*	0.037**	0.046***	0.035***
	Age squared	0.000	0.000	0.000***	0.000***
	Primary education	0.020	−0.090	−0.050	−0.023
	Secondary education	0.059	0.128	0.075	0.150***
	Cap/Bep	0.051	0.110*	0.101**	0.135***
	University	0.196**	0.253***	0.266***	0.235***
	Superior education	0.246***	0.248***	0.351***	0.246***
	Immigrant	−0.027	−0.133**	−0.094**	−0.063**
	Ile-de-France	0.095	0.051	0.066	0.110**
	Public sector	0.101***	0.084*	0.126***	0.011
	White collar	0.342***	0.406***	0.307***	0.406***
	Constant	1.344***	1.200***	0.791***	1.171***
Selection					
	Age	0.121***	0.027	0.073***	−0.001
	Age squared	−0.001***	0.000	−0.001***	0.000
	Primary education	0.552**	−0.055	0.237	−0.042
	Secondary education	0.704***	0.105	0.147	0.476**
	Cap/Bep	0.675***	0.622***	0.303***	0.184
	University	1.231***	0.194	0.686***	0.276**
	Superior education	1.358***	0.423	0.726***	0.468***
	Immigrant	−0.344*	−0.609***	−0.656***	−0.580***
	N. children [0–2]	−0.347	−1.493	−0.459***	−0.206**
	N. children [3–6]	−0.243***	2.961	−0.255***	0.027
	Bad health	−0.977***	−1.297***	−0.425***	−0.764***
	Non-labor incomes	−0.081***	−0.077***	−0.015***	−0.044***
	Ile-de-France	0.498	0.371	−0.086	0.083
	Constant	−1.694*	1.397	−0.238	2.000**
	ρ	−0.018	0.008	0.236***	−0.027

2.3.1 The Arithmetical Model

The first element of SYSIFF 2006 is the arithmetical model which includes a collection of algorithms and parameters that allow to compute, for each family, the amount of direct and local taxes, social contributions, and social benefits received, within a given fiscal system. As discussed in Sect. 2.3.2, singles (women and men) and married women can choose whether not to work, to work 50% part time (18 h per week), 80% part time (28 h per week), or full time (36 h per week). Married men can only choose whether not to work or to work full time (36 h per week). For each labor supply alternative, the arithmetical model computes, for one family at a time, the labor income (which depends on the level of the equilibrium wage determined in the CGE model) and the value of all fiscal instruments.[6] The disposable income, for each labor supply alternative, is given by the labor income minus taxes and contributions, plus the transfers from the government. Consequently, when a shock is introduced in the model, the CGE model determines the effect on the equilibrium wage and the arithmetical model determines, at the family level, the effect on the disposable income for each labor supply alternative. This information is then transmitted to the labor supply behavioral model.

2.3.2 The Labor Supply Behavioral Model

The second element of SYSIFF 2006 is the behavioral model concerning labor supply. This model allows to determine the effect of a shock on the quantity of labor supplied by each family. In particular, this effect depends on the change of the equilibrium wage determined in the CGE model and on the estimated reaction of labor supply to a change in the disposable income. This is why, in this section, we firstly describe how labor supply is estimated and, secondly, how the behavioral model is used to evaluate the effect on labor supply at the family level.

A standard way to estimate labor supply is to consider that individuals choose the optimal number of hours worked in order to maximize their well-being under a time and budget constraint. However, the non-linearity and non-convexity of the budget constraint, due to the characteristics of the tax system, implies the impossibility to derive an explicit solution to this standard utility maximization problem. For this reason, the best option for estimating labor supply behavior is that of discrete choice models *à la* Van Soest (1995). This approach allows to directly estimate the utility function parameters without the need of a Marshallian labor supply function. In particular, discrete choice models have the advantage of capturing behavioral change in corner solution, accounting for market rigidities and avoiding the computational and analytical difficulties arising from non-linear and non-convex budget constraints, since the budget constraint depends on the disposable income which is computed by the arithmetical model and is introduced directly into the utility function.

[6]The list of fiscal instrument modeled in SYSIFF 2006 is reported in Table 3.

The analysis of the distribution of the work alternatives has lead to the choice of four work alternatives for singles and married women (not to work (0 h), 50% part time (18 h), 80% part time (28 h), and full time (36 h)) and two alternatives for married men (not to work (0 h), and full time (36 h)).[7]

The estimates of labor supply are performed on a sub-sample of potential wage earners[8] separately for single men, single women and couples. We consider four alternatives for singles and eight alternatives for couples, and we assume that singles and couples choose the alternative that maximizes their utility. In particular, for single men ($j = 1$), single women ($j = 2$) and couples ($j = 3$), we define a utility function for each labor supply alternative s depending on individual (or family) characteristics $X_{j,k}$ and on the disposable income (provided by the arithmetical model for each alternative) $y_{i,s}$,[9] as follows:

$$U_{j,s} = \sum_k \alpha_{k,s} \cdot X_{j,k} + \beta_j \cdot \ln y_{i,s} + \epsilon_{j,s}$$

With respect to the standard model proposed by Van Soest (1995), which implicitly assumes that non-working people choose not to work, we consider that unemployment may be involuntary, as in Magnac (1991), Bingley and Walker (1997), and Haan and Uhlendorff (2013). Our micro dataset allow us to identify involuntarily unemployed by checking if individuals perceive an unemployment benefit (*allocation chômage*) that is given only to people who are actively searching for a job. In our sample, 19.7% of individuals do not work and 6.3% of the sample is involuntarily unemployed, implying that the unemployment rate is 7.3%. Involuntary unemployment is introduced by randomly assigning (respecting the actual distribution of observed choices) a choice among the work alternatives to involuntarily unemployed. Thus, involuntarily unemployed choose to work but, given the constraints in the labor market, cannot find a job.

The discrete choice labor supply models (for single men, women and couples) are estimated using a Multinomial Logit regression by also taking into account the fictitious choice of the involuntarily unemployed. Tables 5 and 6 report the estimated parameters for singles and couples respectively. The most relevant parameter in these estimates is the income parameter, which is expected to be positive and significant. This is so for single women and couples, while for single men it is not significantly

[7]Clearly, not everybody chose one of these options, so we set-up intervals within which the assigned choice is one of these. 0 h is reserved to non-working people, 50% part time is for people working less than 23 h per week, 80% part time is for people working 23 to 33 h per week, and full-time work is for those working more than 33 h per week.

[8]We exclude from the sample self-employed, retired people, individuals with less than 25 years or over 60 years.

[9]In order to compute the disposable income for the non-observed alternatives it is necessary to generate a potential salary for the unemployed. Potential salaries are generated from the estimation of the wage equation using the Heckman correction model (Heckman 1979). The estimation results are reported in Table 4.

different from zero, probably due to the fact that the vast majority of single men are full-time workers.[10]

We then estimate the probability of not being involuntarily unemployed, using a Probit model. The estimation results are as expected since the probability of not being involuntarily unemployed significantly increases with age and education, while it decreases if the person is an immigrant or if he is in bad health conditions. Living in Paris has a positive but not significant impact on the probability of not being involuntarily unemployed.

After the description of the estimation of labor supply, we now describe how the behavioral model is used to evaluate the effects of a shock on labor supply. As we have already said, when a shock is introduced in the model, the CGE model determines the effect on the equilibrium wage and the arithmetical model determines, at the family level, the effect on the disposable income for each labor supply alternative. The labor supply behavioral model receives this information and determines for each single woman and man and for each couple the optimal labor supply choice, i.e. the alternative that maximizes the utility level. The optimal choice depends on the change in the disposable income for each alternative, on the estimated reaction of labor supply to a change in the disposable income, and on the value of the error terms.[11] However, the optimal choice determined here may be fictitious. In fact, an individual may choose to work but not find a job. The CDF of the predicted probabilities of not being involuntarily unemployed allows to rank individuals and, thus, is used to determine the unemployment status of each individual which depends on the level of unemployment determined at the macro level using the CGE model. Thus, the ranking in the CDF allows to identify the individuals who find a job (if, at the national level, the unemployment rate decreases) or the individuals who lose their job (if, at the national level, the unemployment rate increases).

Next, it is possible to compute, at the family level, the labor income, the value of all fiscal instruments and, thus, the disposable income. Remember that the arithmetical model determines the value of these variables for each labor supply alternative and that an individual, even if his/her optimal choice is to work, may not find a job. Thus, the level of these variables corresponds to the level (determined in the arithmetical model) for the alternative chosen in the behavioral model, and considering whether the individual is or is not involuntarily unemployed. The effect on the disposable income for each family is transmitted to the consumption behavioral model.

Finally, it is possible to aggregate (i) the behavior concerning labor supply, i.e. to compute the total quantity of labor supplied by all families and (ii) the amount of employees' and employers' contributions, of income taxes, and of transfers from the government. The effects on these variables are transmitted to the CGE model.

---

[10]Once the model is estimated the correct prediction is quite large: 88% for single men, 72% for single women and 53% for couples.

[11]For each alternative and for each single and couple, we have generated 300 extreme-value distributed stochastic terms, conditioned on the observed choice. The 300 extractions ensure the statistical properties of labor supply predictions once a shock modifies the disposable income for each labor supply alternative.

Table 5 Discrete choice labor supply estimation for singles

	Single females				Single males			
	Not working	50% part time	80% part time	Full time	Not working	50% part time	80% part time	Full time
Log of dispos. income	Ref. cat.	0.444**	0.444**	0.444**	Ref.cat.	0.012	0.012	0.012
Part time 1		4.248	−8.716	−13.233**		−17.508	−2.692*	−2.620**
Part time 2		0.643	13.905***	−0.846		−28.445	−26.597	−10.203
Domestic worker		−1.630*	−2.571**	−1.833***		1.024	43.158*	46.652**
Baby sitter		−0.590	−0.226	0.222		−4.720	−55.798*	−58.848**
Age		16.179	52.136***	31.865***		0.588	1.261	0.952
Age squared		−21.995	−65.668***	−40.839***		−0.123	0.444	0.603
Primary education		0.956	0.432	1.183*		−0.107	1.463	0.864
Secondary education		1.075	1.811***	1.393**		−0.828	0.423	0.033
Cap/bep		0.412	0.537	0.995**		3.472***	3.873***	3.083**
University		1.937***	1.701***	2.435***		0.457	−0.719	−0.151
Superior education		2.231***	1.953***	2.482***		27.561	24.387	23.462
Immigrant		0.443	0.323	−0.284		6.360	6.020	5.558
N. children [0–2]		−0.078	−0.247	−1.399**		−1.974***	−1.890***	−3.501***

(continued)

Table 5 (continued)

	Single females				Single males			
	Not working	50% part time	80% part time	Full time	Not working	50% part time	80% part time	Full time
N. children [3–6]		−0.247	−0.341	−0.684***		−0.114**	−0.209***	−0.192***
Bad health		−2.096***	−1.828***	−2.421***		0.523	0.064	0.486
Non-labor incomes		−0.024	−0.061**	−0.062***		−1.778	3.909	3.611
Rent		1.176***	1.031***	1.284***		0.122	−0.016	−0.035
Museums		−5.942	−0.237	−0.203		0.078	0.06	0.098
Books		0.091	0.007	0.033		0.870	−6.748	−4.295
Amusements		−0.030	0.101*	0.050		0.000	0.000	0.000
Constant		−3.984	−10.827***	−4.316*		0.000	0.000	0.000

Table 6 Discrete choice labor supply estimation for couples

Husband	Not working				Full time			
Wife	Not working	50% Part time	80% Part time	Full Time	Not working	50% Part time	80% Part time	Full Time
Log of dispos. income	Ref. cat	2.401***	2.401***	2.401***	2.401***	2.401***	2.401***	2.401***
Age wife		108.876	112.073*	76.146**	7.236	20.030	42.130*	46.001**
Age squared wife		−108.164	−133.696*	−91.670**	-5.080	−23.861	−50.852*	−56.638**
Cap/bep wife		−0.147	0.592	0.409	−0.350	−0.014	0.062	−0.231
University wife		−1.509	−15.533	0.332	−0.183	0.209	0.530	0.345
Superior education wife		−1.273	−15.412	0.543	−0.731	−0.473	−0.314	−0.427
Immigrant wife		0.771	−16.678	−0.057	0.398	0.323	−0.436	−0.352
Bad health wife		−0.724	−18.028	−1.139*	−0.298	−1.019*	−0.636	−1.141**
Age husband		−5.975	−47.364	5.005	−26.819	−22.964	−29.254	−35.180
Age squared husband		−13.638	46.666	−7.243	25.663	20.958	28.465	33.523
Cap/bep husband		0.806	−0.309	0.270	−0.730	−0.732	−0.386	−0.489
University husband		0.782	1.047	0.861	−0.501	−0.282	−0.017	0.116
Superior education husband		−0.336	−14.322	−0.130	−0.716	−0.769	−0.428	−0.264
Immigrant husband		−1.091	0.455	0.234	−0.272	−0.180	−0.313	−0.192

(continued)

Table 6 (continued)

Husband	Not working				Full time			
Wife	Not working	50% Part time	80% Part time	Full Time	Not working	50% Part time	80% Part time	Full Time
Bad health husband		0.679	0.716	1.180**	−1.513***	−1.531***	−1.355***	−1.441***
Part-time 1		25.456*	11.706	17.557**	7.210	15.523**	6.507	4.120
Part-time 2		−8.738	−1.448	−6.806	−4.783	−2.856	1.891	−5.125
Non-labour income		0.752**	0.240	0.346	−0.610***	−0.563***	−0.598***	−0.579***
Rent		−0.414	0.353	0.175	0.251	0.253	0.525	0.472
Leisure		0.009	−0.068	0.116	0.119	0.252**	0.215*	0.272**
Number of children		−0.103	−0.807	−0.554*	0.329	−0.070	−0.156	−0.525**
Constant		−25.640*	−13.958	−18.781***	6.585	2.082	−0.897	3.067

2.3.3 The Consumption Behavioral Model

The third element of SYSIFF 2006 is the behavioral model concerning consumption. As for labor supply, we first describe the estimation procedure and, then, how this behavioral model is used to evaluate the effects on consumption.

The estimation of consumption demand is based on the Almost Ideal Demand System proposed by Deaton and Muellbauer (1980) and extended by Banks et al. (1997) with the introduction a quadratic income term in the demand functions that fulfill the necessity of having a higher rank demand system which is useful when Engel curves are non-linear. Along with the quadratic extension, we also introduce demographic heterogeneity through an income translating function, firstly introduced by Gorman (1976). To comply with homogeneity properties required by consumption theory, i.e. to respect linear homogeneity and Slutsky symmetry, the demand system is subject to a set of *a priori* restrictions on the parameters. The system of demand equations is simultaneously estimated by Full Information Maximum Likelihood and using the generalized Heckman procedure to correct for zero expenditures (Shonkwiler and Yen 1999).

The demand of good i, in terms of budget share ω_i, is specified as follows:

$$\omega_i = \alpha_i + \sum_r \tau_{ir} \ln d_r + \sum_j \gamma_{ij} \ln p_j + \beta_i \left(\ln y^* - \ln a(\boldsymbol{p})\right) + \frac{\lambda_i}{b(\boldsymbol{p})} \left(\ln y^* - \ln a(\boldsymbol{p})\right)^2$$

with:

$$\ln y^* = \ln y - \sum_i \sum_r \tau_{ir} \ln d_r \ln p_j$$

$$\ln a(\boldsymbol{p}) = \sum_i \alpha_i \ln p_i + \frac{1}{2} \sum_i \sum_j \gamma_{ij} \ln p_i \ln p_j$$

$$\ln b(\boldsymbol{p}) = \sum_i \beta_i \ln p_i$$

where $\boldsymbol{d}$ is the vector of demographic characteristics, $\boldsymbol{p}$ is the vector of prices, and y is the total expenditure in consumption which is assumed to be proportional to the disposable income. The preference parameters to be estimated are α_i, β_i, γ_{ij}, λ_i and τ_{ir}. To respect linear homogeneity and Slutsky symmetry the following restrictions are imposed:

$$\sum_i \alpha_i = 1; \quad \sum_i \beta_i = 0; \quad \sum_i \lambda_i = 0; \quad \sum_i \tau_{ir} = 0 \text{ for all } r$$

$$\sum_i \gamma_{ij} = 0 \text{ for all } j\,; \quad \sum_j \gamma_{ij} = 0 \text{ for all } i\,; \quad \gamma_{ij} = \gamma_{ji} \text{ for all } i \text{ and } j$$

The dataset used for the estimation is BDF2006. After eliminating a few outliers, families with negative expenditures or negative total expenditure, the sub-sample consists of 10125 families, which is more than 99% of the original sample. To be consistent with the CGE model, consumption goods are aggregated into 11 categories: food, drinks, tobacco, clothing, housing, health care, transport/energy, communication, leisure, food out of home, and other goods. The demographic characteristics included in $\boldsymbol{d}$ are: household size, age of the household head, number of children with less than 3 years, number of children aged between 3 and 6, if living in a city with more than 100 thousands inhabitants, if the household head is married, if the household head is self-employed and if the household head is a manager.

The estimation results, reported in Table 7, show that most parameters of the demand system are significantly different from zero and with expected signs. In addition, self-selection bias due to zero expenditure is detected (and corrected) for almost all goods. The signs of income and uncompensated price elasticities, reported in Table 8 for the average family, are as expected and consistent with consumption theory requirements.

After the description of the estimation of consumption demand, we now describe how the behavioral model is used to evaluate the effect of a shock on consumption. As we have already said, when a shock is introduced in the model, the CGE model determines the effect on the equilibrium prices of the goods and services and the labor supply behavioral model determines, at the family level, the effect on the disposable income. The consumption behavioral model receives this information and determines, at the family level, the optimal level of consumption for each type of good. This optimal choice depends on the change in the disposable income, on the change on the price of each good, and on the estimated value of the elasticity of consumption demand with respect to income and prices.[12]

Finally, it is possible to aggregate the behavior concerning consumption, i.e. to compute the total quantity demanded by all families and for each type of good. The effect on the total quantity of consumption for each type of good is transmitted to the CGE model.

2.4 Micro and Macro Models' Integration

First of all, the macro CGE model is used to simulate a shock or a policy reform. In the CGE model, labor supply, the consumption of goods and services, employees' and employers' contributions, income taxes, and transfers from the government are exogenously fixed at the initial level. The CGE model permits to quantify the macro effects on:

- The equilibrium prices for each type of goods and services.
- The equilibrium wage.
- The unemployment rate.

[12]The consumer reaction is evaluated by considering family specific elasticities rather than average elasticities.

Table 7 FIML estimation of the AIDS system

		Drink	Tobacco	Clothing	Housing	Health	Transport/ energy	Communication	Leisure	Meals	Other	Food
	α	0.165***	0.070***	0.129***	−1.059***	0.166***	0.058***	0.054***	0.356***	0.288***	0.247***	0.526***
Drink		0.012***	0.000	−0.006***	0.011***	−0.002***	−0.002***	0.002***	−0.008***	−0.001***	−0.007***	0.000
Tobacco			0.040***	−0.004***	−0.011***	−0.007***	0.006***	0.013***	−0.012***	−0.007***	−0.008***	−0.011***
Clothing				0.036***	0.044***	−0.020***	−0.001*	−0.003***	−0.031***	−0.007***	−0.003***	−0.005***
Housing					−0.149***	0.024***	−0.003***	−0.002**	0.033***	0.009***	0.012***	0.032***
Health						0.049***	0.000	0.002***	−0.013***	−0.011***	−0.002***	−0.021***
Transport/ Energy							0.035***	0.002***	−0.003***	−0.003***	−0.005***	−0.028***
Communication								0.007***	−0.004***	−0.003***	−0.001***	−0.012***
Leisure									0.075***	−0.013***	−0.009***	−0.016***
Meals										0.055***	−0.008***	−0.012***
Other											0.050***	−0.019***

(continued)

Table 7 (continued)

		Drink	Tobacco	Clothing	Housing	Health	Transport/ energy	Communication	Leisure	Meals	Other	Food
Food												0.093***
	β	−0.018***	0.009***	−0.074***	0.248***	−0.035***	0.000	0.000	−0.053***	−0.018***	−0.019***	−0.041***
	λ	0.001***	−0.003***	0.009***	−0.013***	0.003***	0.000**	0.001***	0.004***	0.000**	−0.001***	0.000**
Hh. size		−0.006***	0.011***	0.003***	−0.007***	−0.006***	−0.001**	−0.004***	−0.006***	−0.004***	0.006***	0.015***
N. children [0–2]		0.002***	−0.023***	0.011***	0.010***	0.026***	-0.002	0.012***	−0.015***	−0.023***	−0.001	0.004
N. children [3–6]		−0.002***	−0.003**	0.005***	0.003***	0.013***	0.005***	0.013***	0.002***	−0.011***	−0.001*	−0.025***
City > 100.000		−0.004***	0.007***	0.009***	−0.012***	−0.006***	−0.006***	0.001*	0.006***	−0.002***	0.003***	0.003
Age of the hh. head		0.002***	−0.006***	−0.004***	0.006***	0.006***	0.005***	0.005***	0.001***	−0.010***	−0.001***	−0.005***
Hh. head is married		0.001***	−0.024***	−0.023***	0.014***	−0.020***	−0.006***	−0.020***	−0.017***	−0.033***	0.012***	0.116***
Hh. head is self-employed		0.003***	0.003*	−0.013***	−0.005***	0.005***	0.006***	0.042***	0.020***	0.007***	−0.019***	−0.050***
Hh. head is manager		−0.005***	−0.011***	−0.005***	−0.017***	−0.006***	−0.013***	−0.011***	−0.007***	0.018***	−0.004***	0.061***
	σ	0.004	0.081***	0.139***	0.305***	0.007*	0.041***	−0.058***	0.017***	0.011***	0.050***	0.000

Table 8 Price and income elasticities from the estimation of the AIDS system

	Drink	Tobacco	Clothing	Housing	Health	Transport/Energy	Communication	Leisure	Meals	Other	Food
Income elasticities	**0.988**	**0.648**	**1.321**	**1.344**	**1.134**	**1.074**	**1.153**	**0.982**	**0.91**	**0.731**	**0.755**
Uncompensated price elasticities											
Drink	**−0.794**	0.045	−0.037	−0.100	0.004	−0.014	0.032	−0.072	−0.019	−0.056	0.021
Tobacco	0.055	**−0.453**	−0.078	−0.089	−0.005	0.074	0.096	−0.084	−0.083	−0.062	−0.019
Clothing	−0.040	−0.057	**−0.484**	−0.123	−0.081	−0.046	−0.036	−0.149	−0.082	−0.066	−0.156
Housing	−0.081	−0.078	−0.140	**−0.270**	−0.105	−0.079	−0.049	−0.166	−0.122	−0.117	−0.137
Health	−0.007	−0.031	−0.063	−0.078	**−0.529**	−0.019	−0.004	−0.084	−0.083	−0.021	−0.213
Transport / Energy	−0.012	0.017	−0.003	−0.027	−0.007	**−0.786**	0.007	−0.027	−0.002	−0.033	−0.200
Communication	0.028	0.116	−0.024	−0.044	0.002	0.014	**−0.741**	−0.110	−0.099	−0.036	−0.260
Leisure	−0.035	−0.092	−0.110	−0.107	−0.056	−0.017	−0.035	**−0.336**	−0.066	−0.053	−0.076
Meals	−0.005	−0.100	−0.045	−0.074	−0.054	0.021	−0.027	−0.063	**−0.466**	−0.041	−0.057
Other	−0.024	−0.106	−0.046	−0.100	0.016	−0.005	0.011	−0.047	−0.035	**−0.355**	−0.041
Food	0.023	−0.051	−0.057	−0.031	−0.088	−0.109	−0.043	−0.026	−0.020	−0.023	**−0.329**

Then, the previous macro effects are introduced into the arithmetical component of SYSIFF 2006 which determines for each labor supply alternative and for each individual:

- The income and local taxes.
- The employees' and employers' contributions.
- The transfers from the government.
- The disposable income.

Then, the effects on the disposable income *for each labor supply alternative* and the effect on the unemployment rate (obtained in the CGE model) are introduced into the labor supply behavioral model which determines for each individual:

- The optimal quantity of labor supplied.
- Whether he/she finds a job or is involuntarily unemployed.
- The value of:
  - The income and local taxes.
  - The employees' and employers' contributions.
  - The transfers from the government.
  - The disposable income.

Then, the effects on the disposable income and on the equilibrium price for each type of goods and services (obtained in the CGE model) are introduced into the consumption behavioral model which determines for each family the optimal demand for each type of goods and services.

Then, the results obtained at the micro level are aggregated in order to compute the percentage change in:

- The total quantity of labor supplied.
- The total demand for each type of goods and services.
- The total amount of employees' and employers' contributions.
- The total amount of income taxes.
- The total amount of transfers from the government.

Then, these aggregate effects obtained at the micro level are introduced into the macro CGE model. The CGE model is solved using these new exogenous levels and determine the new macroeconomic effects, as in the first step.

The previous steps are repeated until a fixed point is reached, i.e. until the changes determined at the macro and micro levels, from one iteration to another, are sufficiently small.

3 Simulation Analysis

In this section we analyze the effects of a real depreciation of the Euro by 10%. We first analyze the macroeconomic effects, both on the whole economy and at the sectoral level, and then the microeconomic effects.

Table 9 Aggregate effects of a 10% real depreciation of the Euro on international trade

		Real terms	Nominal terms
Exports	(Variation in %)	3.3	3.5
Exports (Eurozone)	(Variation in %)	1.3	1.6
Exports (Rest of the world)	(Variation in %)	5.5	5.7
Imports	(Variation in %)	−6.8	−1.9
Imports (Eurozone)	(Variation in %)	2.7	4.3
Imports (Rest of the world)	(Variation in %)	−16.5	−8.2
Current Account / GPD	(Variation in p.p.)	2.7	1.4

3.1 *Macroeconomic Effects*

3.1.1 Macroeconomic Effects on the Whole Economy

The direct effect of a real depreciation of the Euro concerns international trade. In particular, Table 9 shows that exports, at constant prices, increase by 3.3% while imports, at constant prices, decrease by 6.8%. The effects in nominal terms, i.e. in terms of the *numéraire*, are obviously less positive since the Euro's depreciation implies an increase in the price of imports from the non-Eurozone. In nominal terms, imports decrease by 1.9%, while exports increase by 3.5%. The impact on the current account at constant prices is positive and quite important: with respect to GDP, the current account passes from a deficit of 0.5% before the shock to a surplus of 2.2% after the depreciation. Thus, the ratio of the current account to GDP increases, in real terms, by 2.7 p.p.

Table 10 shows the main macroeconomic results. The increase of the current account in real terms stimulates the aggregate demand. Given the macro closure used, the macroeconomic equilibrium between investments and savings is guaranteed by a change in the unemployment rate. In particular: (i) Private consumption at constant prices is negatively affected by the increase in the consumer price index (+0.5%), but positively affected by the reduction in the unemployment rate. The private saving rate increases by 1.3 p.p. (ii) Government savings decrease, at constant prices, and the ratio between the public deficit and GDP increases by 0.7 pp given that (a) the aggregate public expenditure, that is supposed to be proportional to real GDP, increases by 0.7%, (b) total direct taxation decreases by 0.5%,[13] and (c) VAT revenues decrease by 2.3% due to the reduction in imports from the rest of the world. (iii) Savings with respect to the rest of the world are affected by (a) the flow of domestic

[13]This result is explained by the reduction in wages expressed in terms of the *numéraire*. Due to the progressivity of the French fiscal scheme, this negative effect dominates the positive effect related to the increase in employment.

Table 10 Aggregate effects of a 10% real depreciation on the main macroeconomic variables

Real GDP	(Variation in %)	0.7
Unemployment rate	(Variation in p.p.)	−2.0
Labor	(Variation in %)	2.2
Capital	(Variation in %)	−0.8
Real wage	(Variation in %)	−2.0
Real rate of remuneration of capital	(Variation in p.p.)	0.0
Consumer price index	(Variation in %)	0.5
Private consumption	(Variation in %)	0.2
Investments	(Variation in %)	−10.8
Government expenditure	(Variation in %)	0.7
Private saving rate	(Variation in p.p.)	1.3
Public deficit/GDP	(Variation in p.p.)	0.7
Flow of domestic assets to RoW	(Variation in %)	2.4
Flow of foreign assets to France	(Variation in %)	−3.5

assets to the rest of the world (+2.4%) determined by the portfolio decision made by French people to invest in France or abroad and (b) the flow of assets from the rest of the world that equilibrates the balance of payments (−3.5%). (iv) Investments, determined by a specific investment function, are negatively affected by the increase in consumption, in government expenditures and in the current account (−10.8%). Nevertheless, as analyzed in Sect. 4, it is important to note that the crowding-out effect on investments is less important than it could be by using a standard neoclassical closure. The macroeconomic equilibrium between investments and savings requires an important reduction in the unemployment rate, from 8.8 to 6.8% (−2.0 p.p.). Although the quantity of labor that workers want to supply decreases by 0.04% given the reduction in the real wage (−2%), total employment increases by 2.2% thanks to the reduction in the unemployment rate.

Finally, the decrease in the unemployment rate produces a positive effect on the real GDP that increases by 0.7%.

3.1.2 Macroeconomic Effects by Sectors

As we have already said, the direct effect of depreciation concerns imports and exports. Obviously, the effect at the sectoral level depends on its exposure to international trade. Table 11 presents, for each sector, the size of imports (with respect to the total demand of domestic and foreign goods) and the size of exports (with respect to the total production). Seven sectors (energy, mineral products, textile, mechanic

Table 11 Size of imports and exports (total, vs. Eurozone, vs. rest of the world)

		M/(M+X) (%)	MEU/M (%)	MROW/M (%)	E/(E+X) (%)	EEU/E (%)	EROW/E (%)	Y/GDP (%)
1	Food	13.9	63.0	37.0	14.5	66.8	33.2	6.8
2	Beverage	9.0	53.8	46.2	24.5	37.9	62.1	2.2
3	Tobacco	13.1	83.3	16.7	3.3	51.8	48.2	0.4
4	Energy	29.5	27.0	73.0	11.2	56.7	43.3	5.5
5	Mineral products	46.5	57.7	42.3	45.7	48.7	51.3	3.2
6	Textile	45.5	38.5	61.5	37.1	47.5	52.5	1.7
7	Housing	22.5	53.3	46.7	18.9	47.4	52.6	2.8
8	Mechanic industry	34.9	53.7	46.3	36.6	46.7	53.3	3.0
9	Electric industry	54.6	34.5	65.5	51.3	47.7	52.3	2.9
10	Metallurgy	34.0	69.1	30.9	32.4	61.4	38.6	3.2
11	Health	6.5	51.1	48.9	9.9	50.5	49.5	10.1
12	Construction	0.0	0.0	0.0	0.0	0.0	0.0	7.1
13	Transports	40.0	51.1	48.9	43.2	50.5	49.5	7.5
14	Hotels and restaurants	0.0	0.0	0.0	0.0	0.0	0.0	2.3
15	Leisure	17.3	72.7	27.3	18.8	59.2	40.8	7.7
16	Communications	3.0	51.2	48.8	5.1	54.2	45.8	1.8
17	Public administration	0.0	0.0	0.0	0.0	0.0	0.0	4.8
18	Non-financial services and R&D	3.8	51.2	48.8	3.6	54.2	45.8	21.9
19	Financial services	2.8	51.2	48.8	3.9	54.2	45.8	5.0

industry, electric industry, metallurgy, and transports) are exposed to international trade, while three sectors are completely closed to international trade (construction, hotels and restaurants, and public administration).[14] The table also indicates the part of imports and exports with respect to the Eurozone and to the rest of the world. In the last column, we indicate the weight of each sector in terms of production with respect to the production at national level.

Table 12 shows that the real depreciation of the Euro strongly reduces imports of tradable sectors from the rest of the world (energy −12.2%, mineral products −16.3%, textile −19.4%, mechanic industry −23%, electric industry −17.5%, metallurgy −22%, and transports −13.9%), that are only partially replaced by imports from the Eurozone, and stimulates exports toward the rest of the world.

Table 13 indicates other macroeconomic effects at the sectoral level, concerning the production level, labor and capital demand, consumption and investment. In particular, the real depreciation of the Euro induces a significant increase in the production of tradable sectors (energy +2%, mineral products +2.6%, textile +4.7%,

[14] A sector is defined as exposed to international trade if imports represent more than 25% of total demand or exports represent more than 25% of total production.

Table 12 Sectoral effects on international trade (variations in %)

		Exports	Exports (Eurozone)	Exports (Rest of the World)	Imports	Imports (Eurozone)	Imports (Rest of the World)
1	Food	2.3	1.0	4.8	−5.2	2.7	−18.6
2	Beverage	3.5	1.1	5.0	−6.1	0.8	−14.1
3	Tobacco	3.0	0.8	5.2	−1.3	1.2	−13.6
4	Energy	2.4	1.2	4.1	−7.2	6.1	−12.2
5	Mineral products	3.0	1.5	4.4	−5.3	2.7	−16.3
6	Textile	3.6	2.0	5.2	−9.2	6.9	−19.4
7	Housing	3.2	1.3	5.0	−9.1	−0.5	−19.0
8	Mechanic industry	3.2	1.4	4.8	−10.8	−0.3	−23.0
9	Electric industry	3.4	1.9	4.8	−9.9	4.6	−17.5
10	Metallurgy	2.6	1.2	4.7	−5.1	2.4	−22.0
11	Health	3.1	0.9	5.2	−2.8	9.0	−15.1
12	Construction						
13	Transports	2.7	1.4	4.1	−6.5	0.6	−13.9
14	Hotels and restaurants						
15	Leisure	2.7	1.0	5.1	−3.6	2.1	−18.6
16	Communications	2.8	1.0	4.8	−7.3	2.0	−17.0
17	Public administration						
18	Non-financial services and R&D	2.8	0.9	5.0	−7.4	3.4	−18.9
19	Financial services	2.8	0.9	5.1	−7.6	3.1	−18.8

electric industry +3.1%, metallurgy +1.3%). The production level of the construction sector is dramatically reduced (−8.9%) due to the strong fall in investments.[15]

The effects on sectoral prices are reported in Table 14. In particular the domestic price, for each sector, is endogenously determined to guarantee the equilibrium in the domestic market, while the foreign price is computed as the weighted average between the price in the Eurozone and the price in the rest of the world, both affected by the real depreciation of the Euro. We also compute the total effect on the price level for each sector, computed as the weighted average between the domestic and the foreign prices, that is then sent to the microsimulation model and affects the individual consumer behavior. The most important increases in prices are observed in sectors that are more exposed to international trade: energy +2%, mineral products +1.8%, textile +2.3%, electric industry +3.3%, and transports +3.1%.

[15]Note that this negative effect is coherent with the evolution of the construction sector in the early years of the introduction of the Euro. In fact, between 1999 and 2005, the size of the construction sector has significantly increased in the Eurozone. In particular, in this period, the contribution of the construction sector to the total value added has increased by 11% in the Eurozone and, more specifically, by 13% in France, by 39% in Ireland, by 18% in Italy, and by 72% in Spain. In contrast, it has decreased by 27% in Germany. Thus, our simulation results suggest that the increase in the size of the construction sector is partly explained by the real appreciation of the Euro during this period.

Table 13 Sectoral effects on macro variables (variations in %)

		Production	Labor	Capital	Consumption (MS sectors)	Consumption (CGE sectors)	Investments
1	Food	0.7	2.5	0.7	0.0		−10.5
2	Beverage	0.2	2.2	−0.2	−1.7		
3	Tobacco	0.6	2.2	−0.2	0.5		−10.3
4	Energy	2.0	6.3	3.6		0.3	−12.6
5	Mineral products	2.6	5.7	3.3		0.5	−11.1
6	Textile	4.7	6.6	3.9	0.5		
7	Housing	−1.0	1.1	−1.4	−1.4		−10.5
8	Mechanic industry	−1.1	1.4	−1.1		0.5	−11.1
9	Electric industry	3.1	5.8	3.1		0.8	−11.9
10	Metallurgy	1.3	4.1	1.4		−1.0	−10.7
11	Health	0.8	2.5	−0.1	−0.5		−10.2
12	Construction	−8.9	−6.3	−8.8		1.1	−10.6
13	Transports	0.6	5.5	2.4	0.2		−11.9
14	Hotels and restaurants	−0.3	2.1	−0.4	−2.7		
15	Leisure	0.6	2.5	−0.1	−0.1		−10.6
16	Communications	−0.3	2.3	−0.3	−0.2		
17	Public administration	0.8	2.5	−0.1		1.9	
18	Non-financial services and R&D	−0.8	1.3	−1.3		2.9	−10.4
19	Financial services	−0.4	1.4	−1.1	0.4		

3.2 *Microeconomic Results*

The change in real wages affects the disposable income earned by each family and thus its labor market choices. The change in consumer prices and the change in the disposable income earned by each family affect the consumption choices concerning the different goods and services. Moreover, a certain number of involuntary unemployed find a job since, at the macro level, the equilibrium unemployment rate decreases. This implies that a real depreciation of the Euro, which is a pure macroeconomic shock, produces significant effects at the individual level, both in terms of individual choices and of income distribution.

Table 15 reports poverty and inequality measures for the whole population before and after the 10% real depreciation of the Euro.[16] The most notable result is a reduction in the number of the poor[17] of about 1%, accompanied by a similar reduction in the intensity of poverty[18] (−1.2%). Inequality reduction is small: the Gini index

[16]Poverty and inequality analysis is carried out by computing equivalent incomes using the OECD scale.

[17]An individual is defined as poor if his/her equivalent income is below the poverty line corresponding to the 60% of the median equivalent income.

[18]The intensity of poverty is defined as the mean distance separating the population from the poverty line, with the non-poor being given a distance of zero.

Table 14 Sectoral effects on prices (variations in %)

		Domestic prices	Foreign prices			Total
			Eurozone	Rest of the world	Total	
1	Food	0.1	0.7	10.0	3.8	0.4
2	Beverage	−0.2	0.6	10.0	4.8	0.2
3	Tobacco	−0.7	−0.5	10.0	1.2	−0.5
4	Energy	1.6	2.5	10.0	7.8	2.0
5	Mineral products	0.8	2.3	10.0	5.4	1.8
6	Textile	−0.5	2.0	10.0	6.6	2.3
7	Housing	−0.2	0.9	10.0	4.9	0.6
8	Mechanic industry	0.1	1.5	10.0	5.1	1.4
9	Electric industry	0.1	2.4	10.0	7.1	3.3
10	Metallurgy	0.3	1.3	10.0	3.7	1.1
11	Health	−0.7	−0.3	10.0	4.4	−0.6
12	Construction	0.3	0.8	10.0	0.0	0.3
13	Transports	1.5	2.9	10.0	6.3	3.1
14	Hotels and restaurants	0.0	0.5	10.0	0.0	0.0
15	Leisure	−0.5	0.1	10.0	2.6	−0.2
16	Communications	0.1	0.7	10.0	5.0	0.1
17	Public administration	−0.6	−0.2	10.0	0.0	−0.6
18	Non-financial services and R&D	−0.3	0.0	10.0	4.6	−0.3
19	Financial services	−0.4	−0.1	10.0	4.5	−0.4

Table 15 Microeconomic effects on income inequality and income distribution

	Baseline	Shock (%)	Variation (%)
Headcount ratio	9.05%	8.96	−1.00
Poverty gap ratio	2.53%	2.50	−1.20
Gini index	0.2909	0.2899	−0.30
10th percentile	6867	6893	0.40
50th percentile	13363	13345	−0.10
90th percentile	24750	24677	−0.30
90th/10th perc.	3.60	3.58	−0.70

reduces by 0.3%, while the average income increases for the first decile and reduces for the last one, reducing the interdecile ratio by 0.7%.

The reduction in poverty and inequality is substantially explained by the reduction in involuntary unemployment. Tables 16 and 17 report the change in the labor supply for singles and couples. 1.4% of previously unemployed singles finds a full-time job, while 0.3% finds a part-time job. Similarly, for couples, the families in which at least one member find a full-time job are 1.2%, while in 0.4% of families a previously

Table 16 Singles' labor supply reaction

		Shock				
		0	18	24	36	Total
Baseline	0	**14.1**	0.1	0.2	1.4	15.7
	18	0	**6.7**	0	0	6.7
	24	0	0	**8.2**	0	8.2
	36	0	0	0	**69.4**	69.4
	Total	14.1	6.7	8.4	70.8	100

unemployed member finds a part-time job. The number of families that reduce the labor supply due to the decline of the salary is negligible.

These results imply that the number of families that gain from the shock is limited, but their average gain is quite substantial. Table 18 reports the number of winners and losers in terms of disposable income and their average gain and average loss by family type. In general, results confirm that who wins obtains substantial gains but that a rather large part of the population suffers from a moderate loss, due to the slight decrease in equilibrium wages. It is worth noting that the shock benefits mostly the poor. Almost 9% of the poor win and their gain in terms of disposable income represents more than 34%. Another group that benefit from the shock is that composed by singles with children, where about 10% of them gain from the shock although their gain is slightly lower than for the poor. The group that is less affected by the shock is that composed by the elderly. Since there is no behavioral variation in the labor supply and their pension is not affected by the shock, there is no variation in their conditions.[19]

Thus, a first look at the aggregate poverty and inequality measures hides a quite substantial improvement. This is particularly true for poor people, since the reduction in the unemployment rate permits to some individuals to find a job. On the other side, the situation deteriorates—but very slightly—for people who have already a job since the wage level decreases after the shock. At the end, the average gain in terms of disposable income is very important (+24.1%) for the winners (which represent 2.5% of the families), while the average loss is quite limited (−1.1%) for the losers (which represent 36.9% of the families).

Finally, Table 19 reports the effects concerning consumption for each category by family type. Clearly, part of the variation is driven by the increase in prices, especially for tobacco and clothing, that had the sharpest price increase, while large income increases beef up consumption of the poor.

[19]In France pension benefits and subsidies are indexed to inflation, implying that the increase in prices obtained in the macro model has no effect on their real income.

Table 17 Couples' labor supply reaction

		Shock								
		0–0	0–18	0–24	0–36	36–0	36–18	36–24	36–36	Total
Baseline	0–0	**2.8**	0	0	0	0.2	0	0	0	3.1
	0–18	0	**0.8**	0	0	0	0.1	0	0	0.9
	0–24	0	0	**0.7**	0	0	0	0	0	0.7
	0–36	0	0	0	**2.5**	0	0	0	0.4	2.9
	36–0	0	0	0	0	**17.8**	0.2	0.2	0.5	18.7
	36–18	0	0	0	0	0	**9.9**	0	0	9.9
	36–24	0	0	0	0	0	0	**14.5**	0	14.5
	36–36	0	0	0	0	0	0	0	**49.3**	49.4
	Total	2.8	0.8	0.7	2.6	18.1	10.3	14.7	50.2	100

Table 18 Percentage of winners, average percentage gain, percentage of losers and average loss

Family type	Winners (%)	Average gain (%)	Losers (%)	Average loss (%)	Net gain (%)
All families	2.5	24.1	36.9	−1.1	0.2
Poor	8.7	34.2	11.6	−0.3	2.9
Single males	3.9	33.0	48.3	−1.1	0.8
Single females	2.8	14.3	29.2	−1.1	0.1
Singles without children	10.2	18.2	63.6	−0.7	1.4
Couples without children	1.8	28.6	40.3	−1.1	0.1
Couples with 1 child	3.3	18.8	59.4	−1.2	−0.1
Couples with 2 children	2.9	37.9	65.5	−1.1	0.4
Couples with 3 children or more	2.3	20.2	61.3	−0.9	−0.1
Elderly (aged more than 60)	0.0	0.0	0.0	0.0	0.0

4 Sensitivity Analysis

In this section, we present a sensitivity analysis to explore the role of the key element in the simulation of currency devaluation, i.e. the choice of the macro closure rule.

In this sensitivity analysis, we compare the main results with those obtained using different macro closure rules. As discussed in Sect. 2.2.6, the macro closure used in this study can be considered as between the neoclassical and the Keynesian ones. Here we simulate the effects of a 10% real devaluation of the Euro using a neoclassical closure, in which investments are determined by aggregate savings, and a Keynesian closure, in which investments are fixed at a predetermined value.

As Table 20 shows, using a neoclassical closure, a 10% real devaluation of the Euro would produce a negative effect on real GDP (−0.6%). This negative result is explained by the reduction in the stock of capital available in the economy (−0.9%) which, in turns, is related to the choice of French people to invest abroad. Clearly, without this effect due to the optimal asset allocation, the effect on the real GDP would be negligible since, if aggregate capital is constant and the reaction of labor supply to a change in the real wage is small, real GDP is only affected by the reallocation of the production factors across sectors. Thus, in real terms, the positive effect on the current account (exports increase by 2.6% and imports decrease by 8%) is more than

Table 19 Microeconomic effects on income, consumption and prices (variations in %)

Family type	Income	Food	Beverages	Tobacco	Clothing	Housing	Health	Transports	Communic.	Recreation	Hotels	Financial services
All families	0.2	−1.8	0.4	−1.4	−0.7	0.2	−2.7	0.3	−0.2	−0.3	0.0	−0.1
Poor	2.9	1.5	2.5	2.0	4.1	7.3	−0.3	10.6	4.7	2.9	2.1	2.2
Single males	0.8	−1.6	1.5	0.1	−0.5	1.9	−1.9	0.7	1.9	0.2	0.4	0.1
Single females	0.1	−1.8	0.1	−1.5	−0.7	0.1	−2.9	−0.4	−0.4	−0.5	0.0	−0.1
Singles without children	1.4	−0.9	1.6	0.8	0.1	−0.6	−3.0	3.8	0.6	−0.2	0.4	0.5
Couples without children	0.1	−2.0	0.3	−1.7	−0.8	−0.1	−2.8	−0.4	−0.6	−0.2	−0.2	−0.1
Couples with 1 child	−0.1	−2.1	0.0	−2.0	−1.2	−0.6	−3.0	0.0	−0.7	−0.5	−0.5	−0.4
Couples with 2 children	0.4	−1.8	0.1	−2.0	−0.2	1.1	−3.0	2.5	−0.4	−0.5	0.0	0.1
Couples with 3 children or more	−0.1	−2.1	−0.1	−1.5	−1.0	−0.3	−2.5	−0.7	−0.5	0.3	−0.1	−0.2
Elderly (aged more than 60)	0.0	−1.7	0.3	−1.1	−0.6	0.0	−2.6	−0.3	−0.3	0.0	0.0	0.0
Price variation		0.2	−0.5	2.3	0.6	−0.6	3.1	0.0	−0.2	0.1	−0.4	0.4

Table 20 Sensitivity analysis—Macro closures

		Base scenario	Neoclassical closure	Keynesian closure
Real GDP	(Variation in %)	0.7	−0.6	4.6
Unemployment rate	(Variation in p.p.)	−2.0	0.0	−7.9
Labor	(Variation in %)	2.2	0.0	8.6
Capital	(Variation in %)	−0.8	−0.9	−0.8
Real wage	(Variation in %)	−2.0	−1.5	−3.9
Real rate of remuneration of capital	(Variation in p.p.)	0.0	−0.1	0.2
Consumer Price Index	(Variation in %)	0.5	0.5	0.5
Private consumption	(Variation in %)	0.2	−0.5	2.0
Investments	(Variation in %)	−10.8	−14.4	0.0
Government expenditure	(Variation in %)	0.7	−0.6	4.6
Exports	(Variation in %)	3.3	2.6	5.4
Imports	(Variation in %)	−6.8	−8.0	−3.4
Private saving rate	(Variation in p.p.)	1.3	0.9	2.6
Public deficit/GDP	(Variation in p.p.)	0.7	0.8	0.4
Flow of domestic assets to RoW	(Variation in %)	2.4	2.4	2.3
Flow of foreign assets to France	(Variation in %)	−3.5	−3.8	−2.8

compensated by the small decrease in consumption and government expenditures and the strong fall in investments (−14.4%).

Instead, using a Keynesian closure rule, a 10% real devaluation of the Euro produces a strongly positive effect on the real GDP. Even if the stock of capital decreases by 0.8%, real GDP increases by 4.6% thanks to the strong reduction in the unemployment rate (which passes from 8.8% to 0.9%) and, consequently, to the significant increase in total employment (+8.6%) and the real GDP (+4.6%). Concerning the elements of the aggregate demand, the shock produces a positive effect on the current account (exports increase by 5.4% and imports decrease by 3.4%), investments do not change since, with the Keynesian closure rule, they are fixed at the initial level, while consumption and government expenditures increase (+2% and +4.6%, respectively). To resume, we think that both the neoclassical and the Keynesian closures determine an unrealistic effect on real GDP. In particular, with the neoclassical closure the unrealistic effect is related to the excessively high reduction in investments, while with the Keynesian closure the unrealistic effect is related to the excessively high reduction in the unemployment rate. In contrast, in our base scenario, the real devaluation of the Euro produces a quite positive effect on real GDP combined with an important reduction in investments (even if the crowding-out effect on investments is less important than that obtained with the neoclassical closure rule) and with a reduction in the unemployment rate (which is less important than that obtained with the Keynesian closure rule).

5 Conclusions

In this chapter we present a Micro-Macro simulation model applied to the French economy. The Micro-Macro simulation approach is very appealing since it permits to analyze simultaneously both the macro general equilibrium effects and the individual effects of a shock or a policy reform.

We use our Micro-Macro simulation model to evaluate the effects of a pure macroeconomic shock represented by a 10% real depreciation of the Euro, both at the macro and micro level. We find that the real depreciation of the Euro stimulates the aggregate demand by increasing exports and reducing imports. The increase in aggregate demand stimulates the real GDP and reduces the unemployment rate in the economy from 8.8 to 6.8%. At the sectoral level, currency devaluation induces important effects on job creation in tradable sectors (as found, for instance, by Davis and Haltiwanger 2001). Moreover, a positive effect, although less important, can be observed for non-tradable sectors (with the relevant exception of the construction sector) given the interrelations between sectors. Interestingly, currency devaluation does not produce a general reallocation of the workforce from all non-tradable to tradable sectors. In fact, the strong job destruction produced in the construction sector benefits to all the other sectors, especially mineral products, textile and electric industry.

At the individual level, the macroeconomic shock induces a significant alleviation of poverty and a slight reduction of income inequality. The decrease in the equilibrium wage determined in the macro model moderately reduces the disposable income for people who have already a job, while the reduction in unemployment permits to some involuntary unemployed to find a job, substantially increasing their income. The average income of the first decile increases while a rather large part of the population suffers from a moderate loss. About 9% of the poor improve their situation after the shock and their gain in terms of disposable income is more than 34%.

Acknowledgements We are grateful to Philippe Aghion, Luc Arrondel, Francois Bourguignon, and Benjamin Carton for their useful comments. We are especially thankful to an anonymous referee for the comments that have permitted to improve and clarify the paper. The usual disclaimer applies.

References

Álvarez-Martínez, M. T., & Polo, C. (2012). A general equilibrium assessment of external and domestic shocks in Spain. *Economic Modelling, 29*(6), 2486–2493.

Banks, J., Blundell, R., & Lewbel, A. (1997). Quadratic engel curves and consumer demand. *The Review of Economics and Statistics, 79*(4), 527–539.

Bingley, P., & Walker, I. (1997). The labour supply. *Unemployment and Participation of Lone Mothers in In-Work Transfer Programmes, Economic Journal, 107*(444), 1375–1390.

Bourguignon, F. J., Robilliard, A.-S., & Robinson, S. (2008). In *Examining the social impact of the indonesian financial crisis using a macro-micro model*. Open Access publications from Université Paris-Dauphine, Université Paris-Dauphine.

Burgess, S., & Knetter, M. M. (1998). An international comparison of employment adjustment to exchange rate fluctuations. *Review of International Economics, 6*(1), 151–163.

Campa, Jose Manuel, & Goldberg, Linda S. (2001). Employment versus wage adjustment and the U.S. Dollar. *The Review of Economics and Statistics, 83*(3), 477–489.

Cockburn, J. (2004). In *Trade Liberalisation and Poverty in Nepal. A Computable General Equilibrium Micro Simulation Analysis*. Development and Comp Systems, EconWPA.

Davis, S., & Haltiwanger, J. (2001). Sectoral job creation and destruction responses to oil price changes. *Journal of Monetary Economics, 48*(3), 465–512.

Deaton, A. S., & Muellbauer, J. (1980). An almost ideal demand system. *American Economic Review, 70*(3), 312–326.

Haan, Peter, & Uhlendorff, Arne. (2013). Intertemporal labor supply and involuntary unemployment. *Empirical Economics, 44*(2), 661–683.

Hall, R. E. (2009). By how much does gdp rise if the government buys more output? *Brookings Papers on Economic Activity, 40*(2) (Fall), 183–249.

Gorman, W. M. (1976). Tricks with utility functions. In M. J. Artis, & A. R. Nobay (Eds.), *Proceedings of the 1975 AUTE Conference, Essays in Economic Analysis*. Cambridge: Cambridge University Press.

Gourinchas, P.-O. (1999). Exchange rates do matter: French job reallocation and exchange rate turbulence, 1984-1992. *European Economic Review, 43*(7), 1279–1316.

Heckman, J. (1979). Sample selection bias as a specification error. *Econometrica, 47,* 153–161.

Löfgren, H., Harris, R.L., & Robinson, S. (2001). A standard computable general equilibrium (CGE) model in GAMS, No 75, TMD discussion papers, International Food Policy Research Institute (IFPRI).

Magnac, T. (1991). Segmented or competitive labor markets. *Econometrica, 59*(1), 165–187.

Magnani, R. (2015). The solow growth model revisited. Introducing Keynesian Involuntary Unemployment, CEPN Working Papers, HAL.

Magnani, R., & Mercenier, J. (2009). On linking microsimulation and computable general equilibrium models using exact aggregation of heterogeneous discrete-choice making agents. *Economic Modelling, 26*(3), 560–570.

Magnani, R., Piccoli, L., Carré, M., & Spadaro, A. (2017). Would a euro's depreciation improve the French economy? IZA Discussion Paper no. 11094.

Rattso, J. (1982). Different macroclosures of the original Johansen model and their impact on policy evaluation. *Journal of Policy Modeling, 4*(1), 85–97.

Rosensweig, J. A., & Taylor, L. (1990). Devaluation, capital flows, and crowding-out: a cGE model with portfolio choice for Thailand. In L. Taylor (Ed.), *Socially relevant policy analysis*. Cambridge, Massachusetts: MIT Press.

Savard, L. (2003). Poverty and Income Distribution in a CGE-Household Micro-Simulation Model: Top-Down/Bottom Up Approach, Cahiers de recherche, CIRPEE.

Shonkwiler, J. S., & Yen, Steven T. (1999). Two-Step estimation of a censored system of equations. *American Journal of Agricultural Economics, 81*(4), 972–982.

Taylor, L., & Lysy, F. J. (1979). Vanishing income redistributions: Keynesian clues about model surprises in the short run. *Journal of Development Economics, 6*(1), 11–29.

Van Soest, A. (1995). Structural models of family labor supply: a discrete choice approach. *Journal of Human Resources*, *30*(1), 63–88.
Vaqar, A., & O' Donoghue, C. (2007). CGE-Microsimulation modelling: A survey, MPRA Paper, University Library of Munich, Germany.

Green and Blue Dividends and Environmental Tax Reform: Dynamic CGE Model

Francesca Severini, Rosita Pretaroli and Claudio Socci

Abstract The challenge of climate change needs to be tackled with environmental policies carefully designed to achieve environmental benefits and avoid negative economic effects. The introduction of an environmental tax in the economic system can generate a double benefit represented by the attainment of the environmental target (first or green dividend) and other additional benefits (second/third or blue dividends) represented by gains in welfare, employment, consumption etc. In this perspective, the general equilibrium analysis is able to quantify the environmental and welfare direct and indirect effects that an environmental policy generates within the economic system. Since international environmental agreements set clear target deadlines on the reduction of GHG emissions, in this chapter a dynamic CGE model based on a bi-regional SAM framework for Italy is developed.

Keywords Environmental tax reform · SAM · Dynamic CGE model · Double dividend

JEL Classification H23 · D58 · D57

1 Introduction

In recent years, the European Union has promoted initiatives on environmental protection as required by the Kyoto Protocol[1] and the member states are committed to introduce environmental policies for the reduction of greenhouse gas emissions

[1] In 2000, the European Commission launched the European Climate Change Program (ECCP) to identify and develop all the elements necessary to match the Kyoto Protocol. The goal of EU environmental policy for the year 2020 includes the cut of 20% in CO_2 emissions, the increase in renewable energy use of 20% and the increase in energy efficiency by 20% with respect to 1990 levels.

F. Severini (✉) · R. Pretaroli · C. Socci
Department of Economics and Law, University of Macerata, via Crescimbeni 14, 62100 Macerata, Italy
e-mail: francesca.severini@unimc.it

© Springer International Publishing AG, part of Springer Nature 2018
F. Perali and P. L. Scandizzo (eds.), *The New Generation of Computable General Equilibrium Models*, https://doi.org/10.1007/978-3-319-58533-8_10

(GHG). However, the need of structurally reducing the pressure of the human activities on environment cannot be kept apart from the need of stimulating economic growth as suggested by the European Commission (EC 1993). Therefore, the trade-off between environmental and economic targets that often emerges should be taken into consideration when selecting the most appropriate environmental policy instruments to deal with this complex target.

Among the variety of economic measures, more than a few Central Governments adopted taxes on emissions and emission permits trading that are widely known as market-based policy instruments dealing with externalities of pollution (Baumol and Oates 1988; Morris 1999). The environmental taxation, in particular, is considered a powerful tool of pollution control (Parry 2004; Farmer and Steininger 1999) because it discourages the pollutant behaviour and provides public revenue that can be recycled both at state and federal level to enhance environmental and non-environmental benefits (Pearce 1991). As for the potential positive impact of environmental taxation, we can refer to the concept of double or triple dividend (Bovenberg and De Mooij 1994), widely debated in the economic literature. In particular, it is possible to identify as benefits from the environmental taxation the reduction in GHG emissions (first or green dividend) and the positive non-environmental benefit related to economic welfare improvement (second or blue dividend). This latter can be assessed when the tax revenue, collected by Central or Local governments, is used to cut existing distortive taxes (Parry 1995). In the literature, many empirical studies debate around the existence of the double dividend in its weak and strong version, starting from the survey of Goulder (1995b). In general, there is evidence that the weak double dividend hypothesis holds, while the strong version seems to fails, that is to say, the environmental tax revenue can be used to reduce other existing tax distortions but does not easily allow achieving also an income or, more generally, a welfare benefit. For the literature that neglects the presence of a second dividend we can see Goulder (1995a), Bovenberg and Goulder (1996), Bovenberg and Goulder (1997) and Böhringer et al. (1997).

However, when considering the complexity and peculiarities of each economic system, there are many empirical studies that demonstrate the possibility to reach a strong double dividend as a result of an accurate environmental tax revenue recycling scheme. Nonetheless, there is even evidence of a third benefit that might arise with the second dividend effect for certain countries (Schneider 1997; Bovenberg and De Mooij 1998; Manresa and Sancho 2005; Takeda 2007; Glomm et al. 2008; Bor and Huang 2010). These second and third benefits can be represented by better performances of economic variables such as employment, production, consumption, inflation or income (Gimenez and Rodriguez 2010).

As for the Italian economy, the analysis carried out by Pench (2002), Bulckaen and Stampini (2002) and Roson (2002) tries to detect the double dividend effect by introducing different environmental tax reforms in a general equilibrium framework. In particular, they simulate the introduction of a "carbon tax" (Pench 2002) and analyse the impact of the environmental policy in a dynamic framework (Roson 2002). The results of their simulations do not confirm the possibility to get any further advantages apart from the environmental benefit, however the opportunity of getting a double dividend for the Italian economy is still an open question since the introduction of an environmental taxation to realize both environmental and budget objectives is still present in the policy debate. Indeed, it is worth pointing out that the possibility to get a double dividend through an environmental policy strictly depends on the structure of the existing tax system, on the production technology and above all on the structure of tax reform. Following Bovenberg and Goulder (2002), the presence of double dividend depends on a set of different conditions: the structure of primary factors' taxation; the elasticity in primary factors supply; the international immobility of capital; the value of elasticity of substitution between energy and labour (if higher than elasticity of substitution between energy and capital) and the sensibility of real wages to unemployment falls (as a consequence of the reduction of taxes on labour).

From this point of view, in a country characterised by economic differences at regional and social level, the double dividend could differ between regions or it could not occur for all regions where environmental fiscal reform is implemented (Takeda 2007). In this respect, empirical studies on environmental tax reforms and double dividend are typically focused on countries rather than regions and either accept or refuse the hypothesis of double dividend merely observing the effects of the policy on the macroeconomic variables' changes at national level. When the analysis of the environmental tax reform is performed at regional rather than national level, it is possible to figure out the economic and social differences among regions within the same country and let the regional peculiarities in technologies and habits emerge also in terms of ability in generating ecological dividends.

Such regional analysis can be carried out through a set of instrument able to quantify the direct and indirect effects of the environmental policy in a multisectoral and multiregional framework. Computable general equilibrium models (CGE) are widely considered in the literature as suitable instruments of analysis (Radulescu and Stimmelmayr 2010) to quantify the impacts of an exogenous shock on macroeconomic variables along the income circular flow (Ciaschini and Socci 2007a, b; Yeldan 1997). Moreover, since the European Commission sets to the member states clear deadlines (year 2020) to achieve the CO_2 emissions targets, the analysis should be carried out in a long term perspective, moving from the static to a dynamic approach.

In this perspective, this study develops a bi-regional multisectoral dynamic CGE model to verify the compatibility between the environmental taxation and the economic targets in terms of double/triple dividend. In particular we analyse the impact of an environmental fiscal policy for the Italian economy through a dynamic CGE model. Three main aspects inspire the use of a dynamic CGE model: first, static CGE is based upon a single set of equilibrium conditions and leaves aside relationships over time.[2] Second, the vector of prices that solve the static equilibrium does not hold over time and refers to an uncertain time horizon. Finally, even if the assumptions on elasticity of supply and demand can be interpreted as relatively long run adjustments,

---

[2]As an example, producers and consumers, which maximize their utility choosing the optimal allocation of consumes and savings become myopic in the between period decisions (savings and investment).

static models do not account for more than a few factors such as capital accumulation, population growth and technological change (Lau et al. 2002). Therefore, we suggest a disaggregate and regional perspective for the study of environmental policies that can offer a further contribution in the existing debate on double dividend.

In addition, since the introduction of an environmental tax is becoming a prerogative of European countries, we focus on Government taxations from two sides: the side of production and generation of income (that is affected by the introduction of a new environmental tax) and the side of the secondary income distribution (that is affected by different tax revenue recycling schemes). Actually, once the environmental tax rate and tax base are determined, the government action should focus on the best possible use of tax revenues in order to reduce existing distortions in taxation.

The environmental tax reform proposed is characterized by the introduction of a regional environmental tax, which affects the production process according to the level of CO_2 emissions by each commodity. In particular, this tax is designed with a different '*polluter pays*' principle that follows instead, the idea that "who pollutes more should pay more", so that the activities that have a level of emission over a certain limit, pay a tax that is reshaped by and has a progressive structure. The main purpose of this reform is to assess if there is the possibility to obtain, through an environmental tax, a positive effect both on environment and on disposable income.[3] Thus, the corresponding tax revenue is used to cut income taxes and the regional tax on activities' value added. The reduction of income tax has the aims to mitigate the green tax effects on households real disposal income: the price of goods in fact, may be affected by tax shift and the private final consumption may dampen as a consequence of the higher price. On the other side the reduction of the regional tax on activities' value added is applied to face directly the tax shift correlated to the green tax on commodities.

The analysis is carried out on the bi-regional Social Accounting Matrix (SAM) for the Italian economy that allows to quantify both the economic and the environmental effects generated by the environmental fiscal reform in the long run. Furthermore, the aim to identify the convenient green tax reform requires the integration of the SAM with the environmental data set concerning CO_2 emissions by commodity. In this respect, the European Commission suggests the use of the National Accounting Matrix with the integration of the Environmental Accounts (NAMEA) as the basic tool for the integration between environmental and economic flows.[4] This detailed database represents the benchmark for the CGE model that is calibrated on it, and allows to discuss the results of the policy proposed in terms of changes in prices, total output, final demand and total emissions. Moreover, the analysis is integrated with considerations about the changes of the burden index over time. Indeed this

[3]The model assumes that all markets clear, therefore we do not considers any rigidity on wage formation and unintentional unemployment.

[4]The NAMEA integrates the major economic aggregates—total output, value added and final demand—with the GHG emissions data in physical terms according to the input output disaggregation (EC 1994). This approach avoids the difficulties connected to a correct valuation of environmental costs.

indicator is classically considered as a measure of government efficiency in the income distribution act and we assume its performance as a welfare measure.

The next section points out the main features of the dynamic model and describes the tax burden indicator. Section three gives a description of the database and introduces the environmental policy targets for the Italian economy. Than in the fourth section, we suggest a suitable environmental tax reform consistent with the reduction of CO_2 emissions and propose two tax revenue recycling alternatives. The fifth section provides a description of the simulation results emerging from the application for the Italian case, in terms of CO_2 emissions by activities, total output, price trend, gross investment, final consumption and tax burden over time.

2 Dynamic CGE Model Relationships

BiReg17 (Bi-Regional 2017), is a bi-regional and multisectoral dynamic CGE model where the evolution path is a sequence of single period static equilibria linked each other by the capital accumulation condition (Lau et al. 2002). It is a recursive dynamic model that can be illustrated in two phases: the first refers to the description of the single period equilibrium conditions; the second introduces the dynamic rule.

The model considers an open economy with two regions, m commodities, c components of value added, h Institutional Sectors including Households, Firms, Government[5] and Rest of the World. In every time period the demand equals the supply in all commodities and primary factors markets (market clearing conditions) and extra profits are not allowed (no profit conditions) (Pretaroli and Severini 2009).

Bireg17 can be described as an integrated representation of the bi-regional income circular flow (Ciaschini et al. 2012) where the entire process of generation, primary and secondary distribution of income is represented by a system of behavioural equations and income constrains for agents, which are price taker and maximise their utility function. Following the scheme provided by Table 1, the total output ($\mathbf{X_t}$) resulting from the sum of domestic and imported output ($\mathbf{M_t}$)[6] is equal to intermediate demand ($\mathbf{B_t}$), final consumption expenditures ($\mathbf{C_t}$), final consumption expenditure incurred by Government ($\mathbf{CG_t}$), gross fixed capital formation ($\mathbf{I_t}$) and exports ($\mathbf{E_t}$). Likewise, primary factors' endowments correspond to primary factors' demanded by production process ($\mathbf{Y}$) and their markets are perfectly competitive. We do not consider any rigidity on wage formation and thus we assume that there is no unintentional unemployment.[7] Domestic production is formalized as a nested constant return to scale technology. Assuming the Leontief production function,

[5]The Government is represented as a Central Government, that has a national dimension, and as Local Government that is represented together with the other institutional sectors. The assumptions on Institutional Sectors hold also for Central and Local Government.

[6]Following Armington's hypothesis (Armington 1969), imported and domestically produced commodities are not perfect substitutes. This solves the problem that the same kind of good is found to be both exported and imported.

[7]Labor supply (endowment) is exogenous.

domestic output is the combination of intermediate goods (**B**), depending on total output and prices, and value added that is affected by total production and primary factors compensations (**Y**). Then assuming a CES technology, the value added is generated by combining capital and labour that are perfectly mobile across activities.[8]

Following the logic of the Ramsey model, the Institutional Sectors maximise the present value of their intertemporal utility function, which depends on final consumption expenditure ($\mathbf{C_t}$ and $\mathbf{CG_t}$), and gross saving ($\mathbf{S_t}$ and $\mathbf{SG_t}$) subject to the lifetime budget constraint. The budget constrain for Households is verified when the total disposable income (**Rd**) is equal to the sum over time of final consumption expenditures (**C**) and savings (**S**). The primary factor compensations ($\mathbf{R_t}$) plus net transfers from Institutional Sectors ($\mathbf{Tr_t}$), minus income taxes ($\mathbf{Ta_t}$), determine consumers total endowments in every time period. As to Government, public saving (or deficit) ($\mathbf{SG_t}$) is calculate as total tax revenue ($\mathbf{Ta_t}$) minus the sum of final consumption expenditures by Government ($\mathbf{CG_t}$) and transfers to other Institutional Sectors ($\mathbf{Tr_t}$). This description represents the public budget constrain.[9] We distinguish direct income taxes and a set of indirect taxes (tax on products, value-added tax and payroll taxes).

The single period equilibrium regarding the condition on gross capital formation requests that total gross fixed capital formation ($\mathbf{I_t}$) becomes equal to gross savings by Institutional Sectors ($\mathbf{S_t}$ and $\mathbf{SG_t}$). The dynamic component in the model is given by the inter-temporal capital accumulation condition. According to the market clearing condition for capital, any change in gross fixed capital formation must affect the capital yearly growth given a constant rate of capital depreciation (δ).[10]

Then, in the dynamic model, the optimization problem for all the consumers becomes:

$$\max \sum_{t=0}^{\infty} \left(\frac{1}{1+\rho}\right)^t u\left[C_t\left(y_{d_t}, p_t\right)\right] \tag{1}$$

s.t.

$$C_t = f\left(K_t, L_t, M_t, Ta_t\right) - I_t - E_t \tag{2}$$

$$K_{t+1} = (1-\delta)K_t + I_t \tag{3}$$

Every institutional sector maximizes intertemporal utility which depends on consumption, under the constraints represented by two conditions:

---

[8]The elasticity of substitution between labor and capital derives from econometric estimates for Italy (Van der Werf 2007).

[9]The marginal cost of public funds are set equal to zero.

[10]According to the literature on dynamic CGE we employ the term '*depreciation*' in place of the term '*consumption of fixed capital*' used by the SNA. The term 'consumption of fixed capital' refers to the decline, during the course of the accounting period, in the current value of the stock of xed assets owned and used by a producer as a result of physical deterioration, normal obsolescence or normal accidental damage. It is used in the SNA to distinguish it from 'depreciation' as typically measured in business accounts (United Nations 2008).

Table 1 CGE model following SAM framework (our source)

		South-Islands			North-Centre			Central government	Formation of capital	Rest of word
		Commodities	Primary factors	Institutional sectors	Commodities	Primary factors	Institutional sectors			
S-I	*Comm.*	$\mathbf{B}^{ss}(x, p)$		$\mathbf{C}^{ss}(rd, p)$ $\mathbf{GL}^{ss}(rd, p)$	$\mathbf{B}^{sn}(x, p)$		$\mathbf{C}^{sn}(rd, p)$ $\mathbf{GL}^{sn}(rd, p)$	$\mathbf{G}^{sg}(rd, p)$	$\mathbf{I}^{s}(p)$	$\mathbf{E}^{sw}(e, p)$
	Fact.	$\mathbf{Y}^{ss}(x, pi, pk)$			$\mathbf{Y}^{sn}(x, pi, pk)$					
	Ins.Sec.	$\mathbf{T}^{ss}(x)$	$\mathbf{R}^{ss}(y)$	$\mathbf{T}^{ss}(r, t)$ $\mathbf{T}_r^{ss}(r, t)$		$\mathbf{R}^{sn}(y)$	$\mathbf{T}^{sn}(r, t)$ $\mathbf{T}_r^{sn}(r, t)$	$\mathbf{T}_r^{sg}(r, t)$		$\mathbf{t}^{sw}(r)$
N-C	*Comm.*	$\mathbf{B}^{ns}(x, p)$		$\mathbf{C}^{ns}(rd, p)$ $\mathbf{GL}^{ns}(rd, p)$	$\mathbf{B}^{nn}(x, p)$		$\mathbf{C}^{nn}(rd, p)$ $\mathbf{GL}^{nn}(rd, p)$	$\mathbf{G}^{ng}(rd, p)$	$\mathbf{I}^{n}(p)$	$\mathbf{E}^{nw}(e, p)$
	Fact.	$\mathbf{Y}^{ns}(x, pi, pk)$			$\mathbf{Y}^{nn}(x, pi, pk)$					
	Ins.Sec.		$\mathbf{R}^{ns}(y)$	$\mathbf{T}^{ns}(r, t)$ $\mathbf{T}_r^{ns}(r, t)$	$\mathbf{T}^{nn}(x)$	$\mathbf{R}^{nn}(y)$	$\mathbf{T}^{nn}(r, t)$ $\mathbf{T}_r^{nn}(r, t)$	$\mathbf{T}_r^{ng}(r, t)$		$\mathbf{t}^{nw}(r)$
C.G.		$\mathbf{T}^{gs}(x)$	$\mathbf{R}^{gs}(y)$	$\mathbf{T}^{gs}(r, t)$ $\mathbf{T}_r^{gs}(r, t)$	$\mathbf{T}^{gn}(x)$	$\mathbf{R}^{gn}(y)$	$\mathbf{T}^{gn}(r, t)$ $\mathbf{T}_r^{gn}(r, t)$			
F.K.				$\mathbf{S}^{s}(rd)$			$\mathbf{S}^{n}(rd)$	$\mathbf{S}^{g}(rd)$		(+/−)a
RW		$\mathbf{M}^{ws}(x, e)$	$\mathbf{t}^{ws}(y)$	$\mathbf{T}_r^{ws}(r)$	$\mathbf{M}^{wn}(x, e)$	$\mathbf{t}^{wn}(y)$	$\mathbf{T}_r^{wn}(r)$			

(i) total output produced by each commodity $\mathbf{X_t}$ is equal to the sum of intermediate consumption, households consumption expenditures ($\mathbf{C_t}$), government current expenditures ($\mathbf{CG_t}$), gross fixed capital formation $\mathbf{I_t}$ and exports $\mathbf{E_t}$ (market clearing conditions);
(ii) the capital stock in period $t + 1$ is equal to the capital stock in period $t(\mathbf{K_t})$,[11] less depreciation ($\delta\mathbf{K_t}$) plus gross fixed capital formation in period $t(\mathbf{I_t})$[12]. The rate of capital depreciation is exogenous. It depends on the value of steady state interest rate r and growth rate g.[13]

In order to solve the model for a finite number of periods, we approximate the infinite horizon with endogenous capital accumulation condition according to Lau et al. (2002). Thus in order to obtain the terminal period equilibrium we set the terminal gross capital formation growth rate equal to the growth rate of aggregate output (see the Appendix 1).

In BiReg17 two regions are modelled, therefore all equilibrium condition and budget constraint hold for both regions. As for the economic flows between regions, they are not considered as exports or imports, but are modelled as intermediate consumption of commodities associated to the other region (in the production block) and as income transfers between institutional sectors belonging to the other region (in the primary and secondary income distribution). Imports and Exports are determined at national level and include the economic flows of each region only with the rest of the world.

Because there are two regions and a set of Institutional sectors, the model produces a disaggregate set of information on prices, total output by commodity and incomes. However, a welfare measure that allows seeing the overall effects of a policy is represented by the tax burden index that is considered as a measure of government efficiency in income redistribution. Moreover, the computation of this index allows to understand whether there are efficiency gains or only redistribution effects when a policy is implemented. The tax burden index is calculated as the ratio between the sum of all taxes (direct taxes, indirect taxes—tax on products, value-added tax and payroll taxes) and the Gross Domestic Product (GDP). It is set equal to 100pt in the baseline equilibrium and it is recursively calculated each year in order to show its evolution over time.

3 Environmental Accounts in a Social Accounting Scheme

The parameters of BiReg17 are calibrated on the bi-regional SAM for the Italian economy. It describes the production system features and the income circular flow in terms of intra-regional and inter-regional flows (Pretaroli and Severini 2009). A SAM

[11] The capital stock in period t is calibrated on the SAM data following Paltsev (2004).

[12] For the specification of the dynamic model see the appendix Appendix 1.

[13] In our model, we assume $r = 4\%$ (nominal interest rate) and $g = 0.6\%$ (real growth rate). According to the rule for investment on a steady state $I_t = (d + g)K_t$ we calibrate the value of the depreciation rate δ on the SAM data.

scheme of the bi-regional flows is showed in Fig. 1. The flows are split in two macro regions, the North-Centre region and the South-Islands region (Pretaroli and Severini 2008) and the rows and columns of the SAM are headed to 16 commodities [1. Products of agriculture, 2. Energy products, 3. Metal and non metal ore, 4. Non-metallic mineral products, 5. Chemical products, 6. Mechanics, 7. Transport equipment, 8. Food products and beverages, 9. Textile, 10. Other manufacturing products, 11. Construction work, 12. Trade, 13. Transport, 14. Financial services and Insurance, 15. Private services, 16. Government services]; 2 primary factors [Labor and Capital]; 5 Institutional Sectors [I. Households, II. Firms, III. Regional Government, IV. Central Government, V. Rest of the World]. The disaggregation of Institutional Sectors follows the exigency of testing the impacts of policy reforms on public and private balances. For this purpose, different typologies of tax and expenditures are considered. In particular, we distinguish social contributions, regional value added tax, a set of indirect taxes on commodities and income taxes. The average tax rates are calibrated on the SAM data and are fixed at their benchmark level in all scenarios.

Since the paper aims to assess the economic and environmental impacts of a fiscal reform at regional level, the SAM database is integrated with environmental indicators provided by the National Accounting Matrix including Environmental Accounts (NAMEA) developed by ISTAT (2008). We focused on CO_2 emissions by commodity[14] and associated these physical flows to the commodities classification in the SAM. This phase allows to construct a data scheme in which the economic flows related to the 16 commodities in each region (North-Centre and South-Islands) are associated to a specific level of CO_2 emissions. The different polluting power associated to each commodity depends on the technology employed in the production process and is measured by the CO_2 emission coefficient.[15]

4 Fiscal Policy Through an Environmental Tax Reform

According to the Kyoto Protocol, the Italian economy had to reduce the CO_2 emissions by 16.9% with respect to the 1990's level within 2020.[16] From 1990, when the total CO_2 emissions were 360 Mlt, Italy should be reducing the emissions of CO_2 on average by 2.045 Mlt each year, in order to achieve the Kyoto target represented by 300 Mlt in 2020 (ISTAT 2008).[17] Actually, the annual level of CO_2 emissions

[14] We do not consider the CO_2 emissions resulting from final consumption expenditure. The impact on CO_2 emissions is not included in utility function of the Government in order to obtain Environmental Domestic Product.

[15] The emission coefficient by commodity is the ratio between the of CO_2 emission tons by commodity and the total output.

[16] The Kyoto protocol established the reduction of 20% of Italian GHG. CO_2 emissions represent the 85% of total GHG, thus the Kyoto target for Italian CO_2 can be considered as 16.9%.

[17] We do not consider the emissions deriving from final consumption process. Therefore, the levels and the target of emissions considered do not include direct emissions caused by households and firms.

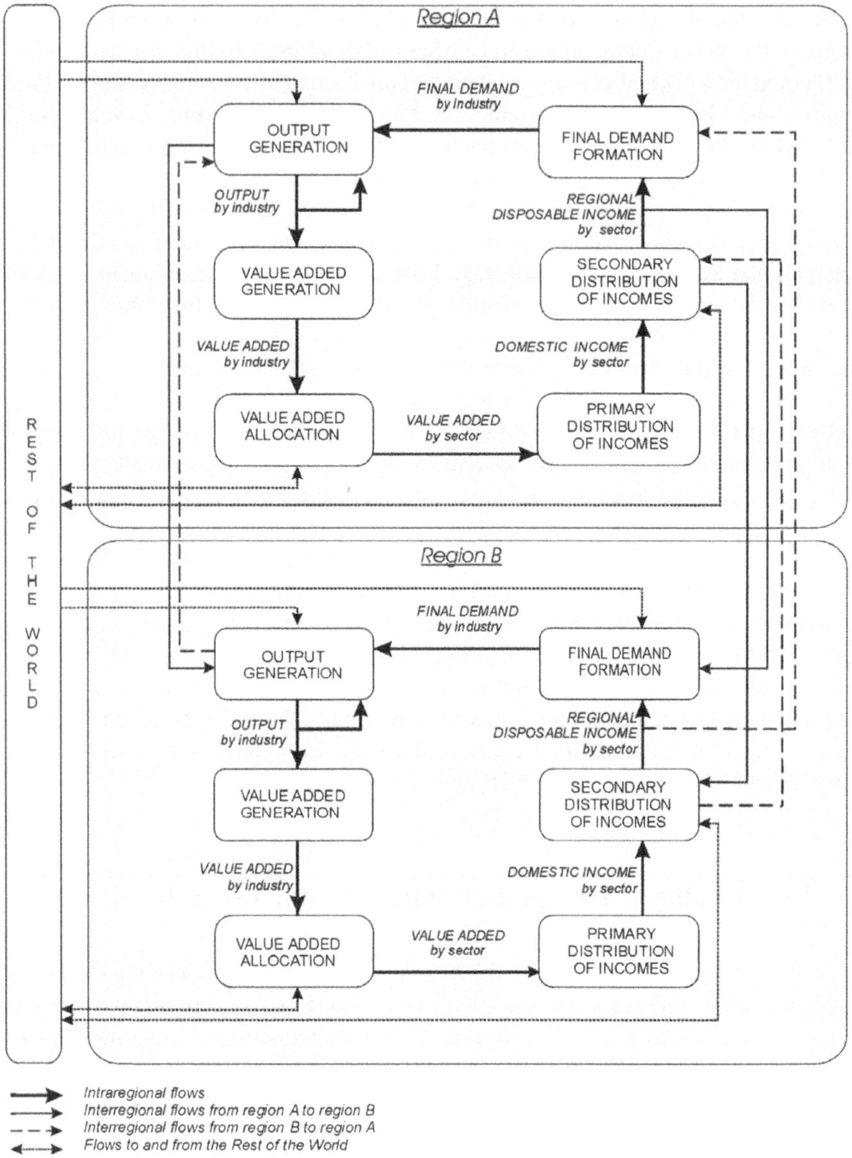

Fig. 1 Bi-regional SAM framework (our source)

is exceeding the hypothetical annual level compatible with the 2020 target and this difference can be easily interpreted as the annual Italian potential debit of CO_2 emissions (Ciaschini et al. 2012, 2014). This trend should reverse or, at least, terminate to approach the 2020 target. For this purpose, we can consider the introduction of an environmental tax that takes into account the above-mentioned polluting license and the polluting power of each commodity represented by the emission coefficient. The average level of emission allowed by commodity can be calculated as the ratio between the total level of CO_2 for Italy in each year and the number of commodities considered in the benchmark.[18] The "no-tax area" therefore, reflects the average level of CO_2 emissions that fulfils the Kyoto Protocol target and the commodities charged by the taxation are those with a level of CO_2 emissions exceeding this level.

In Table 2, we calculate the distance in tons between the actual and the admitted level of emissions in the base year by commodity. Those showing a negative value do not pay any environmental tax since they pollute under the permitted level, on the contrary those with a positive value are burdened by the taxation following the principle "who pollutes more should pay more". This means that they pay a higher marginal tax rate and should have an incentive to reduce their emissions to avoid the taxation, allowing the achievement of the so called green environmental dividend.[19]

Technically, we consider the introduction of a carbon tax on output differentiated by commodity according to CO_2 emissions coefficients. The exemption area is calculated as the ratio between the total level of CO_2 allowed for Italy in the base year and the number of commodities in the benchmark. Thus, the environmental tax is designed with a progressive structure, with 5 classes of taxation and a fixed price per ton of CO_2 emission established in each class. When total emissions by commodity exceeds the cut-off point, the commodity is taxed according to the subsequent class of taxation for the emissions in excess. The structure of the tax for the base year can be described as follow:

from 0 to 10.871.958 t (no-tax area);
from 10.871.958 t to 15.000.000 t (9 euro per CO_2 t);
from 15.000.001 t to 30.000.000 t (16 euro per CO_2 t);
from 30.000.001 t to 50.000.000 t (22 euro per CO_2 t);
over 50.000.001 t (32 euro per CO_2 t).

According to the database, the commodities burdened by the tax in North-Centre region are: 'Energy products', 'Non metallic mineral products', 'Chemical products', 'Mechanics', 'Trade' and 'Transport'. In South-Islands region, the tax is calculated on 'Energy products', 'Non metallic mineral products' and 'Transport'.

The environmental tax revenue can be attributed alternatively to the Central or Local Government and used to cut existing local or central taxes. Depending on the choice, we can identify two hypothetical scenarios: in s1 the tax revenue is allocated to the Regional Government and recycled to reduce the regional tax on value added

[18]There are 32 commodities (16 for North-Centre and 16 for South-Islands regions).

[19]Because we do not know the costs of the environmental damage, we consider the amount of CO_2 emissions as a proxy of the environmental damage and consider its reduction as a positive effect (dividend)

Table 2 Distance from admitted level of CO_2 emissions by commodity tons in base year (our elaboration)

	South-Island	North-Centre
1. Products of agriculture	−8193883	−6481826
2. Energy products	41741701	79085967
3. Metal and non metal ore	−11658202	−10710630
4. Non metallic mineral product	569054	21842393
5. Chemical products	−6322697	22121753
6. Mechanics	−8674561	14339882
7. Transport equipment	−11164576	−9053772
8. Food products and beverages	−9641433	−4622612
9. Textile	−10621875	−1373467
10. Other manufacturing products	−10647291	−1972093
11. Construction work	−10820342	−9343867
12. Trade	−6188216	2824093
13. Transport	−1284255	16843725
14. Financial services and insurance	−11600401	−11099251
15. Private services	−10092737	−7181543
16. Government services	−7543571	−4155521

by activity; in s2 the tax revenue is allocated to the Central Government and recycled to reduce Households income tax.

The reasons that led us to model these two scenarios refers to the opportunity of reducing the indirect effects of the environmental tax on commodity prices and stimulating income generation. Indeed, the tax directly affects the most polluting goods and indirectly leads to higher final prices even for the other commodities, since all the production processes are integrated. For this purpose, by reducing the income tax we attempt to compensate Households for the loss purchasing power. Similarly the reduction of regional tax on value added by activities should reduce the costs of production and compensate for the environmental tax burden on final prices formation. The effectiveness of the policies in this sense is analysed also in terms of tax burden in order to measure the action of the Government in the redistribution of income.

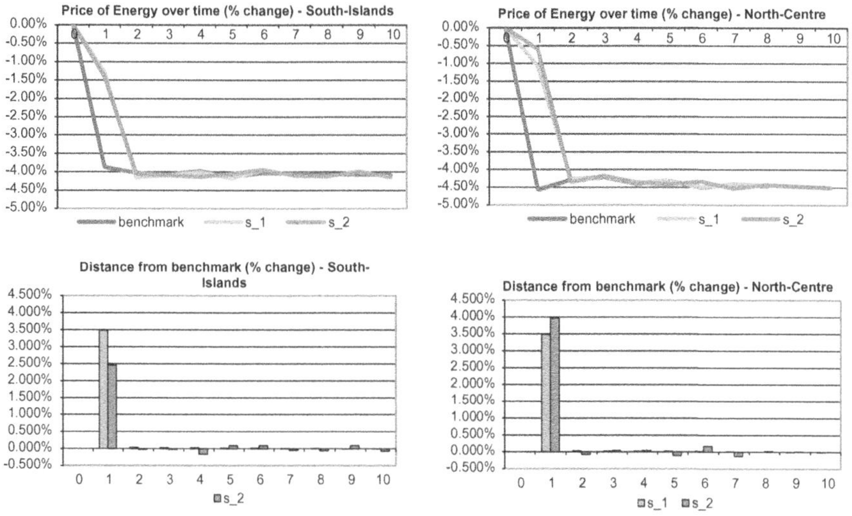

Fig. 2 Effects on energy price—% change

5 Looking for Dividends

5.1 Economic Impact of the Environmental Tax Reform

The simulations compare the baseline equilibrium (or benchmark equilibrium) without any environmental taxation, with the new equilibrium resulting from the environmental policy reform. The distance in every period (year) between the baseline trend and the path after the simulation represents the impact of the policy, that is measured on the main environmental and welfare variables in the long run.

The results of the simulations are discussed starting from the effects on total output, prices, CO_2 emissions, final demand and tax burden index.

In both scenarios, the environmental tax is modeled as a new tax on total output. In particular, the burden is on the commodities whose CO_2 emissions exceed the allowed level (the no-taxed level). Among the other, the Energy commodity is the most pollutant in both regions, thus pays a higher tax (class 4). As a result of the environmental fiscal reform, total output of Energy decreases and the price increases with respect to the benchmark path (Figs. 2 and 3). The impact on prices and outputs is greater in the North-Centre region, where the production of Energy generates a higher level of CO_2 emissions and the tax burden is higher. This effect is more evident when the tax revenue is recycled through a reduction of households' income tax (scenario s2). In the first scenario (s1) in fact, the cut in the regional tax on value added mitigates the pressure of the policy on final prices and reduces the impact on total production.

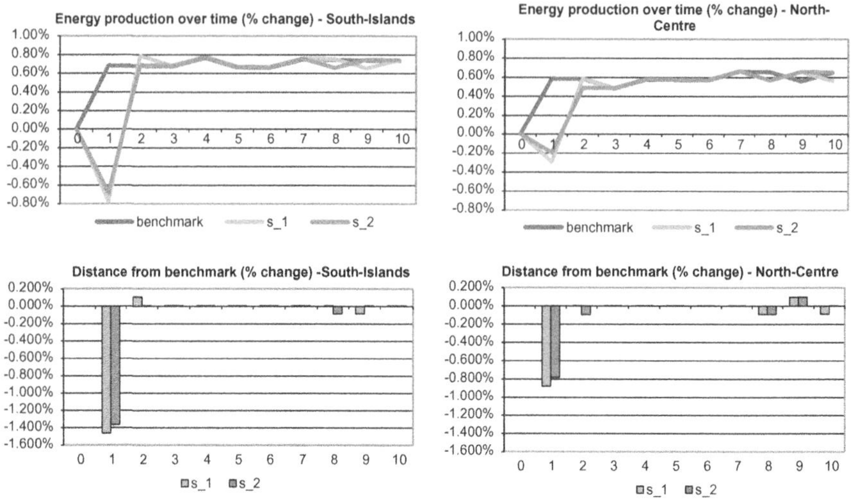

Fig. 3 Effects on energy output—% change

Similarly, total output of all the other goods decreases and prices increase with respect to the benchmark. Nevertheless, the impact of the environmental reform on commodity outputs and final prices is relevant only in the short run. In the long run, changes adjust back to the benchmark trend.

Looking at the general level of CO_2 emissions, in both scenarios the policy allows achieving the expected environmental target as results from the reduction in CO_2 emissions showed in Fig. 4. The decrease in CO_2 emissions is greater in the North-Centre area but, as already observed for the commodities total output, the distance from the benchmark path almost disappears after few periods. It is possible to say that in all scenarios the environmental (green) dividend can be pursued although for a short time and with regional differences. The results allow us to identify the relevance of the second recycling scheme (scenario s2), which provides the reduction of households income tax, in terms of environmental performance in both regions.

The research for further benefits associated to the recycling scheme of the tax revenue, requires the collection of results in terms of income generation and distribution. In particular, to identify a welfare blue dividend, we consider the evolution of final demand formation and distinguish the impact of the reform on gross capital formation and final consumption expenditure. Since the inter-temporal utility depends on the single period utility and the single period utility depends on final consumption expenditures by all institutional sectors, observing the change in final consumption expenditures we derive information on consumers' utility or welfare.

In the South-Islands region we observe a reduction in final consumption in both scenarios for the first year (see Fig. 5) than it gradually follows the benchmark path. In the North-Center region, final consumption in the short run increases with respect to the benchmark in the first scenario (s1) when the environmental tax is recycled

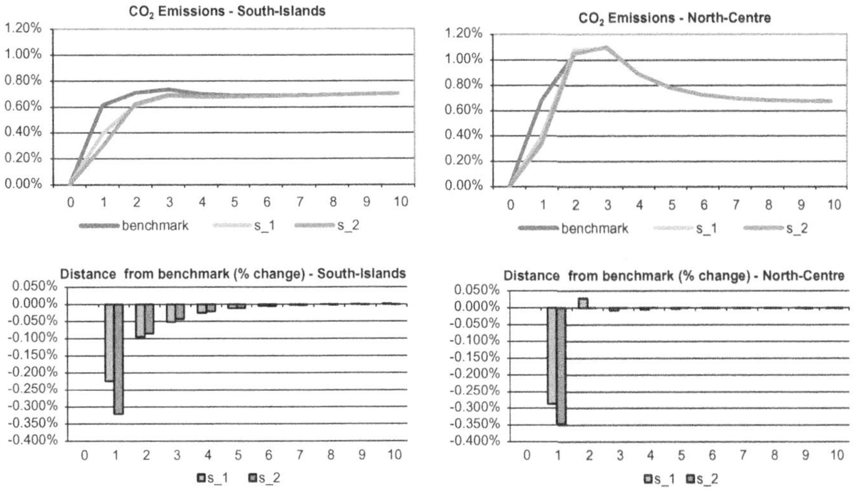

Fig. 4 Effects on total CO_2 emissions—% change

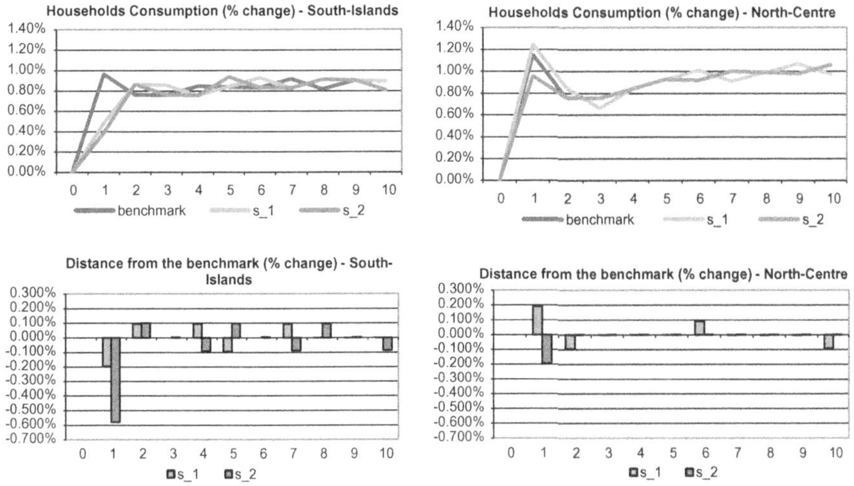

Fig. 5 Effects on households' consumption—% change

by reducing the regional tax on activities. As already observed for the South-Islands region, this rise in final demand disappears in the long run.

Similarly, in South-Islands region it is possible to observe a reduction in gross capital formation with respect to the benchmark path in the short run, regardless to the recycling assumptions. Conversely, in the North-Centre region, this policy does not affect the gross capital formation that almost replicates the same benchmark trend over the time in both scenarios as shown in Fig. 6.

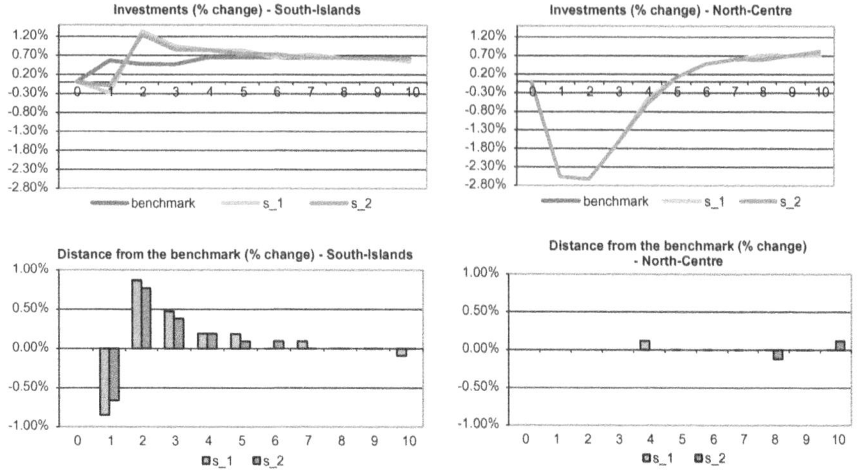

Fig. 6 Effects on gross investment—% change

The combination of the effects on consumption and gross capital determines the performance of the policy reform in terms of final demand (Fig. 7). In general, the introduction of the environmental tax in the economic system generates positive effects on final demand only in the North-Centre region in the first scenario. This result might lead to conclude that recycling the tax revenue through a reduction of income taxes is a less efficient measure than cutting taxes on regional value added in terms of final demand and investment. This result is confirmed also by the performance of the tax burden index. As a measure of government efficiency in income redistribution, this index shows whether the environmental policy generates efficiency gains or only redistribution effects. The value of the index in the benchmark is fixed to 100 pts in the base year and it is recursively calculated in the following years. As showed in Fig. 8, when the tax revenue is recycled according to the first scenario (reducing the regional tax on value added by activities) the tax burden is lower than the benchmark and the second scenario, allowing a gain in welfare. As a result of the previous disaggregate analysis we can assert that probably this overall effect is mostly driven by the effectiveness of the policy more in the North-Centre region than in the South-Islands region.

5.2 *Sensitivity Analysis of the CGE Model Results*

In the economic literature on CGE modeling, the consistency of the results are commonly argued to be strictly dependent on the assumption on exogenous parameters, such as elasticity of substitution in production and utility function (Grassini 2009). Therefore, in order to strengthen the validity of the outcomes already described,

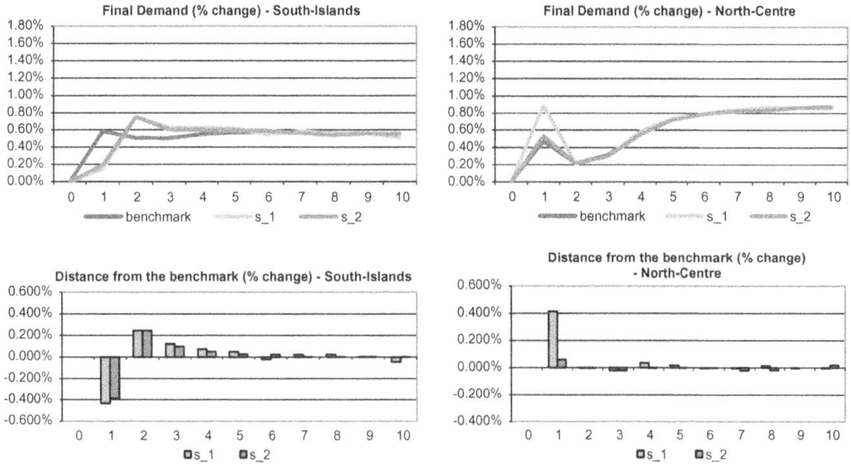

Fig. 7 Effects on final demand—% change

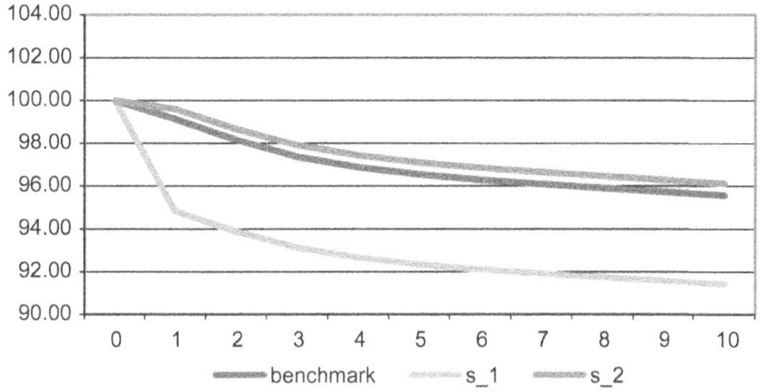

Fig. 8 Tax burden index

we integrate the study with the sensitivity analysis for the elasticity of substitution between primary factors (labor and capital) in the value added aggregate. To be more specific, in the nested production function, when combining labor and capital to generate the value added aggregate, we assumed a CES technology with elasticity of substitution equal to 0.5218 (Van der Werf 2007). The sensitivity analysis is carried out assuming two alternative parameters for the elasticity: sigma1 = 0.6262 (the original parameter increased by 20%) and sigma1 = 0.4174 (the original parameter decreased by 20%).

Then the simulations are run again considering these new parameters. The results of the sensitivity analysis are showed in Tables 3 and 4 in the Appendix 2 and confirm the robustness of the analysis. We compare the outcomes of the new simulations with the results of the simulations s1 and s2 as regard to the changes in Energy commodity output and price, the change in the CO_2 emissions (Table 3), the change in final consumption, gross investment and final demand (Table 4). As showed the results are persistent since only small differences are detected in final consumption, gross investment and final demand change but only in the fifth decimal.

6 Conclusions

The effectiveness of an environmental policy measure can be tested using the multi-sectoral approach that allows determining its direct and indirect (desired and undesired) impacts in a general equilibrium analysis. In particular, the use of the SAM integrated with environmental data on CO_2 emissions allows to calibrate a dynamic CGE model in which the environmental aspects are modelled as dependent to agents behavior over the time.

The environmental tax proposed for the Italian economy is modelled to reduce the pollution power of each activity following the principle "those who pollute more, should pay more" in order to assess the existence of a regional second dividend that integrates the first national dividend related to an improvement of environment through the reduction of CO_2 emissions.

The first step of this analysis consists in the definition of the tax structure. In particular, disaggregated data on CO_2 emissions permits the classification of commodities according to their polluting capacity. They allow to identify the production processes that exceed the level of emissions compatible with International Targets. Then, an environmental tax on output with a progressive structure was introduced to restore the correct level of CO_2 whether the admitted level of emissions is not respected. The most interesting aspect of the policy scenarios is related to the destination of the tax revenue to reduce existing tax burden and provide a better income distribution. Indeed, we focus on the Government action that affects the behavior of economic agents from two sides: the side of production and generation of income (that is affected by the introduction of a new environmental tax) and the side of the secondary income distribution (that is affected by different tax revenue recycling schemes). In particular, two alternative recycling scheme are developed: the first refers to the reduction of income tax, the second concerns the reduction of regional tax on value added generated by activities.

The second step concerns the assessment of the environmental and the social-economic benefits (the first or green and the second or blue dividend). The results show the importance of using a detailed database inthe general equilibrium analysis to detect the impacts of the environmental fiscal reform within the economic system. We ascertain the existence of a green dividend in the economy as a whole, regardless to the recycling scheme, given by the reduction in CO_2 emissions. This effect is particularly strong in the short run and endures in the long run, with a lower trend.

As to the blue dividend, we considered the evolution over the time of the final demand in order to obtain information on the intertemporal utility change, which depends on the consumption over time. The results show that the final demand in the North-Centre region increases in particular when the tax revenue is recycled through

Table 3 Sensitivity analysis results for energy price, energy output and CO_2 emissions

Years	Benchmark		s_1						s_2					
			sigma = 0.5218		sigma = 0.6262		sigma = 0.4174		sigma = 0.5218		sigma = 0.6262		sigma = 0.4174	
	South Islands	North Centre	South Islands	North Centre	South Islands	North Centre	South Islands	North Centre	South Islands	North Centre	South Islands	North Centre	South Islands	North Centre
	Energy production													
0	1.026	1.025	1.027	1.025	1.027	1.025	1.027	1.025	1.027	1.025	1.027	1.025	1.027	1.025
1	1.033	1.031	1.019	1.022	1.019	1.022	1.019	1.022	1.02	1.023	1.02	1.023	1.02	1.023
2	1.04	1.037	1.027	1.028	1.027	1.028	1.027	1.028	1.027	1.028	1.027	1.028	1.027	1.028
3	1.047	1.042	1.034	1.033	1.034	1.033	1.034	1.033	1.034	1.033	1.034	1.033	1.034	1.033
4	1.055	1.048	1.042	1.039	1.042	1.039	1.042	1.039	1.042	1.039	1.042	1.039	1.042	1.039
5	1.062	1.054	1.049	1.045	1.049	1.045	1.049	1.045	1.049	1.045	1.049	1.045	1.049	1.045
6	1.069	1.06	1.056	1.051	1.056	1.051	1.056	1.051	1.056	1.051	1.056	1.051	1.056	1.051
7	1.077	1.067	1.064	1.058	1.064	1.058	1.064	1.058	1.064	1.058	1.064	1.058	1.064	1.058
8	1.085	1.074	1.072	1.064	1.072	1.064	1.072	1.064	1.071	1.064	1.071	1.064	1.071	1.064
9	1.093	1.08	1.079	1.071	1.079	1.071	1.079	1.071	1.079	1.071	1.079	1.071	1.079	1.071
10	1.101	1.087	1.087	1.077	1.087	1.077	1.087	1.077	1.087	1.078	1.087	1.078	1.087	1.078

(continued)

Table 3 (continued)

Years	Benchmark		s_1						s_2					
			sigma = 0.5218		sigma = 0.6262		sigma = 0.4174		sigma = 0.5218		sigma = 0.6262		sigma = 0.4174	
	South Islands	North Centre	South Islands	North Centre	South Islands	North Centre	South Islands	North Centre	South Islands	North Centre	South Islands	North Centre	South Islands	North Centre
	Price of energy													
0	0.852	0.831	0.851	0.831	0.851	0.831	0.851	0.831	0.851	0.831	0.851	0.831	0.851	0.831
1	0.819	0.793	0.840	0.822	0.840	0.822	0.840	0.822	0.839	0.826	0.839	0.826	0.839	0.826
2	0.786	0.759	0.805	0.787	0.805	0.787	0.805	0.787	0.805	0.790	0.805	0.790	0.805	0.790
3	0.754	0.727	0.772	0.754	0.772	0.754	0.772	0.754	0.772	0.757	0.772	0.757	0.772	0.757
4	0.724	0.695	0.741	0.721	0.741	0.721	0.741	0.721	0.740	0.724	0.740	0.724	0.740	0.724
5	0.694	0.665	0.710	0.690	0.710	0.690	0.710	0.690	0.710	0.692	0.710	0.692	0.710	0.692
6	0.666	0.635	0.682	0.659	0.682	0.659	0.682	0.659	0.682	0.662	0.682	0.662	0.682	0.662
7	0.639	0.607	0.654	0.630	0.654	0.630	0.654	0.630	0.654	0.632	0.654	0.632	0.654	0.632
8	0.613	0.580	0.627	0.602	0.627	0.602	0.627	0.602	0.627	0.604	0.627	0.604	0.627	0.604
9	0.588	0.554	0.602	0.575	0.602	0.575	0.602	0.575	0.602	0.577	0.602	0.577	0.602	0.577
10	0.564	0.529	0.577	0.549	0.577	0.549	0.577	0.549	0.577	0.551	0.577	0.551	0.577	0.551

(continued)

Table 3 (continued)

Years	Benchmark		s_1						s_2					
			sigma = 0.5218		sigma = 0.6262		sigma = 0.4174		sigma = 0.5218		sigma = 0.6262		sigma = 0.4174	
	South Islands	North Centre	South Islands	North Centre	South Islands	North Centre	South Islands	North Centre	South Islands	North Centre	South Islands	North Centre	South Islands	North Centre
	CO_2 emissions													
0	107939100	287328500	107691300	287319700	107691400	287319600	107691400	287319700	107740800	287321400	107740800	287321400	107740800	287321400
1	108601900	289279600	108111200	288446800	108111300	288446700	108111300	288446800	108056600	288279800	108056600	288279700	108056600	288279800
2	109371800	292306700	108775000	291545500	108775200	291545800	108775100	291545700	108730600	291296800	108730700	291297100	108730700	291296900
3	110176200	295521800	109518700	294729400	109518800	294729100	109518800	294729200	109483900	294497400	109484000	294497200	109484000	294497300
4	110948200	298151900	110259100	297339200	110259200	297338800	110259200	297339000	110228000	297117200	110228100	297116900	110228000	297117000
5	111713200	300477400	111006800	299650900	111006900	299650600	111006900	299650800	110976300	299433100	110976400	299432800	110976300	299433000
6	112482100	302653400	111765400	301816700	111765500	301816400	111765400	301816500	111734000	301599100	111734100	301598900	111734000	301599000
7	113259200	304761500	112535700	303916500	112535800	303916200	112535700	303916300	112502700	303696700	112502800	303696600	112502700	303696600
8	114046300	306843800	113317600	305991000	113317700	305990800	113317600	305990900	113282500	305767700	113282600	305767600	113282600	305767700
9	114843900	308920800	114110900	308060500	114111000	308060400	114110900	308060500	114073500	307832900	114073600	307832900	114073500	307832900
10	115652200	311002000	114915300	310134200	114915400	310134000	114915400	310134100	114875400	309902000	114875500	309901900	114875400	309901900

Table 4 Sensitivity analysis results for final consumption, gross investment and final demand

Years	Benchmark		s_1						s_2					
			sigma = 0.5218		sigma = 0.6262		sigma = 0.4174		sigma = 0.5218		sigma = 0.6262		sigma = 0.4174	
	South Islands	North Centre	South Islands	North Centre	South Islands	North Centre	South Islands	North Centre	South Islands	North Centre	South Islands	North Centre	South Islands	North Centre
	Final consumption													
0	1.037	1.044	1.038	1.044	1.038	1.044	1.038	1.044	1.038	1.044	1.038	1.044	1.038	1.044
1	1.047	1.056	1.043	1.057	1.043	1.057	1.043	1.057	1.042	1.054	1.042	1.054	1.042	1.054
2	1.055	1.064	1.052	1.066	1.052	1.065	1.052	1.065	1.051	1.062	1.051	1.062	1.051	1.062
3	1.063	1.072	1.061	1.073	1.061	1.073	1.061	1.073	1.059	1.07	1.059	1.07	1.059	1.07
4	1.072	1.081	1.069	1.082	1.069	1.082	1.069	1.082	1.067	1.079	1.067	1.079	1.067	1.079
5	1.081	1.091	1.078	1.092	1.078	1.092	1.078	1.092	1.077	1.089	1.077	1.089	1.077	1.089
6	1.09	1.101	1.088	1.103	1.088	1.103	1.088	1.103	1.086	1.099	1.086	1.099	1.086	1.099
7	1.1	1.112	1.097	1.113	1.097	1.113	1.097	1.113	1.095	1.11	1.095	1.11	1.095	1.11
8	1.109	1.123	1.107	1.124	1.107	1.124	1.107	1.124	1.105	1.121	1.105	1.121	1.105	1.121
9	1.119	1.134	1.117	1.136	1.117	1.136	1.117	1.136	1.115	1.132	1.115	1.132	1.115	1.132
10	1.129	1.146	1.127	1.147	1.127	1.147	1.127	1.147	1.124	1.144	1.124	1.144	1.124	1.144

(continued)

Table 4 (continued)

Years	Benchmark		s_1						s_2					
			sigma = 0.5218		sigma = 0.6262		sigma = 0.4174		sigma = 0.5218		sigma = 0.6262		sigma = 0.4174	
	South Islands	North Centre	South Islands	North Centre	South Islands	North Centre	South Islands	North Centre	South Islands	North Centre	South Islands	North Centre	South Islands	North Centre
	Gross investment													
0	1.059	0.899	1.051	0.899	1.051	0.899	1.051	0.899	1.052	0.899	1.052	0.899	1.052	0.899
1	1.065	0.876	1.048	0.876	1.048	0.876	1.048	0.876	1.051	0.876	1.051	0.876	1.051	0.876
2	1.070	0.853	1.062	0.853	1.062	0.853	1.062	0.853	1.064	0.853	1.064	0.853	1.064	0.853
3	1.075	0.839	1.072	0.839	1.072	0.839	1.072	0.839	1.073	0.839	1.073	0.839	1.073	0.839
4	1.082	0.834	1.081	0.835	1.081	0.835	1.081	0.835	1.082	0.834	1.082	0.834	1.082	0.834
5	1.089	0.835	1.090	0.836	1.090	0.836	1.090	0.836	1.090	0.835	1.090	0.835	1.090	0.835
6	1.096	0.839	1.097	0.840	1.097	0.840	1.097	0.840	1.098	0.839	1.098	0.839	1.098	0.839
7	1.103	0.844	1.105	0.845	1.105	0.845	1.105	0.845	1.105	0.844	1.105	0.844	1.105	0.844
8	1.110	0.850	1.112	0.851	1.112	0.851	1.112	0.851	1.112	0.849	1.112	0.849	1.112	0.849
9	1.117	0.856	1.119	0.857	1.119	0.857	1.119	0.857	1.119	0.855	1.119	0.855	1.119	0.855
10	1.124	0.862	1.125	0.863	1.125	0.863	1.125	0.863	1.126	0.862	1.126	0.862	1.126	0.862

(continued)

Table 4 (continued)

Years	Benchmark		s_1						s_2					
			sigma = 0.5218		sigma = 0.6262		sigma = 0.4174		sigma = 0.5218		sigma = 0.6262		sigma = 0.4174	
	South Islands	North Centre	South Islands	North Centre	South Islands	North Centre	South Islands	North Centre	South Islands	North Centre	South Islands	North Centre	South Islands	North Centre
	Final demand													
0	1.0335	1.0048	1.0323	1.0048	1.0323	1.0048	1.0323	1.0048	1.0325	1.0048	1.0325	1.0048	1.0325	1.0048
1	1.0400	1.0070	1.0338	1.0103	1.0338	1.0103	1.0338	1.0103	1.0348	1.0070	1.0348	1.0070	1.0348	1.0070
2	1.0453	1.0068	1.0415	1.0103	1.0415	1.0100	1.0415	1.0100	1.0423	1.0068	1.0423	1.0068	1.0423	1.0068
3	1.0500	1.0078	1.0483	1.0110	1.0483	1.0110	1.0483	1.0110	1.0485	1.0078	1.0485	1.0078	1.0485	1.0078
4	1.0560	1.0123	1.0545	1.0153	1.0545	1.0153	1.0545	1.0153	1.0548	1.0118	1.0548	1.0118	1.0548	1.0118
5	1.0620	1.0180	1.0610	1.0215	1.0610	1.0215	1.0610	1.0215	1.0613	1.0180	1.0613	1.0180	1.0613	1.0180
6	1.0680	1.0250	1.0668	1.0288	1.0668	1.0288	1.0668	1.0288	1.0675	1.0250	1.0675	1.0250	1.0675	1.0250
7	1.0738	1.0325	1.0730	1.0360	1.0730	1.0360	1.0730	1.0360	1.0730	1.0320	1.0730	1.0320	1.0730	1.0320
8	1.0798	1.0403	1.0788	1.0438	1.0788	1.0438	1.0788	1.0438	1.0793	1.0395	1.0793	1.0395	1.0793	1.0395
9	1.0860	1.0475	1.0850	1.0518	1.0850	1.0518	1.0850	1.0518	1.0850	1.0473	1.0850	1.0473	1.0850	1.0473
10	1.0918	1.0555	1.0905	1.0595	1.0905	1.0595	1.0905	1.0595	1.0910	1.0555	1.0910	1.0555	1.0910	1.0555

the cut of the regional tax on value added by activities. This result is consistent with several studies on double dividend (e.g. Takeda 2007), according to which the combination of environmental taxation and the reduction in capital taxes improves welfare. Thus if we concentrate on the benefits connected with environmental policy, the introduction of a tax on CO_2 emissions by commodity, with a progressive structure and a convenient distribution of the tax revenue, i.e. reduction of income taxes and regional value added taxes, allows for the attainment of both the green and the blue dividends at regional level.

This phenomenon can be considered as an actual second dividend since it is confirmed by the performance of the tax burden index over time. We considered the tax burden on the economic system and derived a measure of the effectiveness of government action within the economic system. This allows the achievement of a second blue dividend in the first scenario both at national and regional level.

The consideration that the result depends at a greater extent on the distinct peculiarities of technology and behavioral habits in the two regions of the national economy encourages further attempts in this direction. However the sensitivity analysis confirms the results.

Appendix 1

Dynamic CGE model specification

The dynamic CGE model developed in this paper is calibrated on the SAM integrated with environmental data. It is solved using the GAMS (General Algebraic Modeling System) software to find the equilibrium prices, quantities and incomes over the time.

Given the structure of the economy described by the SAM, to determine prices and quantities which maximize producers' profits and consumers' utility, we solve the Arrow-Debreu (1954) problem as an optimization problem of the consumer subject to income, technology and feasibility constraints. When programming on GAMS usually, this maximization problem is turned into a Mixed Complimentary Problem (MCP) and solved (solver used MILES) as a system of non-linear equation. In our model, the optimization problem for all the consumers (Böhringer et al. 1997) has been settled as:

$$\max \sum_{t=0}^{T} \left(\frac{1}{1+\rho} \right)^{t} u\left[C_t\right] \tag{4}$$

subject to:

$$C_t = x\left(K_t, L_t, M_t, Ta_t\right) - I_t - E_t \tag{5}$$

$$K_{t+1} = (1-\delta)K_t + I_t \tag{6}$$

The first order conditions deriving from this maximization problem are:

$$P_t = \left(\frac{1}{1+\rho}\right)^t \frac{\delta u\left(C_t\right)}{\delta C_t} \tag{7}$$

$$PK_t = (1-\delta)PK_{t+1} + P_t \frac{\delta x\left(K_t, L_t, M_t, Ta_t\right)}{\delta K_t} \tag{8}$$

$$P_t = PK_{t+1} \tag{9}$$

Than the corresponding mixed complimentary problem can be formulated as a sequence of market clearing, zero profit and budget constraint conditions.

Market clearing conditions holds for all commodities and primary factors markets. Analytically, we can summarize the conditions as follow:

$$X_t \geq B_t, d\left(P_t, RA\right) + I_t + E_t, P_t \geq 0,\ P_t\left(X_t - B_t, d\left(P_t, RA\right) - I_t - E_t\right) = 0 \tag{10}$$

$$\begin{aligned} &L_t \geq X_t \frac{\delta C\left(RK_t, PL_t, PM_t, Ta_t\right)}{\delta PL_t}, \\ &PL_t \geq 0, PL_t\left(L_t - X_t \frac{\delta C\left(RK_t, PL_t, PM_t, Ta_t\right)}{\delta PL_t}\right) = 0 \end{aligned} \tag{11}$$

$$\begin{aligned} &K_t \geq X_t \frac{\delta C\left(RK_t, PL_t, PM_t, Ta_t\right)}{\delta RK_t}, \\ &RK_t \geq 0,\ RK_t\left(K_t - X_t \frac{\delta C\left(RK_t, PL_t, PM_t, Ta_t\right)}{\delta RK_t}\right) = 0 \end{aligned} \tag{12}$$

$$\begin{aligned} &M_t \geq X_t \frac{\delta C\left(RK_t, PL_t, PM_t, Ta_t\right)}{\delta PM_t}, \\ &PM_t \geq 0, PM_t\left(K_t - X_t \frac{\delta C\left(RK_t, PL_t, PM_t, Ta_t\right)}{\delta PM_t}\right) = 0 \end{aligned} \tag{13}$$

Zero profit conditions posits that total supply in each commodity market is determined by the perfect competitive market condition, that is to say, price equals average total cost (profit are zero). In a general equilibrium model, the price that clears the market (demand equals to supply) also equals average total costs for each commodity. Analytically, we can summarize the conditions as follow:

$$P_t \geq PK_{t+1}, I_t \geq 0,\quad I_t\left(P_t - PK_{t+1}\right) = 0 \tag{14}$$

$$PK_t \geq RK_t + (1-\delta)\, PK_{t+1}, K_t \geq 0, K_t(PK_t - RK_t - (1-\delta)\, PK_{t+1}) = 0 \tag{15}$$

$$C\left(RK_t, PL_t, PM_t, Ta_t\right) \geq P_t, X_t \geq 0, X_t\left(C\left(RK_t, PL_t, PM_t, Ta_t\right) - P_t\right) = 0 \tag{16}$$

Income balance conditions derive from the budget constraint:

$$RA \geq PK_0K_0 + \sum_{t=0}^{T}(PL_tL_t + PM_tM_t - Ta_t) - PK_{t+1}K_{t+1},\ RA \geq 0 \tag{17}$$

The variables are:

t	Time periods
T	Terminal period
ρ	Individual time-preference parameter
u	Utility
C_t	Consumption in period t
x	Production function
X_t	Total output in period t
K_t	Capital in period t
L_t	Labour in period t
M_t	Imports in period t
Ta_t	All taxes payed by sectors in period t
I_t	Investment in period t
E_t	Exports in period t
δ	Capital depreciation rate
γ	interest rate
P_t	Price of output in period t
d	Demand function
PK_t	Price of capital in period t
RK_t	Rental of capital in period t
PL_t	Wage in period t
PM_t	Price of imports in period t
RA	Consumer's disposable income

Appendix 2

Results from sensitivity analysis

See Tables 3 and 4.

References

Armington, P. (1969). A theory of demand for products distinguished by place of production. International Monetary Fund Staff Papers.

Baumol, W., & Oates, W. (1988). *The theory of environmental policy*. Cambridge: Cambridge University Press.

Böhringer, C., Pahlke, A., & Rutherford, T. (1997). Environmental tax reforms and the prospect for a double dividend. http://debrue.colorado.edu/.

Bor, Y. J., & Huang, Y. (2010). Energy taxation and the double dividend effect in Taiwan's energy conservation policy. An empirical study using a computable general equilibrium model. *Energy Policy, 38,* 2086–2100.

Bovenberg, A., & De Mooij, R. (1994). Environmental levies and distortionary taxation. *American Economic Review, 94,* 1085–1089.

Bovenberg, A., & De Mooij, R. (1998). Environmental taxes, international capital mobility and inefficient tax systems: Tax burden versus tax shifting. *International tax and Public Finance, 5,* 7–39.

Bovenberg, A., & Goulder, L. (1996). Optimal environmental taxation in the presence of other taxes: General equilibrium analysis. *American Economic Review, 86,* 985–1000.

Bovenberg, A., & Goulder, L. (1997). Costs of environmentally motivated taxes in the presence of other taxes: General equilibrium analysis. *National Tax Journal, 70,* 59–87.

Bovenberg, A., & Goulder, L. (2002). Environmental taxation and regulation. In A. Auerbach & M. Feldstein (Eds.), *Handbook of public economics* (Ch. 23, pp. 1471–1545). Amsterdam, North Holland: Elsevier.

Bulckaen, F., & Stampini, M. (2002). Riforme fiscali verdi e doppio dividendo: una rassegna. In F. Bulckaen, A. Pench, & M. Stampini (Eds.), Riforme fiscali e politiche ambientali (pp. 17–31). Milano: Franco Angeli.

Ciaschini, M., Pretaroli, R., Severini, F., & Socci, C. (2012). Environmental tax reform and double dividend evidence. *Research in Economics, 66,* 273–283.

Ciaschini, M., Pretaroli, R., Severini, F., & Socci, C. (2014). Policies for electricity production from renewable sources: The Italian case. *Journal of Policy Modeling.*

Ciaschini, M., & Socci, C. (2007a). Bi-regional SAM linkages: A modified backward and forward dispersion approach. *Reviews of Urban and Regional Development Studies, 19*(3), 233–254.

Ciaschini, M., & Socci, C. (2007b). Final demand impact on output: A macro multiplier approach. *Journal of Policy Modeling, 29*(1), 115–132.

EC. (1993). Growth, competitiveness and unemployment. Tech. rep., Bruxelles.

EC. (1994). Orientamenti per l'UE in materia di indicatori ambientali e di contabilità verde nazionale - Integrazione di sistemi di informazione ambientale ed economica. Tech. rep., Commissione Europea, 21 Dicembre.

Farmer, K., & Steininger, K. W. (1999). Reducing CO_2 emissions under fiscal retrenchment: A multi-cohort CGE-model for Austria. *Environmental & Resource Economics, 13*(3), 309–340.

Gimenez, E. L., & Rodriguez, M. (2010). Reevaluating the first and the second dividends of environmental tax reforms. *Energy Policy, 38,* 6654–6661.

Glomm, G., Kawaguchi, D., & Sepulveda, F. (2008). Green taxes and double dividends in a dynamic economy. *Journal of Policy Modeling, 30,* 19–32.

Goulder, L. (1995a). Effects of carbon taxes in an economy with prior tax distortions: An intertemporal general equilibrium analysis. *Journal of Environmental Economics and Management, 29,* 271–297.

Goulder, L. (1995b). Environmental taxation and the double dividend: A reader's guide. *International tax and public finance, 2,* 157–183.

Grassini, M. (2009). Rowling along the computable general equilibrium modelling mainstream. *Studies on Russian Economic Development, 20*(2), 134–146.

ISTAT. (2008). Le emissioni atmosferiche delle attività produttive e delle famiglie anni 1990–2005. Annali di Statistica.

Lau, M. I., Pahlke, A., & Rutherford, T. F. (2002). Approximating infinite-horizon models in a complementarity format: A primer in dynamic general equilibrium analysis. *Journal of Economic Dynamic & Control, 26,* 577–609.

Manresa, A., & Sancho, F. (2005). Implementing a double dividend: Recycling eco taxes towards lower labour taxes. *Energy Policy, 33,* 1577–1585.

Morris, G. E. (1999). Integrating environmental taxes on local air pollutants with fiscal reform in Hungary: Simulations with a computable general equilibrium model. *Environment and Development Economics, 4*(4), 537–564.

Paltsev, S. (2004). Moving from static to dynamic general equilibrium economic models. MIT Joint Program on the Science and Policy of Global Change (Technical Note No. 4).
Parry, I. W. (1995). Pollution taxes and revenue recycling. *Journal of Environmental Economics and Management, 29,* 64–77.
Parry, I. W. (2004). Are emission permits regressive? *Journal of Environmental Economics and Management, 47*, 364–387.
Pearce, D. (1991). The role of carbon taxes in adjusting to global warming. *The Economic Journal, 101,* 938–948.
Pench, A. (2002). Riforme fiscali e qualità ambientale: quali criteri per il policy maker. In F. Bulckaen, A. Pench, & M. Stampini (Eds.), Riforme fiscali e politiche ambientali (pp. 151–164). Milano: Franco Angeli.
Pretaroli, R., & Severini, F. (2008). Dalla SAM al modello computazionale CGE. In Ministero dell'Interno (Ed.), Modelli computazionali per la valutazione dell'impatto del Federalismo fiscale sulla finanza locale (Ch. 4, pp. 77–100). Ministero dell'Interno.
Pretaroli, R., & Severini, F. (2009). Assessing the effects of a value added tax policy on the wine sectors. *Enometrica, 2*(2), 41–51.
Radulescu, D., & Stimmelmayr, M. (2010). The impact of the 2008 German corporate tax reform: A dynamic cge analysis. *Economic Modelling, 27*, 454–467.
Roson, R. (2002). Dinamica ed effetti distributivi delle politiche ambientali. In F. Bulckaen, A. Pench, & M. Stampini (Eds.), Riforme fiscali e politiche ambientali (pp. 195–207). Milano: Franco Angeli.
Schneider, K. (1997). Involuntary unemployment and environmental policy: The double dividend hypothesis. *Scandinavian Journal of Economics, 99,* 45–49.
Takeda, S. (2007). The double dividend from carbon regulations in Japan. *The Japanese and International Economies, 21,* 336–364.
United Nations. (2008). System of national accounts. N.U., New York.
Van der Werf, E. (2007). Production function for climate policy modeling: An empirical analysis. Fondazione Eni Enrico Mattei Working Papers 50. http://www.bepress.com/feem/paper50.
Yeldan, A. E. (1997). Financial liberalization and fiscal repression in turkey: Policy analysis in a CGE model with financial markets. *Journal of Policy Modeling, 19*(1), 79–117.

A Sub-national CGE Model for the European Mediterranean Countries

Francesco Bosello and Gabriele Standardi

Abstract This chapter describes the methodology used to develop a Computable General Equilibrium model with sub-national detail for the Euro-Mediterranean area: Italy, France, Spain, Portugal and Greece. The main purpose of this exercise is to perform economic assessments of climate change impacts with a finer spatial resolution compared to that offered by standard CGE models and, in doing so, to increase the comparability of and the possibility to exchange information across economic and physical impact models. Indeed, aiming to represent the high spatial heterogeneity of climate drivers and environmental impacts, both climate models and physical process models (like e.g. land use, crop growth, flood risk models) are spatially detailed. This is not the case for macroeconomic models that typically feature large geo-political blocks or at best the country as the finest investigation units. Accordingly, when physical and economic models are interfaced to produce integrated assessments of climate change impacts, there is an unavoidable loss of richness both of input and output information. Developing a sub-national resolution for the economic analysis thus offers a first useful step to measure more accurately the economic consequences of climate change, to produce an information more relevant for local planners and businesses, and also to better capture the economic feedbacks between regions which can turn to be as important as the international ones. The study addresses conceptual and practical issues related to the regionalization process, and presents simple experiments aimed to test the robustness of the regionalized structure and understand the economic implications in terms of market integration.

Keywords CGE models · Regional economics

JEL CODE C68 · D58 · R11 · R12 · R13

F. Bosello · G. Standardi (✉)
Euro-Mediterranean Centre on Climate Change (CMCC), Venice, Italy
e-mail: gabriele.standardi@cmcc.it

F. Bosello
e-mail: francesco.bosello@cmcc.it

F. Bosello
University of Milan, Milan, Italy

© Springer International Publishing AG, part of Springer Nature 2018

F. Perali and P. L. Scandizzo (eds.), *The New Generation of Computable General Equilibrium Models*, https://doi.org/10.1007/978-3-319-58533-8_11

1 Introduction

Computable General Equilibrium (CGE) models represent a popular tool to assess the economic consequences of different policies and social-economic scenarios. Starting from a neo-classical theoretical structure (Johansen 1974; Shoven and Whalley 1992) they are able to capture all the feedbacks in the economic system in terms of inter-sectoral production reallocation, international trade and investments flows. In the recent years an increasing number of works has seen the application of CGE models in the field of environmental and climate change economics. Examples are the assessment of the implementation of carbon and energy taxes for environmental purposes (see e.g. EC 2008, 2010; Böhringer et al. 2009, 2010, 2012) or the study of economic consequences of climate change impacts (see e.g. Darwin and Tol 2001; Bigano et al. 2008; Aaheim et al. 2010; Eboli et al. 2010; Ciscar et al. 2011; Bosello et al. 2012).

The typical investigation unit of CGE models is the country. Sub national differences are very often overlooked. However, on the one hand climate change impacts can be highly differentiated across different areas within the same country. This raises an immediate interest in quantifying their economic consequences with a similar degree of detail, especially to provide information that can be useful to "local" decision makers. On the other hand, countries also present economic asymmetries within them. These can impact for instance factor mobility and trade as importantly as international dynamics. Accordingly, local specificities constitute important determinants not only of how the economic consequences of climate change spread all over the economic system, but also of climate policy effectiveness. Tracing these sub national effects is thus particularly important to gain a better grasp of the distributional implication of a given policy or impact, and to understand the economic implications related to the different assumptions on market integration and flexibility.

In this chapter we describe the building process of a sub-national CGE model for the Euro-Mediterranean region: Italy, France, Spain, Greece and Portugal. The Mediterranean area has been identified as one of the main climate change hotspots: that is, one of the most responsive areas to climate change (IPCC 2014). The area is populated by over 500 million people, distributed in about 30 countries in Africa, Asia, and Europe. This geographical area is also crucial from an economic and socio-political point of view.

As anticipated, there are few CGE multi-country models also featuring a sub-national detail. This is mainly due to the difficulty to create mutually consistent Social Accounting Matrices (SAMs) and reconstruct all the bilateral trade flows for a large number of sub-national regions. Among these: Peter et al. (1996) developed the MRF (*Multi Regional Forecasting*) model to simulate tax/environmental policy for the Australian economy; Jean and Laborde (2004) developed the DREAM-MIRAGE (*Deep Regional Economic Analysis Model—Modelling International Relationships in Applied General Equilibrium*) model for Europe taking into account 119 NUTS (*Nomenclature of Territorial Units for Statistics*) 1 regions; Canning and Tsigas (2000) built a model for eight macro-regions of the USA; recently EU Joint Research

Centre (JRC) has created RHOMOLO (Regional HOlistic MOdeL) for 267 NUTS2 European regions and 6 macro-sectors (Brandsma et al. 2015; Potters et al. 2014) for analysing the impact of the European Cohesion policy.

Some CGE models exist which present a spatially resolved description of the agricultural sector. Examples of this type are CAPRI-GTAP (*Common Agricultural Policy Regional Impact Analysis—Global Trade Analysis Project*) (Jansson et al. 2009), CAPSIM (*China's Agricultural Policy Simulation Model*) (Yang et al. 2011), GTAP-AEZ (*Global Trade Analysis Project—AgroEcological Zones*) (Hertel et al. 2009; Lee et al. 2009) and the ICES-AEZ (*Intertemporal Computable Equilibrium System—AgroEcological Zones*) (Michetti and Parrado 2012).

The model presented here is an extension of the regionalized model for Italy developed by Standardi et al. (2014). That model has been applied to the economic assessment of flood risk (Carrera et al. 2015; Koks et al. 2015), sea-level rise (Standardi and Eboli 2015) and environmental policies for water saving (Pérez-Blanco et al. 2016).

The chapter is organized as follows. Section 2 presents the database construction and the estimation strategy to obtain trade flows across sub-national regions within countries. Section 3 describes the main theoretical changes made to adapt the standard country-level CGE model to the sub-national framework. Section 4 reports the results of the experiments developed to test the robustness of the model structure and to highlight the economic interactions between sub-national regions. Section 5 concludes and sketches some ideas for future research.

2 Database Development

The starting point is the GTAP 8 database (Narayanan et al. 2012). The 8.1 version consists in a collection of Social Accounting Matrices (SAMs) for 57 economic sectors and 134 countries (or groups of countries) in the world. The calibration/reference year is 2007.

In the case of France, Italy, Spain, Portugal and Greece, which the GTAP dataset already represents as singled-out countries, we further reconstruct a database characterizing the 70 sub-national entities represented in Table 1 and the 57 sectors represented in Table 2. An advantage of using a global database such as GTAP is the already existing rich description of international trade flows. In some sense, this facilitates the subsequent endeavor to further regionalize international trade and keep all the information for the other GTAP countries in the world.

Information sources to substantiate the process were the Eurostat (Economic Accounts for Agriculture 2017; Structural Business Statistics 2017) and the National Statistical Offices. Specifically: for Italy we refer to Istituto Nazionale di Statistica—ISTAT (Conti Economici Regionali, Anni 1995–2009; Agricoltura e Zootecnia; Valore Aggiunto ai Prezzi di Base dell'Agricoltura per Regione, Anni 1980–2011); for France to Institut National de la Statistique et des Etudes Economiques - INSEE (Valeurs Ajoutées régionales), for Spain to Instituto Nacional de Estadistica - INE

Table 1 Sub-national characterization of the Euro-Mediterranean region[a]

France (22 NUTS-2)	Italy (20 NUTS-2)	Spain (19 NUTS-2)	Portugal (5 NUTS-2)	Greece (4 NUTS-1)
1. Île de France	1. Piemonte	1. Galicia	1. Norte	1. Voreia Ellada
2. Champagne-Ardenne	2. Valle d'Aosta	2. Principado de Asturias	2. Algarve	2. Kentriki Ellada
3. Picardie	3. Lombardia	3. Cantabria	3. Centro	3. Attica
4. Haute-Normandie	4. Trentino-Alto-Adige	4. País Vasco	4. Lisboa	4. Nisia-Aigaiou-Kriti
5. Centre	5. Veneto	5. Navarra	5. Alentejo	
6. Basse-Normandie	6. Friuli-Venezia-Giulia	6. La Rioja		
7. Bourgogne	7. Liguria	7. Aragón		
8. Nord -Pas-de-Calais	8. Emilia-Romagna	8. Comunidad de Madrid		
9. Lorraine	9. Toscana	9. Castilla y León		
10. Alsace	10. Umbria	10. Castilla-La Mancha		
11. Franche-Comté	11. Marche	11. Extremadura		
12. Pays de la Loire	12. Lazio	12. Cataluña		
13. Bretagne	13. Abruzzo	13. Comunidad Valenciana		
14. Poitou-Charentes	14. Molise	14. Illes Balears		
15. Aquitaine	15. Campania	15. Andalucía		
16. Midi-Pyrénées	16. Puglia	16. Región de Murcia		
17. Limousin	17. Basilicata	17. Ceuta		
18. Rhône-Alpes	18. Calabria	18. Melilla		
19. Auvergne	19. Sicilia	19. Canarias		
20. Languedoc-Roussillon	20. Sardegna			
21. Provence-Alpes- Côte d'Azur				
22. Corse				

[a]The Nomenclature of territorial units for statistics (NUTS) is a hierarchical system for dividing up the economic territory of the EU (Eurostat 2016)

Table 2 Sectoral detail of the CGE model

Sectors		
1. Paddy rice	20. Meat products nec	39. Transport equipment nec
2. Wheat	21. Vegetable oils and fats	40. Electronic equipment
3. Cereal grains nec	22. Dairy products	41. Machinery and equipment nec
4. Vegetables, fruit, nuts	23. Processed rice	42. Manufactures nec
5. Oil seeds	24. Sugar	43. Electricity
6. Sugar cane, sugar beet	25. Food products nec	44. Gas manufacture, distribution
7. Plant-based fibers	26. Beverages and tobacco products	45. Water
8. Crops nec	27. Textiles	46. Construction
9. Bovine cattle, sheep and goats, horses	28. Wearing apparel	47. Trade
10. Animal products nec	29. Leather products	48. Transport nec
11. Raw milk	30. Wood products	49. Water transport
12. Wool, silk-worm cocoons	31. Paper products, publishing	50. Air transport
13. Forestry	32. Petroleum, coal products	51. Communication
14. Fishing	33. Chemical, rubber, plastic products	52. Financial services nec
15. Coal	34. Mineral products nec	53. Insurance
16. Oil	35. Ferrous metals	54. Business services nec
17. Gas	36. Metals nec	55. Recreational and other services
18. Minerals nec	37. Metal products	56. Public Administration, Defense, Education, Health
19. Bovine meat products	38. Motor vehicles and parts	57. Dwellings

(Contabilidad Regional de España 2000–2016), for Portugal to - Instituto Nacional de Estatística - INE (Gross value added (€) of Enterprises by Geographic localization (NUTS 2002) and Economic activity) and for Greece to the Hellenic Statistical Authority—HSA (Gross value added by industry 2000–2015).

Operationally the sub national development of ICES followed a stepwise procedure. We started from a country, we regionalize its national database and then we moved to regionalize another country starting from the database obtained in the previous step. The country sequence is the following: Italy, Greece, Spain, France and Portugal.

2.1 Splitting the Value Added

As quite typical in CGE models, also our model features an upper level production structure which combines a bundle of primary factors with a bundle of intermediate goods by means of a Leontief technology. The two composites of production factors are thus perfect complement. The value of primary factors (in the model: labour, capital, land, natural resources) which coincides with their remuneration, constitutes total value added.

The first step for the sub national development consists thus in detailing the value added, originally available at the country level, to the new regional scope.

To do this, first, we match the sectors of the GTAP database with those of our data sources. Then, for each sector, the regional shares of value added, and accordingly of labour, capital, land and natural resources are computed using the sub-national data. Finally, these shares are used to distribute original country-level data across sub-national units.

National Statistical Offices of Italy and Spain provide information on both capital and labor at the sectoral level. Other countries are not that data rich. In these cases we use the weight of value added to split the national value of all primary factors. For some manufacturing activities we referred to Structural Business Statistics (SBS) of Eurostat because they have a more detailed description of these sectors. To regionalize the agricultural economic components of value added we mainly rely on Economic Accounts for Agriculture of Eurostat because of the rich and already standardized information across EU regions.

2.2 The Derivation of Sub-national Demand for Domestic and Imported Goods: Simple Location Quotients (SLQs)

One of the most challenging tasks in the database construction is the derivation of the sub-national domestic demand and trade patterns with other regions within and outside the country. This is because these data are often missing and need to be reconstructed using different techniques. The derivation of intra-national trade is particularly important. In our case we rely on the so-called Simple Locations Quotients (SLQs) method (Miller and Blair 1985; Bonfiglio and Chelli 2008; Bonfiglio 2008).[1] The formula for the SLQs is the following:

[1]SLQs are so called "non-survey" techniques to derive input-output tables and SAMs. SLQs are not the most precise across non survey coefficients known in the literature (Bonfiglio and Chelli 2008). However they offer the great advantage to be of particularly easy application when large global databases are involved as in our case.

$$SLQ_{i,r} = \frac{X_{i,r}/X_r}{X_{i,c}/X_c} \tag{1}$$

where i is the sector and X the output, r and c represent the regional and national indexes, respectively. SLQ gives a measure of the regional specialization in the economic activity. Two extreme cases are possible: absence of the economic activity i, and perfect specialization. In the first case we have $X_{i,r} = 0$ which implies $SLQ_{i,r} = 0$. This means that the region will need to import i, whether intermediate and final goods, from other regions.

At the other extreme we have $X_{i,r} = X_{i,c}$ and $SLQ_{i,r} = X_c/X_r$. This means that the sectoral regional value added coincides with the national one and that region will tend to export the good for intermediate or final consumption.

Finally in the case of $X_{i,r}/X_r = X_{i,c}/X_c$ the sub-national demand structure will follow exactly the national one and the share of domestic and imported demand will be the same.

Obviously in almost all the cases the SLQ values will be in between the two extreme cases and not equal to the last one. The sub-country shares of domestic and imported demand will be given by multiplying the national shares times SLQs and then normalizing these shares, as illustrated in the following equations:

$$ShrDom_{i,r} = ShrDom_{i,c} \cdot SLQ_{i,r} \tag{2}$$

$$ShrImp_{i,r} = ShrImp_{i,c} \cdot (1/SLQ_{i,r}) \tag{3}$$

$$ShrDom^*_{i,r} = ShrDom_{i,r}/(ShrDom_{i,r} + ShrImp_{i,r}) \tag{4}$$

$$ShrImp^*_{i,r} = ShrImp_{i,r}/(ShrDom_{i,r} + ShrImp_{i,r}) \tag{5}$$

where ShrImp and ShrDom are the not normalized shares of domestic and imported demand and ShrImp* and ShrDom* the normalized ones. In the extreme case of no economic activity we put ShrImp = 1 and ShrDom = 0 to avoid the infinite numbers.

2.3 *Estimation of Bilateral Trade Flows Between Sub-national Regions*

The second step consists in the determination of the bilateral trade flows across sub-national regions. These data are very often missing. To overcome the problem the procedure usually adopted is the so-called gravitational approach as in Horridge and Wittwer (2010) and Dixon et al. (2012). By this method, the bilateral intra-country trade flows are estimated using a gravity equation as in the Newtonian physics. It accounts for the sectoral production in the origin region and sectoral demand in the destination regions as attractors and the distance between them as friction.

Table 3 Components of matrix Π

	North	Centre	South	Tot
North	π_{11}	π_{12}	π_{13}	$\Pi_{1.}$
Centre	π_{21}	π_{22}	π_{23}	$\Pi_{2.}$
South	π_{31}	π_{32}	π_{33}	$\Pi_{3.}$
Tot	$\Pi_{.1}$	$\Pi_{.2}$	$\Pi_{.3}$	1

Some alternative approaches exist. For example, Chintrakarn and Millimet (2006) and Canning and Tsigas (2000) use transport data for United States to obtain trade flows across member States. Dubé and Lemelin (2005) also use transport data to estimate the trade flows across three sub-national regions of Quebec. In addition, they integrate this information with economic data about aggregate sub-national exports and imports and apply a cross-entropy optimisation method to make the two types of information consistent.

We follow the gravitational approach adjusting the trade flows across sub-national regions by the RAS statistical method (Deming and Stephan 1940; Bacharach 1970) to make consistent the intra-national trade obtained through the SLQs and the application of the gravitational approach for the bi-lateral trade flows between sub-national regions.[2]

In practice, the procedure is the following. Consider the share matrix **Π** represented in Table 3. Afterwards, vectors and matrices are in bold type. For simplicity, Table 3 features just three hypothetical sub-country regions: North, Centre and South.

In matrix **Π**, the rows represent the origin, and the columns the destination sub-national regions. Its general element π_{od}, where $0 \leq \pi_{od} \leq 1$, is computed through the gravitational criterion, that is the kilometric distance between the capital cities of the origin and destination regions. As our procedure is valid for all the sectors, for sake of algebraic simplicity we do not consider a sector index in the rest of the section (Table 4).

Denoting Y_{NAT} the Italian sectoral production sold countrywide that is the value of sectoral production sold domestically, D the sub-national demand (excluded demand for foreign goods), EXP the sub-national exports towards the other sub-national regions, IMP the sub-national imports from the other sub-national regions, EXPAG the aggregate sub-national exports towards the rest of country and IMPAG the aggregate sub-national imports from the rest of country, we compute these variables for, say, sub-national region Centre, applying the following formulas:

[2]The RAS abbreviation stems from the names of the vectors (R and S) and matrix (A) used by Bacharach in the original formulation of the algorithm. According to McDougall (1999) RAS is a type of cross-entropy optimization method and it should be preferred in the absence of information about variation in column structure or row structure of the matrix.

Table 4 GDP % changes wrt database

	1st Scenario: reference	2nd Scenario: bigger products substitution	3rd Scenario: lab/cap mobility	4th Scenario: both components
Norte	−9.86	−9.90	−17.47	−20.13
Algarve	−0.01	0.02	4.77	7.51
Centro	−0.01	0.00	3.92	5.32
Lisboa	−0.03	−0.01	2.94	3.80
Alentejo	0.00	0.03	4.65	6.37
Portugal	−2.89	−2.89	−2.64	−2.56

$$\begin{aligned}
(\pi_{12}+\pi_{22}+\pi_{32})\cdot Y_{ITA} &= D_{Centre}\\
\pi_{21}\cdot Y_{ITA} &= EXP_{Centre,\,North}\\
\pi_{23}\cdot Y_{ITA} &= EXP_{Centre,\,South}\\
(\pi_{21}+\pi_{23})\cdot Y_{ITA} &= EXPAG_{Centre}\\
\pi_{12}\cdot Y_{ITA} &= IMP_{North,\,Centre}\\
\pi_{32}\cdot Y_{ITA} &= IMP_{South,\,Centre}\\
(\pi_{12}+\pi_{32})\cdot Y_{ITA} &= IMPAG_{Centre}
\end{aligned} \tag{6}$$

We apply the same procedure for each sub-national region.

Now, it well may happen that the regional production and demand that can be inferred for a given sector by applying the SLQs are not consistent with the aggregate imports and exports obtained by the gravitation approach:

$$\begin{aligned}
Y_{North} &= D_{North} + EXPAG_{North} - IMPAG_{North}\\
Y_{Centre} &= D_{Centre} + EXPAG_{Centre} - IMPAG_{Centre}\\
Y_{South} &= D_{South} + EXPAG_{South} - IMPAG_{South}
\end{aligned} \tag{7}$$

The required adjustment takes place through the bi-proportional RAS method. Consider the bilateral trade matrix:

$$\mathbf{A} = \boldsymbol{\Pi}\, Y_{ITA}$$

of size 3 × 3, where we put $\pi_{11} = \pi_{22} = \pi_{33} = 0$. In matrix **A**, the general element is a_{od} where row *o* represents the origin and column *d* the destination sub-national region respectively. We also have a target vector of row totals **E** (aggregate sub-national exports to the rest of country, size 3 × 1) and a target vector of column totals **M** (aggregate sub-national imports from the rest of country, size 3 × 1). Targets are computed using the National and Eurostat statistical information about economic production (Y_{North}, Y_{Center} and Y_{South}) according to the following equations:

$$\begin{aligned}
E_{North} &= Y_{North} - D_{North} + IMPAG_{North} \\
E_{Centre} &= Y_{Centre} - D_{Centre} + IMPAG_{Centre} \\
E_{South} &= Y_{South} - D_{South} + IMPAG_{South} \\
M_{North} &= D_{North} + EXPAG_{North} - Y_{North} \\
M_{Centre} &= D_{Centre} + EXPAG_{Centre} - Y_{Centre} \\
M_{South} &= D_{South} + EXPAG_{South} - Y_{South}
\end{aligned} \tag{8}$$

The RAS method attempts to find a new matrix **B** such that:

$$\sum_{o} b_{od} = M_d$$

$$\sum_{d} b_{od} = E_o$$

where b_{od}, e_o and m_d are, respectively, the general element of matrix **B**, vector **E** and vector **M**.

The new matrix **B** is related to the original **A** via the iterative procedure:

$$b_{od} = (rm)_o \cdot (cm)_d \cdot a_{od}$$

where $(rm)_o$ is the multiplier of row o and $(cm)_d$ is the multiplier of column d.

For this initial application, we split the national exports and imports using the sectoral sub-national share of value added for exports and a combination of sub-national GDP share and SLQs for imports.

3 Changes in the Model Structure

Regionalization implies two work phases: one on the database, and another on the model structure. The first phase has been described in the previous section.

The second phase requires modifying the functional structure of the model especially to introduce a different degree of factors and goods mobility for the sub-national regions respect to the national ones. In fact, either goods or factors are expected to move more easily within the same country or the same political and economic union such as EU than between different macro-regions such as Europe and Asia for example.

In our original CGE model primary factors of production like labour and capital are imperfectly mobile across sectors, within the country or the aggregated macro-region, and implicitly also perfectly mobile "spatially" within the country or the aggregated macro-region. They are not mobile across countries. GTAP also includes land among primary factors. Land does not move physically, but can be used for different purposes, namely to grow different crops. It is a "sluggish" factor of production as there are constraints in land uses captured by an elasticity of transformation parameter which determines the land supply in each agricultural sector. This sectoral mobility of primary input is clearly technological/sectoral rather than spatial. The issue is slightly different for intermediates and final consumption goods. Both can be imported and thus are "mobile" across countries. However, in the CGE framework, to prevent unrealistic specialization phenomena and trade overflows that could warp the results of the model, the Armington assumption (1969) is introduced. It postulates imperfect substitutability between homologue domestic and imported goods. The values of the Armington elasticity are set by econometric estimations, which are carried out at the national level.

When, as in our case, the spatial detail of the CGE model is increased, it would be unrealistic to simply transfer to sub national entities the same parameterization used in the national model.

Both intra national primary factor mobility and goods' and intermediates' substitution require additional assumptions.

As to the first point it is reasonable to assume some, but not perfect, degree of factor mobility across sub-national regions within the same country or EU but this depends also on the policy scenario of the experiment (short, medium or long run).

As to the second point some imperfect substitution between goods produced in different sub-national regions must be introduced. If not, unrealistic full specialization or trade flows could be observed also at the sub-national levels. Following the empirical evidence that trade is bigger within than between countries given the same distance—the so-called border effect (McCallum 1995)—these Armington elasticities should be higher intra than inter country.

3.1 Mobility in Factors Market: The CET Approach

The value added in the standard GTAP model originates from five primary factors: land, natural resources, unskilled labour, skilled labour and capital. All the sectors use labour and capital while only some use land and natural resources (agriculture and mining-related sectors, respectively). Land and natural resources supply is sluggish across sectors while labour and capital are perfectly mobile. All the primary factors are spatially immobile. For our sub-national context, we assume the following:

(1) Primary factors sectoral mobility does not change.
(2) Land and natural resources remain spatially immobile at the sub-national level.
(3) Sub-national unskilled labour, skilled labour and capital supply is still immobile with respect to the rest of the world but can be geographically sluggish within the country or the EU depending on the type of experiment and the aim of the research.

The third assumption is new with respect to the standard GTAP model. It is implemented through a CET (*Constant Elasticity of Transformation*) function: as a result, workers and capital can move outside the sub-national region they belong to in response to economic shocks. It is worth noting that the model allows a flexible aggregation scheme. If the focus is on a specific European Mediterranean country, there is the possibility to increase the mobility just within this country.

First order conditions of the CET supply function and the formula to determine the EU price of the endowment (shadow price) are given in the Eqs. 9–14, where QL, QH, QK, PL, PH, and PK represent, respectively, the quantity of supplied unskilled labour, skilled labour, capital and the associated prices. EU and r are, respectively, the European aggregate index and the sub-national index. The parameters σ_L, σ_H and σ_K are the elasticity of substitution of the endowment supply, they are a measure of geographical mobility. Increasing the absolute value of these parameters means increasing the factors mobility within EU.

$$QL_r = QL_{EU}\left(\frac{PL_{EU}}{PL_r}\right)^{\sigma_L} \quad \text{with}\ \sigma_L < 0 \tag{9}$$

$$\sum_r QL_r PL_r = QL_{EU} PL_{EU} \tag{10}$$

$$QH_r = QH_{EU}\left(\frac{PH_{EU}}{PH_r}\right)^{\sigma_H} \quad \text{with}\ \sigma_H < 0 \tag{11}$$

$$\sum_r QH_r PH_r = QH_{ITA} PH_{ITA} \tag{12}$$

$$QK_r = QK_{EU}\left(\frac{PK_{EU}}{PK_r}\right)^{\sigma_K} \quad \text{with}\ \sigma_K < 0 \tag{13}$$

$$\sum_r QK_r PK_r = QK_{EU} PK_{EU} \tag{14}$$

3.2 The Trade Structure of the Sub-national Regions: The CRESH Approach

In the standard GTAP model the demand side is composed by private consumption, government spending and intermediate goods. The demand tree follows a double nest (Fig. 1). The first nest links domestic demand and aggregate foreign imports of a specific commodity (irrespective of origin country) for each agent (households, government, firms). The second nest differentiates foreign imports according to the geographical origin.

The second model improvement thus consists in modifying that tree in order to make sub-national products closer substitutes among them within EU.

To achieve this goal we modify the second nest in Fig. 2 by using the CRESH (Constant Ratios of Elasticities of Substitution, Homothetic) function (see Hanoch 1971; Pant 2007; Cai and Arora 2015). The CRESH function is very flexible compared to the standard CES function because for every good it allows for differing levels of substitution between any pair of countries/sub-national regions.

The Eq. 15 shows the demand function behind the new trade structure in Fig. 2:

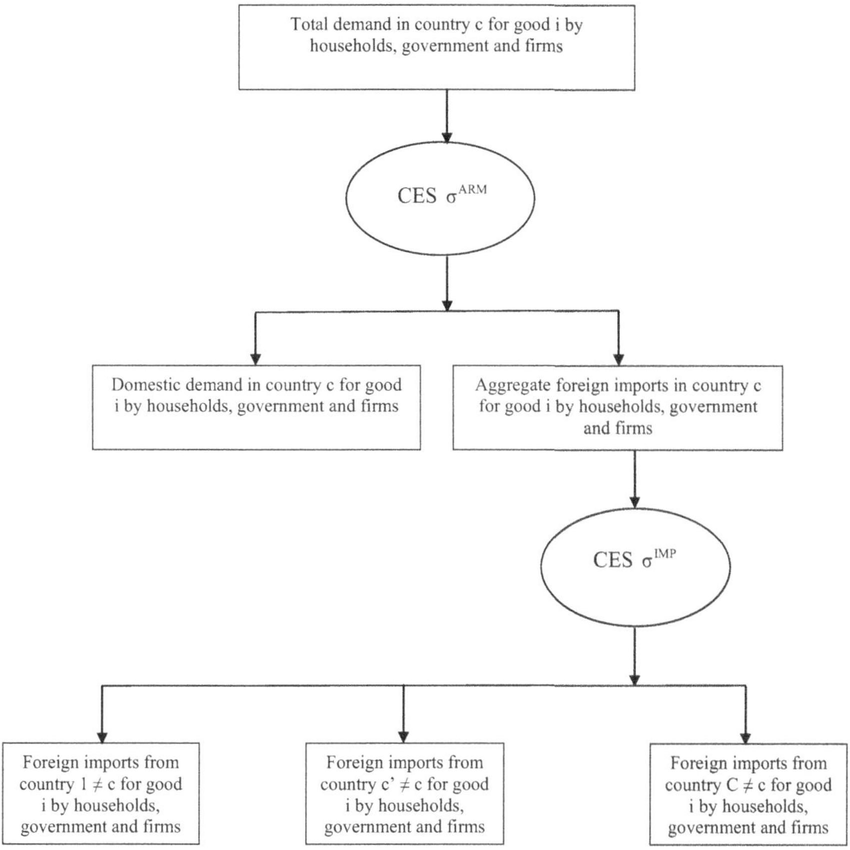

Fig. 1 GTAP standard CES demand structure

$$\sum_{r}\left(\frac{Q_{r,s}}{QIMP_{s}}\right)^{d_{r,s}}\frac{D_{r,s}}{d_{r,s}}=\kappa_{s} \tag{15}$$

In this equation $Q_{r,s}$ is the trade flow between region r and s and QIMP is total amount of imports in regions s, d is a parameter with a value less than 1 but not equal to zero, each D parameter associated with a particular good is positive, and the values of D and κ are normalized. In the special case when $d_i = d$ for all i the CRESH function collapses to a CES function.

Each agent (household, government and firm) minimizes the expenditure subject to (15). The first order conditions become:

$$Q_{r,s}=\left(\frac{P_{r,s}}{P_{s}}\right)^{\sigma_{r,s}^{IMP}} QIMP_{s} \tag{16}$$

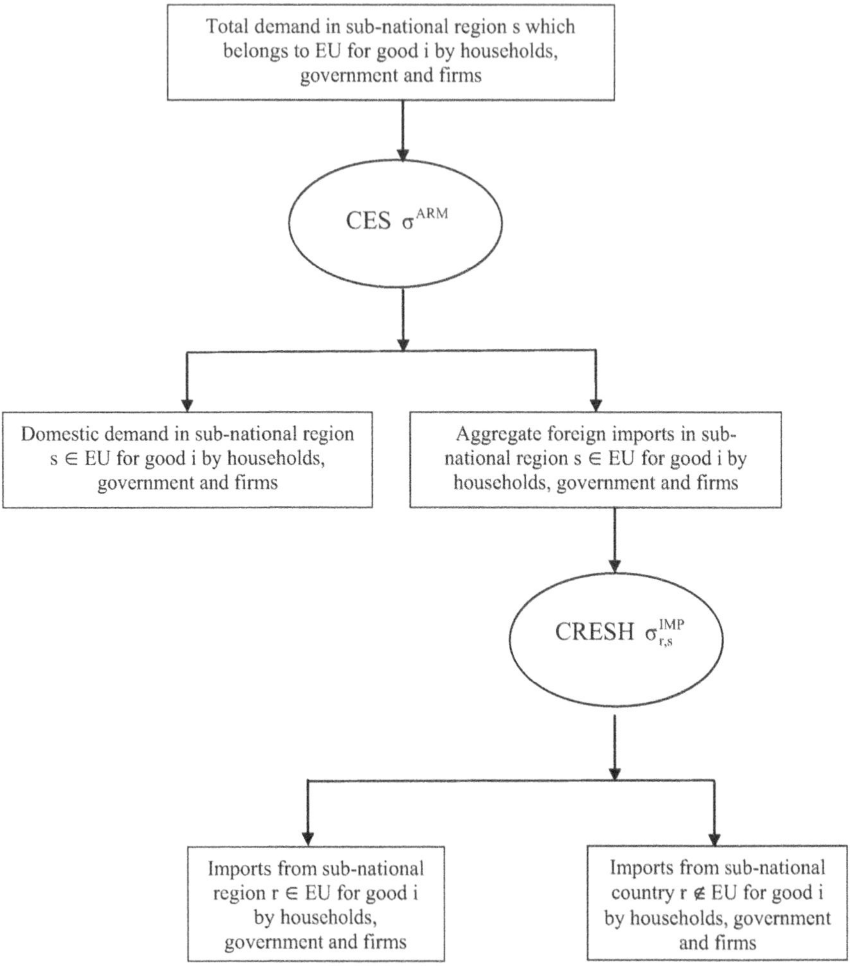

Fig. 2 CRESH sub-national demand structure

where

$$\sigma_{r,s}^{IMP} = \frac{1}{1 - d_{r,s}}, \quad P_s = \sum_r \sigma_{r,s}^{IMP} S_{r,s} P_{r,s}, \quad S_{r,s} = \frac{P_{r,s} Q_{r,s}}{\sum_r P_{r,s} Q_{r,s}}$$

The good point of the CRESH approach is that compared to the standard CES Armington elasticities in the second nest, σ^{IMP}, the CRESH Armington elasticities, $\sigma_{r,s}^{IMP}$ have two geographical indexes specifying the origin and the destination of the trade flow. This means that we can set different values for each pair of sub-national regions and/or countries with the maximum level of flexibility.

4 Testing the Model

This section tests the performance of our sub-national model first for a uniform productivity climate shock across the sub-national regions and then for an asymmetric one. In the first case our aim is to verify the robustness of the economic effects for different assumptions on primary factors and goods mobility within the EU. In the second case we want to measure and understand the distributional economic impacts following an asymmetric climate shock.

4.1 Symmetric Shock

In this section, we conduct a sensitivity analysis on the Armington elasticities for trade and the CET elasticities for labour and capital mobility at the sub-national level. These two parameters are fundamental drivers of the model results. Moreover, there is limited quantitative support to their econometric estimation. This is a further motivation to justify a sensitivity test.

Our aggregation scheme is the 70 sub-national regions already cited in Table 1 plus rest of the EU and rest of the world. We keep the sectoral aggregation as simple as possible to make easier the computation of the economic general equilibrium. Three sectors are considered:

(1) agriculture, forestry and fishing
(2) manufactures and extraction
(3) services.

The shock is a uniform 20% decrease in the primary factors productivity of all regions: capital, labour, land and natural resources.

We start with the Armington elasticities. Factor mobility is kept at the reference case ($\sigma_{FAC} = 0$) which corresponds to the case of capital and labour immobility at the sub-national level. Then we progressively increase the substitution across products (i.e., the Armington elasticities $\sigma_{r,s}^{IMP}$) within the European Union. The top graph of Fig. 3 represents GDP % changes of EU under three different assumptions on products mobility, implemented varying the elasticity of substitution σ^{IMP} of Eq. 16. The formulas below represent, respectively, low, medium and high mobility in the goods market):

$$\text{arm\_1} \rightarrow \sigma_{r,s}^{IMP} = \sigma^{IMP}\ \forall r,s$$
$$\text{arm\_2} \rightarrow \sigma_{r,s}^{IMP} = 2\sigma^{IMP}\ \forall r,s \in EU$$
$$\text{arm\_3} \rightarrow \sigma_{r,s}^{IMP} = 3\sigma^{IMP}\ \forall r,s \in EU$$

They are depicted on the horizontal axis.

In the central part of Fig. 3 we modify the value of σ_L, σ_H and σ_K for the primary factors supply (Eqs. 9–14). For sake of simplicity we assume that $\sigma_L = \sigma_H = \sigma_K = \sigma_{FAC}$. Armington elasticity is kept at the reference case, $\sigma_{r,s}^{IMP} = \sigma^{IMP} \forall r,s$. Factor mobility is increased according to the following scheme:

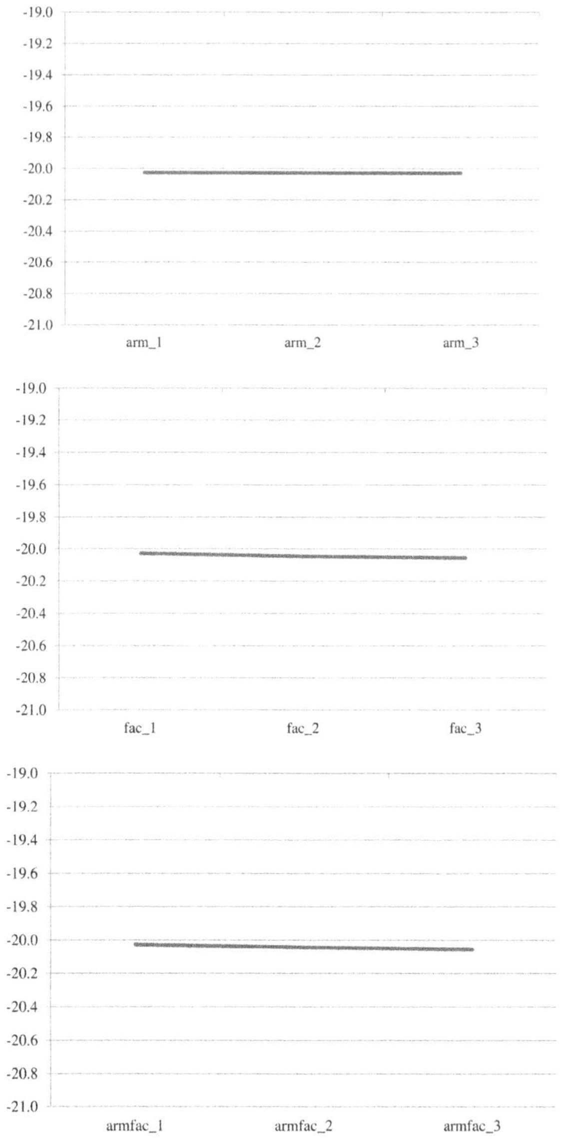

Fig. 3 EU real GDP % changes wrt the database: Armington (top), primary factors (center) and both components (bottom)

fac_1 $\rightarrow \sigma_{FAC} = 0$
fac_2 $\rightarrow \sigma_{FAC} = -5$
fac_3 $\rightarrow \sigma_{FAC} = -20$

where fac_1 represents no factor mobility case and fac_3 the highest level of factors mobility.

Finally, we look at the interaction between the two components:

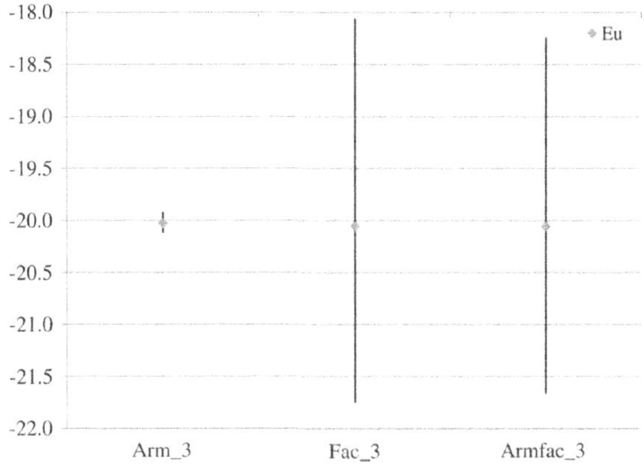

Fig. 4 Segments representing the lowest and highest GDP % Ch. across the sub-national regions (Armington elasticities, primary factors and both components)

armfac_1 → fac_1, arm_1
armfac_2 → fac_2, arm_2
armfac_3 → fac_3, arm_3

Figure 3 clearly shows that results of EU GDP are very stable for all the scenarios considered. Increasing the economic integration in the goods market, labour market and both does not change substantially the results for EU. The regionalized structure of the model is thus robust to productivity shocks at the aggregate level.

From Fig. 4 we can understand the sub-national economic dynamics. Results for all the sub-national regions are reported in the Table 5 of the Appendix. A greater substitutability of sub-national products has almost no effect on the GDP of the 70 sub-national regions while a bigger mobility of primary factors (labour and capital) causes a divergence process across the sub-national regions in a range of less than ±2%. Interestingly the flexibility in the goods market slightly reduces this range. The fact that the assumptions on primary factors mobility have deeper consequences than assumptions on goods market integration is not very surprising. In fact Armington elasticities can change the distribution of trade across sectors and regions but they cannot affect the stock of capital and labour in the region. Introducing primary factors mobility across the regions makes this possible and has a more direct and stronger impact on the regional GDP.

4.2 Asymmetric Shock

The above-mentioned dynamics depends also on the type of shock analysed. Affecting uniformly all the primary factors for all the European regions, as we did, is not very realistic for climate change impacts and risk to underestimate the economic potential of labour and capital mobility to re-distribute production and consumption across the sub-country regions.

Table 5 GDP % changes wrt the database in all scenarios

	arm 1	arm_2	arm 3	fac_1	fac_2	fac_3	armfac_1	armfac_2	armfac_3
Row	−0.06	−0.05	−0.05	−0.06	−0.06	−0.06	−0.06	−0.05	−0.05
REU	−20.02	−20.02	−20.02	−20.02	−20.01	−19.99	−20.02	−19.95	−19.84
Ile de France (France)	−20.09	−20.10	−20.10	−20.09	−19.93	−19.87	−20.09	−20.08	−20.26
Champagne-Ardenne (France)	−20.10	−20.11	−20.12	−20.10	−20.48	−20.63	−20.10	−20.65	−21.28
Picardie (France)	−20.09	−20.10	−20.10	−20.09	−20.36	−20.47	−20.09	−20.50	−20.97
Haute Normandie (France)	−20.03	−20.01	−19.99	−20.03	−19.38	−19.09	−20.03	−19.18	−18.24
Centre (France)	−20.08	−20.09	−20.09	−20.08	−20.17	−20.20	−20.08	−20.35	−20.74
Basse Normandie (France)	−20.07	−20.07	−20.06	−20.07	−19.93	−19.85	−20.07	−19.92	−19.74
Bourgogne (France)	−20.09	−20.10	−20.10	−20.09	−20.28	−20.35	−20.09	−20.44	−20.89
Nord Pas de Calais (France)	−20.07	−20.08	−20.08	−20.07	−20.19	−20.23	−20.07	−20.32	−20.66
Lorraine (France)	−20.08	−20.08	−20.09	−20.08	−20.38	−20.51	−20.08	−20.51	−21.01

(continued)

Table 5 (continued)

	arm 1	arm_2	arm 3	fac_1	fac_2	fac_3	armfac_1	armfac_2	armfac_3
Alsace (France)	−20.07	−20.08	−20.09	−20.07	−20.26	−20.34	−20.07	−20.41	−20.86
Franche-Comte (France)	−20.08	−20.09	−20.09	−20.08	−20.44	−20.59	−20.08	−20.56	−21.10
Pays de la Loire (France)	−20.10	−20.11	−20.11	−20.10	−20.18	−20.20	−20.10	−20.32	−20.66
Bretagne (France)	−20.09	−20.11	−20.12	−20.09	−20.08	−20.06	−20.09	−20.33	−20.76
Poitou-Charentes (France)	−20.09	−20.10	−20.11	−20.09	−20.16	−20.17	−20.09	−20.33	−20.72
Aquitaine (France)	−20.08	−20.09	−20.09	−20.08	−20.08	−20.08	−20.08	−20.23	−20.54
Midi-Pyrenees (France)	−20.09	−20.10	−20.11	−20.09	−20.10	−20.09	−20.09	−20.22	−20.49
Limousin (France)	−20.10	−20.11	−20.12	−20.10	−20.28	−20.35	−20.10	−20.47	−20.97
Rhone-Alpes (France)	−20.08	−20.09	−20.09	−20.08	−20.13	−20.13	−20.08	−20.24	−20.49
Auvergne (France)	−20.10	−20.10	−20.11	−20.10	−20.56	−20.75	−20.10	−20.68	−21.29

(continued)

Table 5 (continued)

	arm 1	arm_2	arm 3	fac_1	fac_2	fac_3	armfac_1	armfac_2	armfac_3
Languedoc-Roussillon (France)	−20.04	−20.05	−20.05	−20.04	−19.89	−19.82	−20.04	−20.01	−20.10
Provence (France)	−20.06	−20.07	−20.07	−20.06	−19.50	−19.29	−20.06	−19.68	−19.63
Corse (France)	−20.03	−20.07	−20.08	−20.03	−19.00	−18.61	−20.03	−19.33	−19.15
Voreia (Greece)	−20.03	−20.04	−20.04	−20.03	−20.86	−21.25	−20.03	−20.91	−21.50
Kentriki (Greece)	−20.01	−20.01	−20.01	−20.01	−20.83	−21.22	−20.01	−20.82	−21.29
Attiki (Greece)	−19.98	−19.98	−19.98	−19.98	−20.87	−21.25	−19.98	−20.66	−20.86
Nisia (Greece)	−20.06	−20.09	−20.11	−20.06	−19.35	−19.21	−20.06	−19.43	−19.12
Piemonte (Italy)	−20.07	−20.07	−20.07	−20.07	−20.72	−21.10	−20.07	−20.62	−21.05
Valle d'Aosta (Italy)	−20.05	−20.05	−20.05	−20.05	−20.97	−21.49	−20.05	−20.83	−21.40
Lombardia (Italy)	−20.07	−20.07	−20.08	−20.07	−20.72	−21.11	−20.07	−20.65	−21.15
Trentino Alto Adige (Italy)	−20.05	−20.05	−20.06	−20.05	−20.79	−21.20	−20.05	−20.71	−21.23
Veneto (Italy)	−20.07	−20.08	−20.08	−20.07	−20.59	−20.91	−20.07	−20.57	−21.06

(continued)

Table 5 (continued)

	arm_1	arm_2	arm_3	fac_1	fac_2	fac_3	armfac_1	armfac_2	armfac_3
Friuli Venezia Giulia (Italy)	−20.05	−20.05	−20.05	−20.05	−21.03	−21.57	−20.05	−20.91	−21.59
Liguria (Italy)	−20.05	−20.05	−20.05	−20.05	−20.53	−20.82	−20.05	−20.40	−20.60
Emilia Romagna (Italy)	−20.05	−20.06	−20.06	−20.05	−20.58	−20.91	−20.05	−20.58	−21.08

(continued)

Table 5 (continued)

	arm 1	arm_2	arm 3	fac_1	fac_2	fac_3	armfac_1	armfac_2	armfac_3
Toscana (Italy)	−20.07	−20.07	−20.07	−20.07	−20.71	−21.08	−20.07	−20.63	−21.09
Umbria (Italy)	−20.05	−20.06	−20.06	−20.05	−20.83	−21.28	−20.05	−20.77	−21.40
Marche (Italy)	−20.07	−20.07	−20.07	−20.07	−20.79	−21.21	−20.07	−20.69	−21.20
Lazio (Italy)	−20.06	−20.05	−20.05	−20.06	−21.21	−21.75	−20.06	−20.96	−21.47
Abruzzo (Italy)	−20.05	−20.06	−20.06	−20.05	−20.74	−21.16	−20.05	−20.69	−21.27
Molise (Italy)	−20.03	−20.03	−20.03	−20.03	−20.86	−21.35	−20.03	−20.77	−21.37
Campania (Italy)	−20.06	−20.06	−20.06	−20.06	−20.98	−21.45	−20.06	−20.82	−21.32
Puglia (Italy)	−20.05	−20.04	−20.04	−20.05	−20.86	−21.29	−20.05	−20.72	−21.18

(continued)

Table 5 (continued)

	arm 1	arm_2	arm 3	fac_1	fac_2	fac_3	armfac_1	armfac_2	armfac_3
Basilicata (Italy)	−20.09	−20.10	−20.10	−20.09	−21.01	−21.48	−20.09	−21.01	-21.66
Calabria (Italy)	−20.06	−20.06	−20.05	−20.06	−21.16	−21.68	−20.06	−20.97	−21.48
Sicilia (Italy)	−20.00	−19.98	−19.97	−20.00	−20.62	−20.96	−20.00	−20.30	−20.23
Sardegna (Italy)	−20.04	−20.04	−20.03	−20.04	−20.40	−20.64	−20.04	−20.16	−20.02
Galicia (Spain)	−19.97	−19.98	−19.98	−19.97	−19.12	−18.71	−19.97	−19.42	−19.19
Asturias (Spain)	−19.97	−19.98	−19.99	−19.97	−19.30	−18.92	−19.97	−19.60	−19.51
Cantabria (Spain)	−19.98	−19.98	−19.99	−19.98	−19.19	−18.78	−19.98	−19.47	−19.26
Pais Vasco (Spain)	−19.99	−19.99	−20.00	−19.99	−19.46	−19.12	−19.99	−19.70	−19.58
Navarra (Spain)	−20.00	−20.01	−20.01	−20.00	−19.62	−19.35	−20.00	−19.86	−19.86
La Rioja (Spain)	−19.97	−19.98	−19.99	−19.97	−19.64	−19.41	−19.97	−19.96	−20.18
Aragon (Spain)	−19.98	−20.00	−20.01	−19.98	−18.87	−18.33	−19.98	−19.22	−18.88

(continued)

Table 5 (continued)

	arm 1	arm_2	arm 3	fac_1	fac_2	fac_3	armfac_1	armfac_2	armfac_3
Madrid (Spain)	−19.94	−19.95	−19.95	−19.94	−19.02	−18.59	−19.94	−19.28	−18.99
Castilla y Leon (Spain)	−19.98	−19.99	−20.00	−19.98	−19.26	−18.90	−19.98	−19.57	−19.47
Castilla-La Mancha (Spain)	−19.97	−19.99	−19.99	−19.97	−19.29	−18.92	−19.97	−19.64	−19.63
Extremadura (Spain)	−19.96	−19.98	−20.00	−19.96	−19.12	−18.72	−19.96	−19.48	−19.41
Cataluna (Spain)	−19.97	−19.98	−19.98	−19.97	−19.13	−18.69	−19.97	−19.36	−19.01
Valencia (Spain)	−19.96	−19.97	−19.98	−19.96	−19.05	−18.59	−19.96	−19.37	−19.11
Balears (Spain)	−19.92	−19.94	−19.95	−19.92	−18.62	−18.07	−19.92	−18.95	−18.47
Andalucia (Spain)	−19.95	−19.96	−19.96	−19.95	−18.98	−18.54	−19.95	−19.28	−19.02
Murcia (Spain)	−19.96	−19.97	−19.98	−19.96	−18.98	−18.52	−19.96	−19.32	−19.10
Ceuta (Spain)	−19.94	−19.95	−19.95	−19.94	−19.74	−19.66	−19.94	−19.86	−19.92

(continued)

Table 5 (continued)

	arm_1	arm_2	arm_3	fac_1	fac_2	fac_3	armfac_1	armfac_2	armfac_3
Melilla (Spain)	−19.94	−19.95	−19.95	−19.94	−19.53	−19.36	−19.94	−19.72	−19.74
Canarias (Spain)	−19.90	−19.91	−19.92	−19.90	−18.60	−18.06	−19.90	−18.92	−18.48
Norte (Portugal)	−20.00	−20.01	−20.01	−20.00	−20.27	−20.39	−20.00	−20.38	−20.80
Algarve (Portugal)	−20.06	−20.08	−20.08	−20.06	−20.04	−20.05	−20.06	−20.19	−20.44
Centro (Portugal)	−19.99	−20.00	−20.00	−19.99	−19.91	−19.89	−19.99	−20.01	−20.16
Lisboa (Portugal)	−19.98	−19.98	−19.98	−19.98	−20.21	−20.30	−19.98	−20.19	−20.38
Alentejo (Portugal)	−20.03	−20.06	−20.07	−20.03	−19.85	−19.79	−20.03	−20.10	−20.38
EU	−20.03	−20.03	−20.03	−20.03	−20.04	−20.05	−20.03	−20.04	−20.06

To test this we impose an asymmetric shock on primary factors productivity. In order to simplify the presentation of the results we take only Portugal (PT) as example for the experiment. A 10% productivity reduction of all primary factors is assumed in the Portuguese Norte region while the other four regions (Algarve, Centro, Lisboa and Alentejo) are not affected. Four scenarios are considered:

(1) in the first one labour and capital are immobile at the sub-national level ($\sigma_{FAC} = 0$) and Armington elasticities are kept at the reference case ($\sigma^{IMP}_{r,s} = \sigma^{IMP} \forall r,s$).
(2) in the second scenario labour and capital are immobile at the sub-national level ($\sigma_{FAC} = 0$) and Armington elasticities are doubled compared to the reference case ($\sigma^{IMP}_{r,s} = 2\sigma^{IMP} \forall r,s \in PT$).
(3) in the third one labour and capital are mobile within Portugal ($\sigma_{FAC} = -5$) and Armington elasticities are kept at the reference case ($\sigma^{IMP}_{r,s} = \sigma^{IMP} \forall r,s$).
(4) in the last scenario labour and capital are mobile within Portugal ($\sigma_{FAC} = -5$) and Armington elasticities are doubled compared to the reference case ($\sigma^{IMP}_{r,s} = 2\sigma^{IMP} \forall r,s \in PT$).

This also confirms the flexibility of the model in the choice of the regional aggregation and experiment. Results are displayed in Table 4.

Also in the case of an asymmetric shock, varying the Armington elasticities does not affect substantially the results (comparison between second and third column in Table 4). However when capital and labour mobility is introduced within Portugal we observe a huge re-allocation of these primary factors across the Portuguese sub-national regions and large GDP changes (fourth column). Mobility of labour and capital improves the GDP performance for Portugal. This is because the negative productivity shock in the Norte region triggers a substitution process reducing the demand of goods produced in the Norte and consequently the demand for labour and capital in this region. This means that remunerations of capital and labour go down and these two factors move towards not negatively affected regions where remunerations are higher. This in turn determines the exacerbation of the looser/winner economic dynamics but also a GDP gain at the aggregate level for Portugal because primary factors are moving toward more productive regions. Differently from the previous symmetric shock the Armington component amplifies the GDP divergences across regions (fifth column).

5 Conclusions and Further Research

This chapter describes and applies a methodology to develop a sub-national CGE model starting from a global model and database. Eventually a CGE model is built for 70 sub-national regions in the Mediterranean European area. Empirical and theoretical issues are discussed through the paper. The model allows for intra-national trade and factor mobility within each country or EU.

We run a number of simulations to test the robustness of our regionalized structure and understand the potential economic mechanisms behind climate impacts in the context of different levels of market integration. In the case of a symmetric shock on productivity, results for GDP are very stable at the aggregate level for the EU. Diverging patterns of GDP can be observed at the sub-national level when interregional mobility is introduced in the factors market, while different degrees of substitutability in consumption of goods from different sub-national regions play a minor role.

When the shock is asymmetric GDP divergences strongly amplify, exacerbating the winner/looser economic dynamics but we also notice a GDP gain at the aggregate level. Inter-regional factor mobility is crucial in explaining this outcome.

Further research involves the extension of this first version to the sub-national regions of some African and Asian Mediterranean countries. In particular ongoing work is focusing on three countries: Morocco, Turkey and Egypt. Not surprisingly data constraints become more stringent for countries in the Southern Mediterranean coast.

The model should be tested to evaluate economic consequences and policies to cope with climate change impacts in the Euro-Mediterranean area at the sub-national level: sea level rise, water management, change in the agriculture yields, mitigation and adaptation policies.

Acknowledgements The research leading to these results has received funding from the Italian Ministry of Education, University and Research and the Italian Ministry of Environment, Land and Sea under the GEMINA project.
The authors are the only responsible for errors and omissions in this work.

References

Aaheim, A., Dokken, T., Hochrainer, S., Hof, A., Jochem, E., Mechler, R., et al. (2010). National responsibilities for adaptation strategies: Lessons from four modelling frameworks. In M. Hulme & H. Neufeld (Eds.), *Making climate change work for us: European perspectives on adaptation and mitigation strategies*. Cambridge: Cambridge University Press.

Armington, P. (1969). A theory of demand for products distinguished by place of production. *IMF Staff Papers, 16*(1), 159–178.

Bacharach, M. (1970). *Biproportional matrices & input-output change*. Number 16 in University of Cambridge Department of Applied Economics Monographs. Cambridge University Press.

Bigano, A., Bosello, F., Roson, R., & Tol, R. (2008). Economy-wide impacts of climate change: A joint analysis for sea level rise and tourism. *Mitigation and Adaptation Strategies for Global Change, 13*(8).

Böhringer, C., Rutherford, T. F., & Tol, R. (2009). The EU 20/20/20 targets: An overview of the EMF22 assessment. *Energy Economics, 31,* 268–273.

Böhringer, C., Fisher, C., & Rosendahl, K. E. (2010). The global effects of subglobal climate policies. Resources for the Future Discussion Paper, pp. 10–48.

Böhringer, C., Balistreri, E. J., & Rutherford, T. F. (2012). The role of border carbon adjustment in unilateral climate policy: Overview of an Energy Modeling Forum study (EMF 29). *Energy Economics, 34,* S97–S110.

Bonfiglio, A. (2008). Evaluating implications of agricultural policies in a rural region through a CGE analysis, No 328, Working Papers, Universita' Politecnica delle Marche, Dipartimento di Scienze Economiche e Sociali.
Bonfiglio, A., & Chelli, F. (2008). Assessing the behaviour of non-survey methods for constructing regional input-output tables through a Monte Carlo simulation. *Economic Systems Research, 20*(3), 243–258.
Bosello, F., Nicholls, R., Richards, J., Roson, R., & Tol, R. (2012). Economic impacts of climate change in Europe: Sea-level rise. *Climatic Change, 112*(1), 63–81. (Springer)
Brandsma, A., Kancs, D., Monfort, P., & Rillaers, A. (2015). RHOMOLO: A dynamic spatial general equilibrium model for assessing the impact of cohesion policy. Papers in *Regional Science* 94. https://doi.org/10.1111/pirs.12162.
Cai, Y., & Arora, V. (2015). Disaggregating electricity generation technologies in CGE models: A revised technology bundle approach with an application to the U.S. Clean Power Plan. *Applied Energy, 154*(C), 543–555.
Canning, P., & Tsigas, M. (2000). Regionalism, federalism, and taxation: A food and farm perspective. Technical Bulletin No. 1882, Economic Research Services, U.S. Department of agriculture.
Carrera, L., Standardi, G., Bosello, F., & Mysiak, Y. (2015). Assessing direct and indirect economic impacts of a flood event through the integration of spatial and computable general equilibrium modelling. *Environmental Modelling and Software, 63,* 109–122.
Chintrakarn, P., & Millimet, D. L. (2006). The environmental consequences of trade: Evidence from subnational trade flows. *Journal of Environmental Economics and Management, 52*(1), 430–453. (Elsevier).
Ciscar, J.-C., Iglesias, A., Feyen, L., Szabó, L., Van Regemorter, D., Amelung, B., et al. (2011). Physical and economic consequences of climate change in Europe. *Proceedings of the National Academy of Sciences of the United States of America, 108*(7), 2678–2683.
Darwin, R., & Tol, R. (2001). Estimates of the economic effects of sea level rise. *Environmental & Resource Economics (European Association of Environmental and Resource Economists), 19*(2), 113–129.
Deming, W. E., & Stephan, F. F. (1940). On a least-squares adjustment of a sampled frequency table when the expected marginal totals are known. *Annals of Mathematical Statistics, 11,* 427–444.
Dixon, P., Rimmer, M., & Wittwer, G. (2012). USAGE-R51, a State-level Multi-regional CGE Model of the US Economy. https://www.gtap.agecon.purdue.edu/resources/download/5933.pdf.
Dubé, J., & Lemelin, A. (2005). Estimation Expérimentale des Flux d'Echanges Interrégionaux par la Méthode de Minimisation de l'Entropie Croisée. *Revue Canadienne des Sciences Régionales/Canadian Journal of Regional Science, 28*(3), 513–534.
European Commission. (2008). Package of implementation measures for the EU's objectives on climate change and renewable energy for 2020. Commission Staff working document SEC (2008) 85 II.
European Commission. (2010). Analysis of options to move beyond 20% greenhouse gas emission reductions and assessing the risk of carbon leakage. Commission Staff working document SEC (2010) 650.
Eurostat. (2016). Eurostat database. http://ec.europa.eu/eurostat/web/nuts/overview.
Eurostat. (2017). Economic accounts for agriculture. http://appsso.eurostat.ec.europa.eu/nui/show.do.
Eurostat. (2017). Structural business statistics. http://ec.europa.eu/eurostat/web/structural-business-statistics/data/database.
Eboli, F., Parrado, R., & Roson, R. (2010). Climate change feedback on economic growth: Explorations with a dynamic general equilibrium model. *Environment and Development Economics, 15*(5), 515–533.
Hanoch, G. (1971). CRESH production functions. *Econometrica, 39,* 695–712.
Hellenic Statistical Authority. Gross value added by industry 2000–2015. http://www.statistics.gr/en/statistics/-/publication/SEL45/-.

Hertel, T. W. (Ed.). (1997). *Global trade analysis: Modeling and applications*. Cambridge and New York: Cambridge University Press.
Hertel, T. W., Lee, H., Rose, S., & Sohngen, B. (2009). Modelling land use related greenhouse gas sources and sinks and their mitigation potential. In T. W. Hertel, S. Rose, & R. Tol (Eds.), *Economic analysis of land use in global climate change policy (Chapter 6)*. London and New York: Routledge.
Horridge, M., & Wittwer, G. (2010). Bringing regional detail to a CGE model using census data. *Spatial Economic Analysis, 5*(2), 229–255.
INE Spain. Contabilidad Regional de España 2000–2016. http://www.ine.es/dynt3/inebase/index.htm?type=pcaxis&path=/t35/p010/base2008&file=pcaxis.
INE Portugal. Gross value added (€) of Enterprises by Geographic localization (NUTS - 2002) and Economic activity (Subclass - CAE Rev. 3); Annual—Statistics Portugal, Integrated business accounts. http://censos.ine.pt/xportal/xmain?xpid=INE&xpgid=ine_indicadores&indOcorrCod=0006560&contexto=pti&selTab=tab10.
INSEE. Valeurs Ajoutées régionales. http://www.insee.fr/fr/themes/detail.asp?reg_id=99&ref_id=pib-va-reg-base-2005.
IPCC. (2014). In V. R. Barros, C. B. Field, D. J. Dokken, M. D. Mastrandrea, K. J. Mach, T. E. Bilir, M. Chatterjee, K. L. Ebi, Y. O. Estrada, R. C. Genova, B. Girma, E. S. Kissel, A. N. Levy, S. MacCracken, P. R. Mastrandrea, & L. L. White (Eds.), *Climate change 2014: Impacts, adaptation, and vulnerability. Part B: Regional aspects. Contribution of working group II to the fifth assessment report of the intergovernmental panel on climate change* (688 pp). Cambridge, United Kingdom and New York, NY, USA: Cambridge University Press.
ISTAT. Conti Economici Regionali. Anni 1995–2009. http://www.istat.it/it/archivio/12718.
ISTAT. Agricoltura e Zootecnia. http://agri.istat.it/sag_is_pdwout/index.jsp.
ISTAT. Valore Aggiunto ai Prezzi di Base dell'Agricoltura per Regione. Anni 1980–2011. http://www.istat.it/it/archivio/66513.
Jansson, T. G., Kuiper, M. H., & Adenäuer, M. (2009). Linking CAPRI and GTAP. SEAMLESS report no. 39.
Jean, S., & Laborde, D. (2004). The impact of multilateral liberalisation on European regions: A CGE assessment. CEPII Working Paper, No. 2004-20.
Johansen, L. (1974). *A multi-sectoral study of economic growth* (2nd ed.). Amsterdam: North-Holland Publishing Company.
Koks, E. E., Carrera, L., Jonkeren, O., Aerts, J. C. J. H., Husby, T. G., Thissen, M., et al. (2015). Regional disaster impact analysis: Comparing input-output and computable general equilibrium models. *Nat. Hazards Earth Syst. Sci. Discuss., 3*, 7053–7088.
Lee, H., Hertel, T. W., Rose, S., & Avetisyan, M. (2009). An integrated global land use database for CGE analysis of climate policy options. In T. W. Hertel, S. Rose, & R. Tol (Eds.), Economic analysis of land use in global climate change policy (Chapter 4). London and New York: Routledge.
McCallum, J. (1995). National borders matter: Canada-U.S. regional trade patterns. *American Economic Review, 85*(3), 615–623.
McDougall, R. (1999). Entropy theory and RAS are friends. GTAP Working Papers 300, Center for Global Trade Analysis, Department of Agricultural Economics, Purdue University.
Michetti, M., & Parrado, R. (2012). Improving land-use modelling within CGE to assess forest-based mitigation potential and costs. Working Paper 2012.19, Fondazione Eni Enrico Mattei.
Miller, R. E., & Blair, P. D. (1985). *Input-output analysis: Foundations and extensions*. Englewood Cliffs, New Jersey: Prentice-Hall Inc.
Narayanan, B., Aguiar, A., & McDougall, R. (2012) Global trade, assistance, and production: The GTAP 8 Data Base. Center for Global Trade Analysis, Purdue University.
Pant, H. (2007). GTEM: Global trade and environment model. ABARE Technical Report. Canberra: Australian Bureau of Agricultural and Resource Economics and Sciences.

Pérez-Blanco, C. D., Standardi, G., Mysiak, J., Parrado, R., & Gutiérrez-Martín, C. (2016). Incremental water charging in agriculture. A case study of the Regione Emilia Romagna in Italy. *Environmental Modelling & Software* (forthcoming).

Peter, M. W., Horridge, M., Meagher, G. A., Naqvi, F., & Parmenter, P. R. (1996). The theoretical structure of MONASH-MRF. Centre of Policy Studies/IMPACT Centre Working Papers op-85, Monash University, Centre of Policy Studies/IMPACT Centre.

Potters, L., Conte, A., Kancs, D., & Thissen, M. (2014). Data needs for regional modelling. A description of the data used in support of RHOMOLO. JRC Publication No: JRC80845.

Shoven, J. B., & Whalley, J. (1992). *Applying general equilibrium*. Cambridge University Press.

Standardi, G., & Eboli F. (2015). Sea level rise in the Italian regions: A macro-economic assessment, Research Papers CMCC, RP0251.

Standardi, G., Bosello, F., & Eboli, F. (2014). A sub-national CGE model for Italy, Working Papers 2014.04, Fondazione Eni Enrico Mattei.

Yang, J., Huang, J., Li, N., Rozelle, S., & Martin, W. (2011). The impact of the Doha trade proposals on farmers' incomes in China. *Journal of Policy Modeling, 33*(3), 439–452.

A Regional Dynamic General Equilibrium Model with Historical Calibration: A Counterfactual Exercise

Stefania Lovo, Riccardo Magnani and Federico Perali

Abstract This chapter develops a regional dynamic general equilibrium model calibrated using two regional SAMs for the Italian region Valle D'Aosta for the years 1963 and 2002. A historical calibration procedure is performed over the 40 years period and ensures that the modelled tendencies perfectly reproduce the actual observed growth patterns of the main regional macroeconomic variables. The dynamic general equilibrium model provides an original and powerful tool for historical counterfactual analysis not available using standard dynamic general equilibrium models. The model is used to compare the growth path followed by the region during the period of interest with a counterfactual scenario intended to evaluate how the region would have performed in the case of a contraction of the transfers from the national government to the regional government and the families.

Keywords Regional dynamic general equilibrium model · Historical calibration Historical counterfactual analysis

JEL Classification C68 · R13

1 Introduction

Understanding the historical paths of institutional and economic development is of central importance in understanding the current differences in economic performances across regions and countries (Engerman and Sokoloff 2000; Acemoglu and Robinson 2000, 2012; Acemoglu et al. 2001, 2005). The choice of a specific

S. Lovo
London School of Economics, Grantham Research Institute, London, UK

R. Magnani
Department of Economics, Université Paris 13, Sorbonne Paris Cité, CEPN, Paris, France

F. Perali (✉)
Department of Economics, University of Verona, Via Cantarane 24, 37129 Verona, Italy
e-mail: federico.perali@univr.it

© Springer International Publishing AG, part of Springer Nature 2018
F. Perali and P. L. Scandizzo (eds.), *The New Generation of Computable General Equilibrium Models*, https://doi.org/10.1007/978-3-319-58533-8_12

development and institutional path, in a certain point in time, in fact, can trigger alternative growth trajectories and lead to different levels of efficiency. The use of economic theory and statistical technique to analyse economic history is becoming a popular exercise. This historical approach, which has been mainly carried out using econometric techniques, is here extended in a general equilibrium context.

In this study, we reproduce the economic development path undertaken by Valle D'Aosta during the last 4 decades using a regional dynamic general equilibrium model built on two regional Social Accounting Matrices (SAMs) constructed for the years 1963 and 2002. The availability of two regional SAMs, equal in the structure and referred to two different periods in time, offers an extraordinary opportunity to perform a dynamic calibration procedure based on the knowledge of the initial conditions and of the current economic circumstances. In particular, the calibration procedure adopted in this paper permits to perfectly reproduce the two SAMs and, thus, ensures that the modelled tendencies closely approximate the actual observed growth patterns of the main regional macroeconomic variables. The calibrated dynamic model provides an original and powerful tool for counterfactual analysis in which the path actually followed by the regional economy can be compared to alternative policy scenarios to draw lessons for future policy recommendations.

Computable general equilibrium (CGE) models are widespread tools for policy analysis in spite of criticisms because of weak validation and inferential power. Conducting a validation exercise can help assessing the model limitation and predictive capacity. Validation, which consists in verifying the matching between modelled and historical tendencies over a chosen period, is a major concern for all simulation and operations research models (Gass 1983; Kleijnen 1999). None of these studies exploit the comprehensive range of information that can be obtained from the use of two SAMs constructed for two different points in time. Thurlow (2004) presents a dynamic general equilibrium model for South Africa which involves the use of two SAMs for the years 1993 and 2000, although the model accounts for a between period component, no explicit attention is given to whether the model replicates 2000 actual figures. In this paper, we develop a regional dynamic general equilibrium model calibrated on historical data capable to reproduce the regional economic structure at the final 2002 year starting from the regional SAM for the year 1963.

The model represents a unique experimental setting for a historical counterfactual analysis of alternative policy scenarios which is not available using standard dynamic CGE models. This exercise is backward looking and compares the model generated outcomes with alternative scenarios obtained by introducing shocks throughout the observed period. By comparing the modelled trend for the period 1963 to 2002 with alternative tendencies obtained by shocking the model throughout the period, we can implement a historical counterfactual exercise that cannot be applied using standard dynamic general equilibrium techniques.

The paper proceeds as follows. Section 2 describes some of the salient economic features of the region during the post-war period of interest. Section 3 describes the structure of the regional dynamic general equilibrium model, while Sect. 4 describes the historical calibration procedures. In Sect. 5 the results of the historical counterfactual analysis are discussed. Section 6 concludes.

2 Some Distinctive Features of the Valle D'Aosta Economy in Retrospect

Valle D'Aosta is one of the Italian regions along with the Trentino Alto Adige, Friuli-Venezia Giulia in the North, and Sardinia and Sicily regions in the South enjoying a high level of governing, financial and legislative autonomy which have been fully implemented since 1981. These regions keep for local use almost all state taxes. The northern regions with a special statute do not have to contribute resources to the state solidarity fund for the less developed regions of Italy. The per capita level of public expenditure in these regions is more than twice as large as the expenditure in regions with ordinary autonomy. These regions enjoy freedom to spend and waste. Institutions are transformed into jobs and consensus creating factories. Unemployment rates are in the northern autonomous regions very low as compared even to the other ordinary northern regions. At the beginning of the 21st century the unemployment rate in Valle d'Aosta was about 3% mainly thanks to the abnormal growth of the public administration sector and associated activities (Noto and Meneghelli 2008). Further, in the post-war period, Valle d'Aosta has been the target of a steady migration flow from the Southern region of Calabria. Nowadays, residents of Calabria origin are about one fifth of the population. This aspect may also have contributed to sustain the pro-public sector growth.

Table 1 shows the geography of public employment in Italy. In Valle d'Aosta there are about 79 public workers per 1000 inhabitant, 23 more than in the Lazio region where the central government resides. The Valle d'Aosta region clearly outperforms when compared to the other autonomous regions of Trentino Alto Adige, Sicily and Sardinia (79 vs. 55, 50 and 48 respectively). Veneto and Lombardy are the most parsimonious regions with 32 and 31 public workers to be compared with an Italian average of 41 workers.

Table 2 reports the distribution of regional value added at factor costs across sectors and reflects the level of sector disaggregation used later in the model. In general, we observe a large reduction in the contribution of the secondary sector, including the manufacturing industries and construction, to the formation of the gross domestic product. While in 1963 the manufacturing, construction and service sectors were equally contributing to the regional GDP, in 2002 there is an evident overtaking by the private and public services showing a shift out of the manufacturing to the services of the regional economy. The construction sector appeared to be very vital during the first decade of the analysis mainly due to the large demand for infrastructures and public works. In the 70s and 80s the vigour of the sector was ensured by the demand for tourism construction. This positive tendency is, however, interrupted by a downfall in the second half of the 90s caused mainly by the introduction of new European regulation on public procurement rules and the more severe limits imposed on national government public expenditure.

The regional administration has notably increased the contribution to the formation of regional domestic product (Table 3). During the period of interest, public services almost doubled in terms of percentage of value added. In 2002, the

Table 1 List of sectors in the CGE model

1	Agriculture
2	Extractive industry
3	Metallurgic industry
4	Mechanic industry
5	Chemical industry
6	Agri-food and textile
7	Other manufacturing industries
8	Construction
9	Electricity
10	Wholesale trade
11	Hotels
12	Transports and communication
13	Credit and insurance
14	Public administration

public services account for 23% of GDP in comparison to 8.8% of the manufacturing sector. The comparison with national figures and with the nearby north-western Italian regions reveals the abnormal size of the public administration. The comparison with another mountainous and autonomous region in the North of Italy, Trentino Alto Adige,[1] where the contribution to value added of the public sector is around 15% (Istat 2004a, b), reinforces the relevant role played by the public sector in Valle D'Aosta. Public services generate 35% of total labour income and employs about 30% of the labour forces. They acted as a social damper during the slowdown of the regional economy to alleviate unemployment.

The composition of the regional gross domestic product reported in Table 4 shows a significant change in the overall structure of GDP during the considered 4 decades. The largest components of GDP in 1963 are exports and imports. This is an aspect that typically characterises small regions such as Valle D'Aosta that is the smallest region in Italy with a population of 125,000. In 2002 the economy is much less export oriented. Imports almost tripled to become about 1.4 times exports. There is a notable increase in the role of private domestic consumption. This may reflect a loss of competitiveness with respect to the nearby territories. On the other side, the share of imports as a percentage of GDP has fallen by only 9 percentage points. The share of fixed investment and the tax burden have slightly increased.

The traditional dynamic general equilibrium analysis reproduces an initial economic situation, as recent as data availability permits, and forecast the future development path of the economy. In this study, using a novel dynamic calibration technique, we are interested to reproduce the main features of the economy at both the initial and final economic situation in order to implement a what if analysis to learn whether

[1]In Trentino Alto Adige, the composition of value added is the following: manufacturing 13.8%, construction 10.3%, private services 55.9% and public services 14.6%.

Table 2 SAM 1963 in millions of euro at constant price 2002

	1	2	3	4	5	6	7	8	9	10	11	12
1	38.96	0.78	0.13	0.00	0.37	65.33	1.21	0.25	0.01	0.12	3.92	0.12
2	0.00	0.24	32.06	0.00	16.20	0.06	0.97	2.22	3.58	0.03	0.59	0.19
3	0.13	0.88	499.66	3.94	0.51	0.21	0.27	14.19	0.35	0.00	0.00	0.07
4	0.62	1.19	4.19	5.47	1.04	2.20	0.56	7.78	2.69	1.87	3.53	4.92
5	4.75	2.32	39.86	0.55	24.66	6.76	3.78	6.59	4.23	4.14	11.72	7.28
6	6.65	0.00	0.00	0.10	1.88	55.03	0.48	0.11	0.00	0.57	28.95	0.47
7	0.52	0.10	10.02	1.11	2.34	4.11	7.20	50.26	0.47	3.55	10.88	1.87
8	0.05	0.00	2.29	0.04	0.21	0.19	0.06	0.21	1.11	0.54	6.28	0.95
9	1.22	2.15	36.07	0.33	3.07	1.97	0.92	1.15	1.51	1.88	7.85	0.77
10	2.93	0.45	1.92	0.50	0.93	4.61	0.41	5.10	0.41	6.16	8.39	1.29
11	0.04	0.86	1.99	0.35	1.23	1.02	0.75	1.70	2.10	4.23	0.77	4.86
12	0.69	1.18	25.54	0.72	2.35	3.42	0.98	7.68	0.52	3.62	14.67	9.98
13	2.65	4.02	102.68	0.81	3.52	3.55	1.00	15.23	2.13	5.21	15.81	3.72
14	0.01	0.00	0.17	0.00	0.02	0.02	0.00	0.14	0.00	0.12	0.00	0.05
High	0.00	0.64	4.99	0.24	0.77	1.04	0.68	31.75	8.61	0.56	0.92	0.31
Medium	0.00	2.41	0.97	0.42	2.82	1.92	0.69	20.56	8.32	0.81	1.35	0.45
Low	8.50	4.69	77.29	7.92	1.08	6.80	4.30	27.03	0.37	12.35	19.80	4.39
K	27.60	31.61	191.85	1.82	6.91	10.48	13.15	190.28	107.38	62.81	36.24	31.48
H												
Gov	0.63	1.22	24.22	0.14	22.80	20.69	1.07	6.85	7.11	8.20	0.25	4.22
Sav												
RoW	55.78	39.78	87.29	79.03	89.54	62.84	106.40	1.92	20.78	90.52	0.53	46.54
Tot	151.73	94.52	1 143.20	103.49	182.24	252.25	144.88	391.01	171.68	207.28	172.44	123.93

(continued)

Table 2 (continued)

	13	14	High	Medium	Low	K	H	Gov	Inv	RoW	Tot
1	0.08	0.17					35.3	0.89	3.74	0.36	151.73
2	0.02	0.07					0.4		2.60	35.34	94.52
3	0.01	0.16					0.0		7.52	615.30	1 143.20
4	1.46	3.87					39.4		7.04	15.70	103.49
5	2.55	1.10					29.4		0.60	31.92	182.24
6	0.33	0.03					118.80	2.66	1.68	34.51	252.25
7	3.59	1.52					25.9	3.19	1.95	16.31	144.88
8	39.79	2.77					0.4		254.56	81.61	391.01
9	1.05	0.70					7.6		0.00	103.42	171.68
10	0.98	0.69					121.6		3.21	47.66	207.28
11	7.88	5.60					40.4	65.96		32.66	172.44
12	1.66	1.27					14.4	4.61	1.19	29.49	123.93
13	4.63	6.83					21.3	38.12		83.77	314.96
14	0.02	1.14					7.6	138.54		43.46	191.31
High	2.38	11.50									64.39
Medium	4.66	44.85									90.23
Low	16.60	95.47									286.58
K	128.32	9.49									849.42
H			30.32	51.12	140.29	368.03		99.99		40.0	729.72
Gov	5.55	0.50				2.18	103.18			156.4	365.25
Sav							164.11	11.29		108.69	284.09
RoW	93.40	3.58	34.07	39.11	146.29	479.21					1 476.61
Tot	314.96	191.31	64.39	90.23	286.58	849.42	729.72	365.25	284.09	1 476.61	

Table 3 SAM 2002 in millions of euro at current price 2002

	1	2	3	4	5	6	7	8	9	10	11	12
1	8.81		0.08	0.01	0.06	47.47	0.88	0.05	0.03	0.04	6.38	0.05
2	0.01	2.10	5.92	0.01	0.15	0.12	1.95	2.59	24.80	0.03	0.01	0.01
3	0.10	0.09	82.37	19.66	0.17	1.68	10.19	28.64	2.22	2.82	0.31	1.02
4	0.06	0.43	3.16	10.44	0.10	0.92	1.84	2.98	1.58	1.96	0.52	3.18
5	2.28	0.58	8.77	0.85	5.39	1.57	20.54	2.83	1.53	1.68	1.62	0.82
6	7.47	0.00	0.40	0.06	0.54	61.59	1.75	0.29	0.02	0.53	78.73	0.88
7	1.15	2.85	35.34	9.35	1.74	11.37	131.51	77.60	4.85	31.70	6.97	62.43
8	0.02	0.05	2.24	0.20	0.05	0.27	1.25	38.03	4.42	1.53	1.16	5.68
9	1.38	1.26	20.05	1.24	0.85	4.36	8.97	1.89	109.14	8.79	7.48	5.41
10	2.73	1.06	21.26	3.23	0.90	19.72	14.45	5.91	1.55	13.45	12.59	14.59
11	0.01	0.05	1.97	0.56	0.19	0.33	1.93	2.45	0.71	2.92	1.16	10.26
12	1.39	1.10	23.79	5.30	1.45	11.82	16.73	11.62	2.70	21.84	7.43	89.57
13	2.03	2.03	27.77	8.19	1.45	11.15	26.33	24.16	7.04	94.60	16.01	59.65
14	0.30	0.07	2.22	0.55	0.26	2.53	1.66	1.52	0.94	6.02	3.84	3.52
High	3.59	1.21	11.98	2.76	0.59	4.64	8.38	19.55	6.88	22.16	27.16	68.11
Medium	3.44	4.51	50.11	11.55	2.46	19.41	35.06	39.38	27.13	46.49	79.39	84.52
Low	2.11	0.60	3.35	0.77	0.16	1.30	2.34	21.31	2.05	10.56	34.08	7.15
K	28.96	5.60	11.55	3.77	2.33	43.67	50.35	55.76	120.74	176.30	115.07	130.73
H												
Gov	0.09	0.01	1.69	2.57	4.78	65.21	113.15	21.51	15.59	60.44	21.62	18.58
Sav												
RoW	78.75	35.30	183.91	160.76	109.81	342.11	817.21	11.66	12.55	135.03	2.77	173.52
Tot	144.68	58.90	497.94	241.83	133.43	651.24	1 266.48	369.73	346.47	638.89	424.31	739.68

(continued)

Table 3 (continued)

	13	14	High	Medium	Low	K	H	Gov	Inv	RoW	Tot
1	0.90	0.84					26.2	0.12	0.11	52.64	144.68
2	0.02	0.01					0.0	0.00		21.17	58.90
3	3.28	3.15					2.4	0.00	31.60	308.21	497.94
4	2.54	4.71					13.7	0.00	119.41	74.26	241.83
5	3.43	36.21					30.7	0.00		14.63	133.43
6	4.55	5.21					281.35	1.03		206.84	651.24
7	32.53	40.65					295.7	0.00	199.93	320.82	1 266.48
8	24.40	10.78					4.3	0.00	264.89	10.46	369.73
9	12.35	13.39					72.1	0.31		77.53	346.47
10	8.76	12.74					397.4	0.00	55.82	52.70	638.89
11	13.21	5.87					379.5	0.45		2.74	424.31
12	45.38	31.72					209.9	6.23	17.91	233.79	739.68
13	249.31	95.84					436.6	167.76	161.85	133.03	1 524.79
14	39.83	134.80					165.8	718.72	4.83	58.68	1 146.08
High	182.14	284.30									643.47
Medium	30.78	117.45									551.67
Low	12.00	20.85									118.65
K	680.68	278.79									1 704.30
H			584.19	497.94	106.67	1 395.98		843.28		297.2	3 725.23
Gov	49.89	25.74				112.323	1 005.25			346.6	1 865.00
Sav							404.27	127.10		324.98	856.35
RoW	128.81	23.02	59.29	53.74	11.99	196.00					2 536.22
Tot	1 524.79	1 146.08	643.47	551.67	118.65	1 704.30	3 725.23	1 865.00	856.35	2 536.22	

Table 4 Value added by sector in 1963 and 2002

	1963		2002	
	Value	In %	Value	In %
Agriculture	36.1	2.8	38.1	1.3
Extractive industry	39.4	3.0	11.9	0.4
Metallurgic industry	275.1	21.3	77.0	2.6
Mechanic industry	10.4	0.8	18.9	0.6
Chemical industry	11.6	0.9	5.5	0.2
Agri-food and textile	20.2	1.6	69.0	2.3
Other manufacturing industries	18.8	1.5	96.1	3.2
Construction	269.6	20.9	136.0	4.5
Electricity	124.7	9.7	156.8	5.2
Wholesale trade	76.5	5.9	255.5	8.5
Hotels	58.3	4.5	255.7	8.5
Transport and communication	36.6	2.8	290.5	9.6
Credit and insurance	152.0	11.8	905.6	30.0
Public administration	161.3	12.5	701.4	23.2
Total	1290.6	100.0	3018.1	100.0

the economy could have improved its performance given different policy scenarios by comparing the actual final situation with the counterfactual final situation. The next section describes the main characteristics of the regional dynamic model.

3 A Dynamic Regional General Equilibrium Model

The model is recursive dynamic which means that agents' behaviour depends on current and past states of the economy and do not form expectations about future events.

3.1 Data

The regional dynamic model is calibrated using two regional social accounting matrices for the region Valle D'Aosta constructed for the years 1963 and 2002. The original matrix accounts have been aggregated to obtain a reduced SAM as required by the structure of the applied dynamic model. The original matrices are reported in Lovo et al. (2008) together with the description of their content and of the procedures and sources adopted in the construction.

The aggregated SAMs are reported in Tables 2 and 3. The matrices include 14 sectors, 4 factors of production (high-skill labour, medium-skill labour, low-skill labour and capital), one private institution account incorporating households and enterprises, the regional government and the rest of the world. The rest of the world sector is a simplified account that includes three main trading partners: the rest of Italy (including the national government), the European Union and the others non-European countries. The 1963 SAM has been converted to constant prices 2002 using the price index reported by the national institute of statistics, Istat (2005).

The two regional SAMs adopt the same structure and are therefore fully comparable. The rest of this section is devoted to the presentation of the main characteristics of the regional dynamic CGE model.

3.2 The Static Specification

In our model, we consider 14 sectors noted by i and j listed in Table 1, one representative household, the regional government (which is supposed to be completely independent with respect to the national government), and the rest of the world (which includes the rest of Italy, other European countries having adopted the euro, and the other countries which have a different currency).

Producers maximize profits under perfect competition given the technological constraint. Following a standard procedure (Löfgren et al. 2001), production is related to intermediate inputs and primary factors according to a nested constant elasticity of substitution (CES) function.

Interregional trade is modelled in an aggregated way, namely, considering a single trade partner that includes the three main trading regions: the rest of Italy, the rest of Europe and non-European countries. Output is sold domestically and outside the region subject to imperfect substitutability between exports and domestic sales represented by a constant elasticity of transformation (CET) function. Domestic demand matches the supply of a composite commodity obtained by an Armington aggregation of imports and domestic sales which reproduces the imperfect substitutability between the two.

The peculiar characteristics of the region help to add few simplifications to the modelling of the government account. Under the current constitution, Valle D'Aosta is one of the five Italian regions enjoying a high level of governing and financial autonomy. This confers to the region, for analytical purposes, the condition of a "small country" within the country. Given the financial independence, government revenues and expenditures can be seen as occurring within the regional borders. The regional government obtains revenues from production and income taxes which are imposed on the regional sectors and institutions. The government budget includes also the net transfers from the national government. Total public expenditure is assumed to be a fixed percentage of the regional GDP and is allocated according to a CES function.

Households, enterprises and the regional government earn factor incomes in proportion to the owned share of factor stocks. Government and non-government institutions receive transfers from the rest of the economy and from the other institutions. Households use their income, net of direct taxes, to consume and save. Household consumption is allocated according to a CES utility function.

Capital and the three types of labour are assumed to be mobile across sectors. However, only capital is assumed to be fully employed. In our model, we consider a non-standard macroeconomic closure (Magnani 2015) which permits to endogenize unemployment by introducing a particular investment function. In this way, the unemployment rate depends on the level of the aggregate demand.

3.2.1 Firms

For each sector i, we use a multi-level production function. In the first stage, the quantity produced $Xs_{t,i}$ depends, according to a Leontief technology, on two aggregates: the value added $VA_{t,i}$ and the total intermediate good $Int_{t,i}^{tot}$. The optimal quantities are chosen by each sector i in order to maximize its profit given the technological constraint. The first order conditions for profit maximization are:

$$VA_{t,i} = \alpha_{t,i}^{VA} \cdot Xs_{t,i}$$
$$Int_{t,i}^{tot} = \alpha_{t,i}^{Int} \cdot Xs_{t,i}$$
$$P_{t,i} \cdot \left(1 - \tau_{t,i}^{Xs}\right) \cdot Xs_{t,i} = P_{t,i}^{VA} \cdot VA_{t,i} + P_{t,i}^{Int} \cdot Int_{t,i}^{tot}$$

where $\alpha_{t,i}^{VA}$ and $\alpha_{t,i}^{Int}$ are parameters to be calibrated, $P_{t,i}$ is the gross producer price, $\tau_{t,i}^{Xs}$ is the tax rate on production, $P_{t,i}^{VA}$ is the aggregate price of the value added and $P_{t,i}^{Int}$ is the aggregate price of intermediate goods.

In the second stage, each sector i chooses, on the one hand, the optimal demand for labour $L_{t,i,s}$ (expressed in efficiency units) for each skill type s (high skilled, medium skilled and low skilled) and capital $K_{t,i}$ which constitute the value added $VA_{t,i}$ according to a CES technology, and, on the other hand, the optimal demand for intermediate goods produced by sector j $Int_{t,j,i}$ which constitute the total intermediate good $Int_{t,i}^{tot}$ according to a CES technology. The first order conditions for cost minimization are:

$$L_{t,i,s} = \alpha_{t,i,s}^{L} \cdot \left(\frac{P_{t,i}^{VA}}{W_{t,s}}\right)^{\sigma_i^{VA}} \cdot VA_{t,i}$$

$$K_{t,i} = \alpha_{t,i}^{K} \cdot \left(\frac{P_{t,i}^{VA}}{R_t}\right)^{\sigma_i^{VA}} \cdot VA_{t,i}$$

$$P_{t,i}^{VA} \cdot VA_{t,i} = \sum_s W_{t,s} \cdot L_{t,i,s} + R_t \cdot K_{t,i}$$

$$Int_{t,j,i} = \alpha_{t,j,i}^{Int} \cdot \left(\frac{P_{t,i}^{Int}}{P_{t,j}^{X}}\right)^{\sigma_i^{Int}} \cdot Int_{t,i}^{tot}$$

$$P_{t,i}^{Int} \cdot Int_{t,i}^{tot} = \sum_j P_{t,j}^{X} \cdot Int_{t,j,i}$$

where $\alpha_{t,i,s}^{L}$, $\alpha_{t,i}^{K}$ and $\alpha_{t,j,i}^{Int}$ are parameters to be calibrated, σ_i^{VA} and σ_i^{Int} are the elasticities of substitution whose values come from the literature, $W_{t,s}$ is the remuneration per unit of efficient labour with skills s, R_t is the remuneration rate of capital, and $P_{t,i}^{X}$ is the sales price of commodity i.

3.2.2 Households

The revenue earned by the representative household is given by the remuneration of primary factors, the exogenous transfers from the government of Valle d'Aosta $Transf_t^G$ and the exogenous transfers from the rest of the world $Transf_t^{RoW}$ (in particular from the Italian government).

$$Y_t = (1 - \tau_t) \cdot \left[\sum_s W_{t,s} \cdot A_{t,s} \cdot N_{t,s}^{H} \cdot (1 - u_{t,s}) + R_t \cdot K_t^{H}\right] + Transf_t^G + Transf_t^{RoW}$$

In particular, labour incomes for each skill level s depend on the wage per unit of effective labour $W_{t,s}$, on productivity $A_{t,s}$, on the quantity of labour supplied $N_{t,s}^{H}$, and on the (endogenous) unemployment rate $u_{t,s}$. Capital incomes depend on the remuneration rate of capital R_t and on the quantity of capital supplied K_t^{H}.

The representative household pays income taxes on the basis of a tax rate τ_t and he saves a fraction s_t of his disposable income. Thus, the budget constraint is:

$$P_t^C \cdot C_t = (1 - s_t) \cdot Y_t$$

where C_t is aggregate consumption and P_t^C is the consumer price index.

Then, the representative household chooses the optimal demand $Xd_{t,i}$ for each commodity i in order to maximize his well-being (described by CES preferences), given the previous budget constraint. The first order conditions are:

$$Xd_{t,i} = \alpha^C_{t,i} \cdot \left(\frac{P^C_t}{P^X_{t,i} \cdot (1 + \tau^C_{t,i})} \right)^{\sigma^C} \cdot C_t$$

$$P^C_t \cdot C_{t,i} = \sum_i P^X_{t,i} \cdot (1 + \tau^C_{t,i}) \cdot Xd_{t,i}$$

where $\alpha^C_{t,i}$ is a parameter to be calibrated, σ^C is the elasticity of substitution whose value comes from the literature, $\tau^C_{t,i}$ is the VAT tax rate and $P^X_{t,i} \cdot (1 + \tau^C_{t,i})$ is the price paid by households to buy one unit of commodity i.

3.2.3 Regional Government

We assume that the regional government has a complete autonomy with respect to the national government in the sense that it collects all taxes on the incomes produced in Valle d'Aosta.

The public surplus is given by the difference between public revenues and public expenditures:

$$S^G_t = \sum_i \tau^{Xs}_{t,i} \cdot P_{t,i} \cdot Xs_{t,i} + \sum_i \tau^C_{t,i} \cdot P^X_{t,i} \cdot Xd_{t,i}$$
$$+ \tau_t \cdot \left[\sum_s W_{t,s} \cdot A_{t,s} \cdot N^H_{t,s} \cdot (1 - u_{t,s}) + R_t \cdot K^H_t \right] + R_t \cdot K^G_t + Tax^{RoW}_t$$
$$- \left(G^{tot}_t + Transf^G_t \right)$$

In particular, public revenues are given by indirect taxes on production and consumption, direct taxes on incomes, capital revenues (where K^G_t is the capital owned by the regional government) and transfers Tax^{RoW}_t perceived from the rest of the world (in particular, from the national government). Public expenditures are given by the expenditures to buy goods and services G^{tot}_t and transfers to families $transf^G_t$.

We assume that an exogenous fraction α^{Gtot}_t of GDP is devoted to public expenditures where GDP is defined as the total value added created by the economic sectors. Thus:

$$G^{tot}_t = \alpha^{Gtot}_t \cdot GDP_t$$

$$GDP_t = \sum_i P^{VA}_{t,i} \cdot VA_{t,i}$$

Finally, we assume that the regional government chooses the optimal demand $G_{t,i}$ for each commodity i in order to maximize his well-being (described by Cobb-Douglas preferences), given its budget constraint. The first order conditions are:

$$P^X_{t,i} \cdot G_{t,i} = \alpha^G_{t,i} \cdot G^{tot}_t$$

where $\alpha^G_{t,i}$ is a parameter to be calibrated and G^{tot}_t represents total public expenditures which are supposed to be equal to an exogenous fraction of GDP.

3.2.4 Investments and Savings

Commodities can also be used as investment goods. Total invtment I^{tot}_t is decomposed into the elements $I_{t,i}$ according to a CES technology in order to minimize the total cost. The first order conditions are:

$$I_{t,i} = \alpha^I_{t,i} \cdot \left(\frac{P^I_t}{P^X_{t,i}} \right)^{\sigma^I} \cdot I^{tot}_t$$

$$P^I_t \cdot I^{tot}_t = \sum_i P^X_{t,i} \cdot I_{t,i}$$

where $\alpha^I_{t,i}$ is a parameter to be calibrated and P^I_t is the aggregate investment price.

The macroeconomic equilibrium condition states that aggregate investment must be equal to aggregate savings given by private savings, public savings and savings with respect to the rest of the world:

$$P^I_t \cdot I^{tot}_t = s_t \cdot Y_t + S^G_t + S^{RoW}_t$$

3.2.5 International Trade

In our paper, imports are modelled using the Armington hypothesis according to which domestic products and foreign products are perceived as imperfect substitutable (implying that their prices will be different) because of the different origin. We use a CES technology to model this imperfect substitutability. The first order conditions for cost minimization are:

$$D_{t,i} = \alpha^D_{t,i} \cdot \left(\frac{P^X_{t,i}}{P^D_{t,i}} \right)^{\sigma^M_i} \cdot X_{t,i}$$

$$M_{t,i} = \alpha^M_{t,i} \cdot \left(\frac{P^X_{t,i}}{P^{RoW}_{t,i} \cdot e_t} \right)^{\sigma^M_i} \cdot X_{t,i}$$

$$P_{t,i}^X \cdot X_{t,i} = P_{t,i}^D \cdot D_{t,i} + P_{t,i}^{RoW} \cdot e_t \cdot M_{t,i}$$

where $\alpha_{t,i}^D$ and $\alpha_{t,i}^M$ are parameters to be calibrated and σ_i^M is the elasticity of substitution between domestic and foreign products. $D_{t,i}$ represents the domestic demand for domestic products, $M_{t,i}$ represents imports, and $X_{t,i}$ represents the composite good (given by a CES function of domestic sales and imports). $P_{t,i}^D$ is the price of domestic sales, $P_{t,i}^{RoW} \cdot e_t$ is the price of imported goods where e_t is the exchange rate, and $P_{t,i}^X$ is the weighted average between the two previous prices. The composite good $X_{t,i}$ is sold in the market to firms as intermediate good, to households, to the government or used as investment good. Thus:

$$X_{t,i} = \sum_j Int_{t,j,i} + Xd_{t,i} + I_{t,i} + G_{t,i}$$

Domestic output $Xs_{t,i}$ can be sold in the domestic market or exported. We assume that domestic sales $DD_{t,i}$ and exports $E_{t,i}$ are not perfectly substitutable, i.e. the quality of goods that are sold in the domestic market is different with respect to the quality of goods that are exported. We use a CET function to model this imperfect substitutability. The first order conditions for profit maximization are:

$$E_{t,i} = \alpha_{t,i}^E \cdot \left(\frac{P_{t,i}^{RoW} \cdot e_t}{P_{t,i}} \right)^{\sigma_i^E} \cdot Xs_{t,i}$$

$$DD_{t,i} = \alpha_{t,i}^{DD} \cdot \left(\frac{P_{t,i}^D}{P_{t,i}} \right)^{\sigma_i^E} \cdot Xs_{t,i}$$

$$P_{t,i} \cdot Xs_{t,i} = P_{t,i}^D \cdot DD_{t,i} + P_{t,i}^{RoW} \cdot e_t \cdot E_{t,i}$$

where $\alpha_{t,i}^E$ and $\alpha_{t,i}^{DD}$ are parameters to be calibrated and σ_i^E is the elasticity of transformation between domestic sales and exports. $P_{t,i}^D$ is the price of domestic sales, $P_{t,i}^{RoW} \cdot e_t$ is the price of exports, and $P_{t,i}$ is the average sales price.

The equilibrium of the balance of payments states that net capital inflows are equal to the difference between income outflows and income inflows.

$$S_t^{RoW} = \left(\sum_i P_{t,i}^{RoW} \cdot e_t \cdot M_{t,i} + \sum_s W_{t,s} \cdot A_{t,s} \cdot N_{t,s}^{RoW} + R_t \cdot K_t^{RoW} \right) - \left(\sum_i P_{t,i}^{RoW} \cdot e_t \cdot E_{t,i} + Tax_t^{RoW} + Transf_t^{RoW} \right)$$

In particular, income outflows are given by imports and labour and capital incomes perceived in Valle d'Aosta by the non-residents, while income inflows are given by exports, transfers perceived by the regional government from the rest of the world Tax_t^{RoW} (in particular, from the national government), and transfers perceived by the representative household from the rest of the world $Transf_t^{RoW}$ (in particular from the Italian government). The exchange rate is considered as exogenous while net capital inflows adjust in order to guarantee the equilibrium of the balance of payments.

3.2.6 Equilibrium Conditions

In the labour market, for each skill type s, the total demand of labour (in efficiency units) must be equal to the quantity supplied by the representative household and the non-residents.

$$\sum_i L_{t,i,s} = A_{t,s} \cdot N_{t,s}^H \cdot (1 - u_{t,s}) + A_{t,s} \cdot N_{t,s}^{RoW}$$

This equation determines the equilibrium wage per unit of effective labour $W_{t,s}$ for each skill type.

In the capital market, the total demand by firms must be equal to the quantity supplied by the representative household, the regional government and the non-residents:

$$\sum_i K_{t,i} = K_{t,K}^H + K_t^G + K_t^{RoW}$$

This equation determines the equilibrium remuneration rate of capital R_t.

In the market of goods and services, for each commodity i, domestic sales $DD_{t,i}$ (i.e. the quantity of output that is not exported) must be equal to the domestic demand for the domestic good $D_{t,i}$:

$$DD_{t,i} = D_{t,i}$$

This equation determines the equilibrium domestic price for each commodity $P_{t,i}^D$. The numeraire of the model is the consumer price index P_t^C.

3.3 *The Dynamic Specification*

The model is solved forward in a dynamically recursive fashion, in which the solution depends only on current and past variables.

The evolution of capital, owned by the representative household, the regional government and the non-residents, depends on capital depreciation according to a depreciation rate δ and savings expressed in real terms, i.e. deflated using the aggregate investment price:

$$K_{t+1}^{H} = K_{t}^{H} \cdot (1-\delta) + \frac{s_t \cdot Y_t}{P_t^I}$$

$$K_{t+1}^{G} = K_{t}^{G} \cdot (1-\delta) + \frac{S_t^G}{P_t^I}$$

$$K_{t+1}^{RoW} = K_{t}^{RoW} \cdot (1-\delta) + \frac{S_t^{RoW}}{P_t^I}$$

The quantity of labour supplied for each skill type s by the representative household and by the non-residents varies over time according to an exogenous growth rate differentiated by skill type:

$$N_{t+1,s}^{H} = N_{t,s}^{H} \cdot \left(1 + n_{t+1,s}^{H}\right)$$
$$N_{t+1,s}^{RoW} = N_{t,s}^{RoW} \cdot \left(1 + n_{t+1,s}^{RoW}\right)$$

The productivity $A_{t,s}$ for each skill type s increases over time according to an exogenous rate differentiated by skill type:

$$A_{t+1,s} = A_{t,s} \cdot \left(1 + g_{t+1,s}\right).$$

3.4 Macro Closure and Unemployment

In this study, we do not use the standard neoclassical closure according to which investments are savings-driven. This macro closure implies that if one element of the aggregate demand increases because of a shock or an economic policy, the level of investments reduces by the same amount. Thus, real GDP remains unaffected[2] when the shock is produced and the long-term effect is negative because of the lower capital accumulation. Moreover, we do not consider the Keynesian closure according to which investments are fixed at an exogenous level. In fact, in this case, a shock that stimulates aggregate demand produces an implausible increase in real GDP and reduction in unemployment. Instead, as described in Magnani (2015), we introduce the following investment function which permits to create a class of intermediate models between the neoclassical model and the Keynesian model:

$$I_t^{tot} = \overline{I}_t^{tot} + \beta \cdot \left[I_t^{tot} \cdot \left(\frac{1-\bar{u}_t}{1-u_t}\right)^{1-\alpha_L} - \overline{I}_t^{tot} \right]$$

where $\overline{I}_t^{tot}$ represents the pre-shock level of total investments, $\bar{u}_t$ the pre-shock level of the regional unemployment rate, α_L is the ratio between the wage bill and the regional GDP and β is an exogenous parameter that lies between zero and one and

[2]Real GDP can be affected because of the reallocation of the production factors across sectors.

that measures the crowding-in/crowding-out effect on investments. If the parameter β is equal to one the model coincides with the neoclassical model, while if it is equal to zero the model reduces to the Keynesian model. This parameter is set to 0.8 in line with estimates presented by Magnani (2015) on seven OECD countries.

The investment function determines the level of aggregate investments while the equilibrium condition between savings and investments determines the equilibrium regional unemployment rate.

In order to determine the unemployment rate differentiated by skill level s, we consider that the unemployment rate is equal to a weighted average between the unemployment rate differentiated by skill level $u_{t,s}$ and we assume that the ratio between the number of unemployed for each skill level and the total number of unemployed is exogenous:

$$u_t = \frac{\sum_s u_{t,s} \cdot N_{t,s}^H}{\sum_s N_{t,s}^H}$$

$$\alpha_{t,s}^U = \frac{u_{t,s} \cdot N_{t,s}^H}{u_t \cdot \sum_s N_{t,s}^H}.$$

4 Historical Calibration

This section describes the procedure followed in the historical calibration of the regional dynamic general equilibrium model.

In the economic literature, few studies have attempted to carry out an historical calibration procedure for dynamic general equilibrium models. For instance, Dixon and Rimmer (1999) employ a historical closure in which some of the variables normally not explained by CGE models, for example tax rates, technology and preferences, are considered endogenous to reproduce the observed movements in the main endogenous variables. More recently Arndt et al. (2002) suggested a maximum entropy approach to estimate the behavioural parameters of a static CGE model. This method, besides making use of historical records provides also statistical tests for the estimates. The historical targets considered include GDP, sales, imports, exports, investment, consumption by commodities and household types.

The peculiar experimental design of this study allows us to use a calibration procedure which differs from the standard dynamic calibration methods because we do not impose exogenous growth rates, normally borrowed from external data sources, on key variables, such as total factor productivity (Marrocu et al. 2000; Leonida et al. 2004). The new calibration approach introduced in this study exploits the available historical information contained in the initial and final input/output matrices that allows us to perfectly match model generated outcomes, such as sector value added, with actual data in two specific points in time. The calibration of the dynamic CGE model is carried out in four steps.

In the first step, we calibrate two static CGE models in order to reproduce the 1963 SAM and the 2002 SAM. The static models are calibrated following a standard procedure commonly used in computable general equilibrium models. We assume that the economy is in equilibrium and calibrate the value of all parameters and exogenous variables, while initializing all endogenous variables, so that we can perfectly reproduce the initial and final SAMs. In the calibration process, we make use of external information for the value of the elasticities adopted in the production and trade functional specifications and for the stock of endowments. When times series or cross section data are not available, elasticities cannot be estimated and are commonly taken from external sources. For instance, from similar contexts or as a sort of average tendency of estimates in the general equilibrium literature. In this paper, we adopted the elasticities used in Finizia et al. (2005) where a general equilibrium model employing similar functional forms for production, investment and foreign trade is applied to the whole Italian economy.

In the second step, we run the model for all periods between 1963 and 2002, where the value of all parameters and exogenous variables is determined by a linear interpolation between the beginning and end of the period. It is important to bear in mind that in this step we do not introduce the dynamics of productivity, capital and labour described in Sect. 3.3. Both SAMs are still perfectly reproduced.

In the third step, we introduce the dynamics as described in Sect. 3.3 that produces dynamic effects on relative prices and optimal quantities. Consequently, the 1963 SAM is still perfectly reproduced because we maintain the parameters and exogenous variables determined using the linear interpolation, while the final 2002 SAM is the outcome of the dynamic rules.

In the last step, we endogenize most of the parameters of the model in order to perfectly reproduce both SAMs by setting the new value of the parameters equal to the previous level multiplied by an adjustment parameter as if the regional economy were partially adapting to the endogenous dynamic changes. In the historical calibration, while elasticities are kept constant, the structural parameters are updated assuming a constant change over time to match 2002 structural relations and policy changes as explained above.

This procedure allows us to reproduce the two SAMs in a consistent way that is coherent with the general equilibrium evolution of all endogenous variables, including the equilibrium relative prices.

Figures 1, 2, 3, 4, 5, 6, 7 and 8 describe the evolution of the Val d'Aosta Region as reconstructed using the historical calibration procedure starting from the economic situation captured by the initial and final SAMs. Figure 1 shows the sharp decline of the metallurgic and construction industries. The value added from these sectors accounted for about 20% of regional value added at the beginning of the period. In 2002, the economic weight of these two sectors was less than 5%. The public administration and credit and insurance services were initially contributing about 25% of the regional value added and then escalated to about 55% at the end of the period.

Figure 2 describes the evolution of the components of consumption, investments, government expenditure and net trade regional GDP. In line with expectations, the

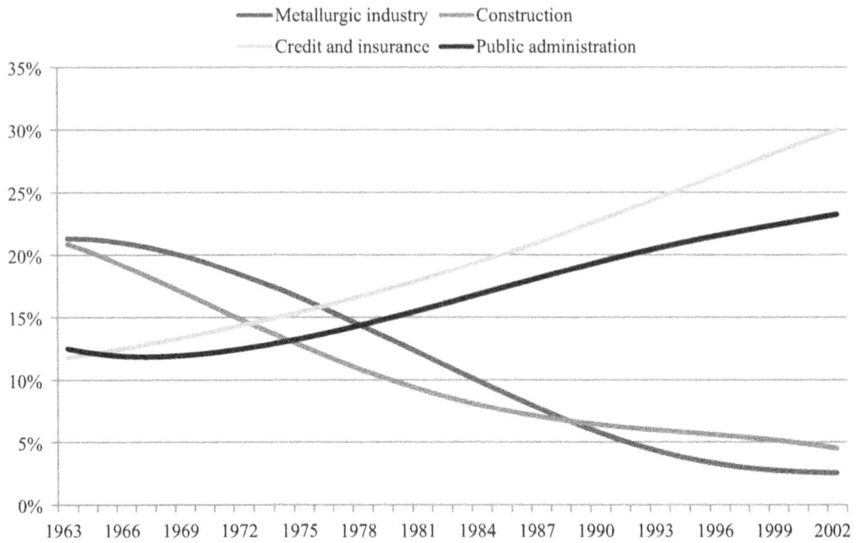

Fig. 1 Evolution (generated by the CGE model) of the value added of four sectors

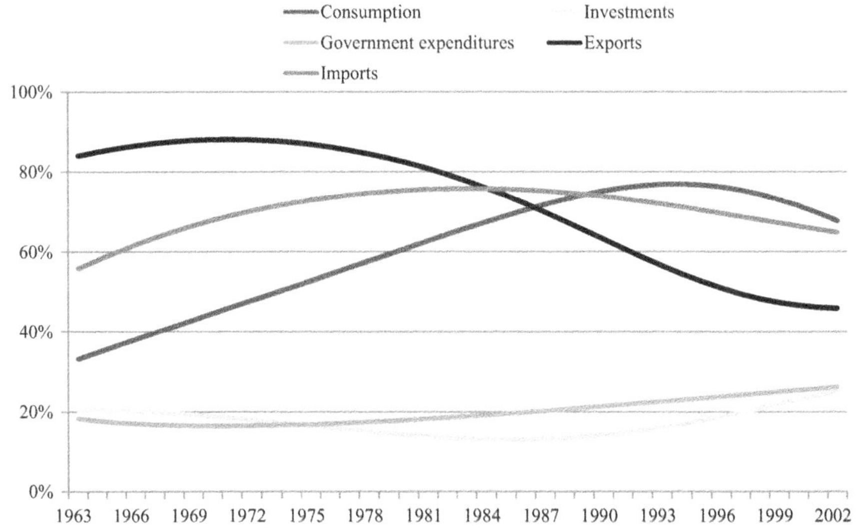

Fig. 2 Evolution (generated by the CGE model) of the components of the regional GDP

growth of the public and service sector was accompanied by a sharp increase in the level of household consumption that passed from a share of about 40% at the beginning of the period to a peak of about 80% in the mid-90s to decrease to about 70% at the end of the period. The growth in consumption also spurred the growth of imports from an initial value of 60% to reach a maximum by the end of the 1980s of about

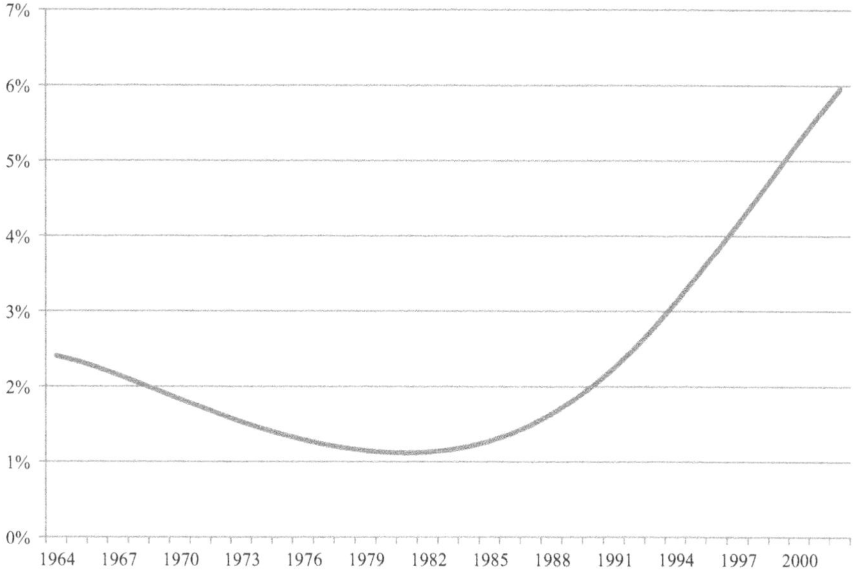

Fig. 3 Evolution (generated by the CGE model) of the growth rate of the regional GDP

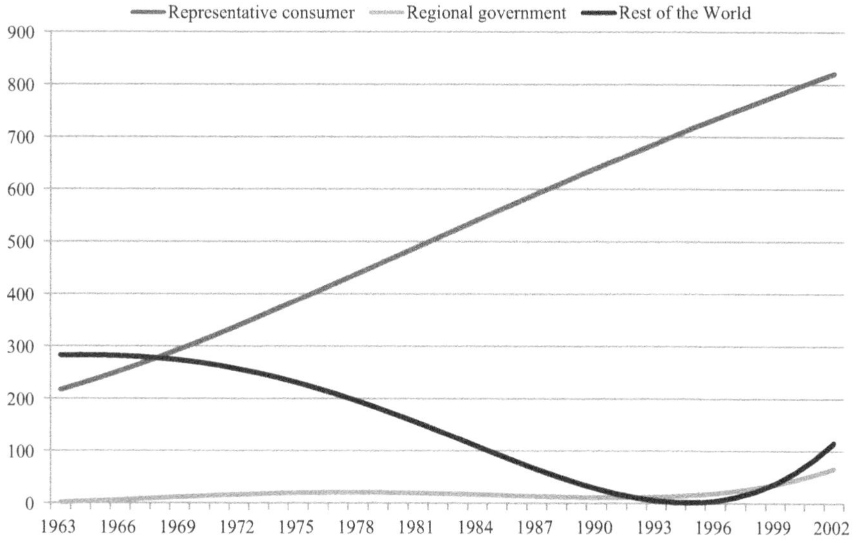

Fig. 4 Evolution (generated by the CGE model) of capital

80% when the balance of regional trade approached zero and become increasingly negative by the end of the period. The level of government expenditures and private investments remained more or less constant throughout the period. This configura-

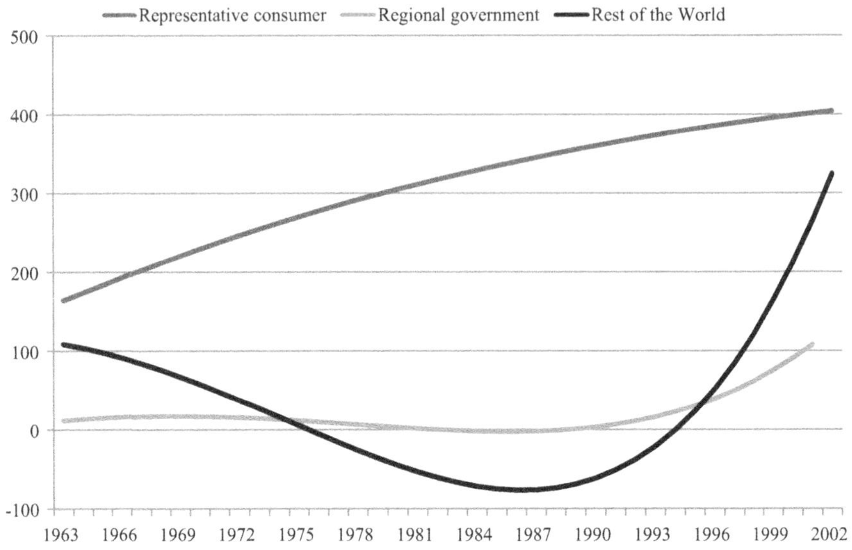

Fig. 5 Evolution (generated by the CGE model) of savings

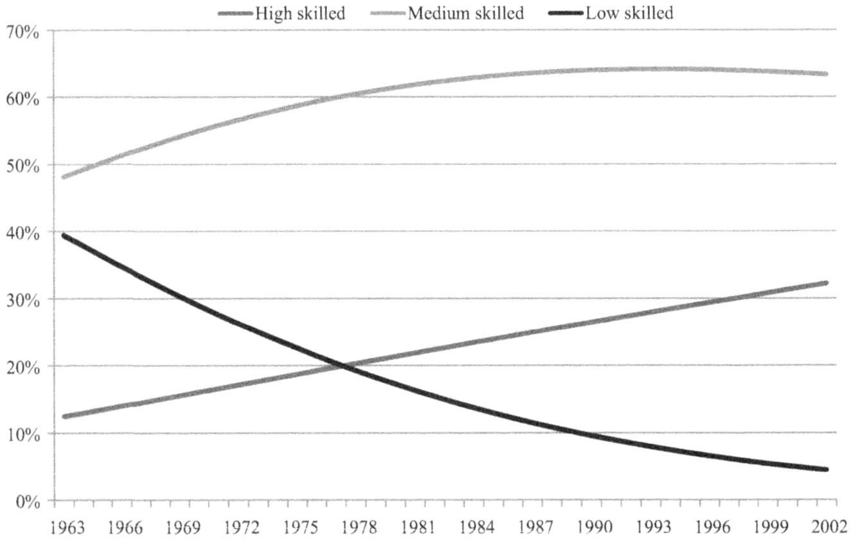

Fig. 6 Evolution (generated by the CGE model) of the number of workers by skill level (in % of the total workers)

tion of the budget balance equation, along with the changing sectorial structure of the regional economy, explains the evolution of regional GDP growth presented in Fig. 3. The rate of regional growth initially decreased from 2 to 1 percentage points

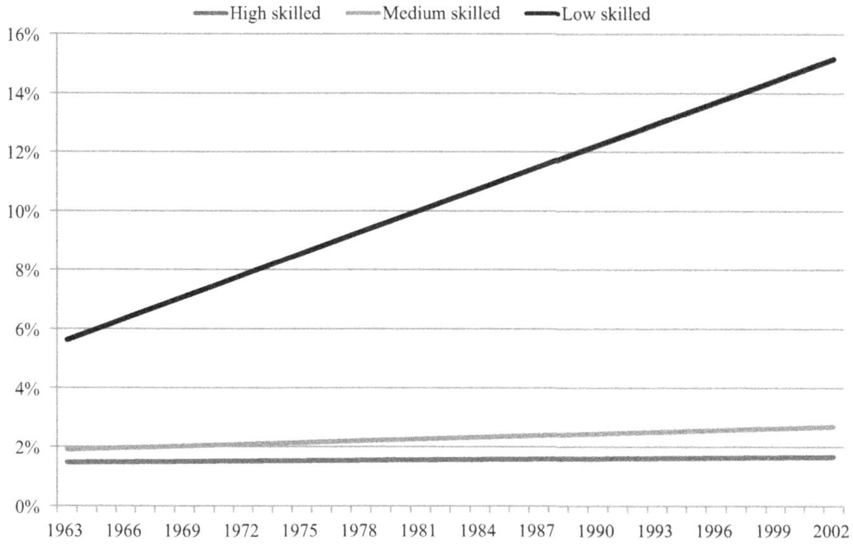

Fig. 7 Evolution (generated by the CGE model) of the unemployment rate by skill level

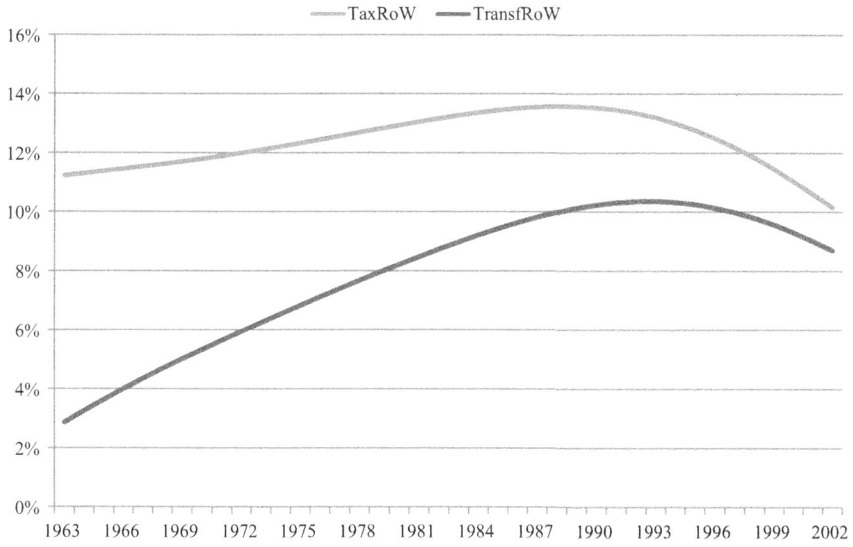

Fig. 8 Evolution (generated by the CGE model) of the transfers from the rest of the world to the regional government (TaxRoW) and to families (TransfRow)

until the beginning of the '80s when the economy rebounded to steadily grow up to a 6% growth rate at the end of the period. The GDP growth was sustained by a fourfold increase in the levels of household capital accumulation (Fig. 4) and a twofold increase in household savings (Fig. 5).

As shown in Figs. 6 and 7, the workers endowed with low skills were especially exposed to the risk of unemployment. Because of the large contraction of mainly the mining, manufacturing, and construction sectors traditionally employing low skilled people and the increasing general adoption of labour-saving technologies, the composition of the labour force passed from a 50-40-10% share of low, medium and high skill workers at the beginning of the period to a composition of 5-63-32% at the end of the period. Among the low skilled workers, the unemployment rate increased from 6% in the 60s to about 15% around the change of the millennium. The rate of unemployment for the medium and high skilled workers remained stable throughout the period around a physiological rate of 2% thanks also to the parallel increase of the private and public services.

Inspection of Fig. 8 reveals the degree of insulation of the regional economy. The economic importance of transfers from the State[3] to the regional government levelled around 10% throughout the period. As a result of the special statute, the Val d'Aosta region is granted an almost complete financial autonomy based on the control of about 90% of the taxes levied within the region and full jurisdiction over the redistribution of the tributes across regional activities and institutions. The economic choices of the regional government can therefore be considered the main determinants of the evolution of the regional economy. Despite the relatively higher resources available to the 120,000 inhabitants of the region as compared to people living in non-autonomous Italian regions, Val d'Aosta families received an increasing amount of resources from the rest of the world[4] reaching almost 10% of the regional budget by the end of the period. These transfers were meant to help the Val d'Aosta families coping with the social costs associated with the major industrial restructuring taking place in both the Valley and the Piedmont region in the second half of the twentieth century.

It is then sensible to ask how the Val d'Aosta economy and its labour market would have performed if the region had to face a contraction of the transfers to both the regional government and the families, mainly unemployed, needing greater social protection. This is indeed the path that we chose to alter in the counterfactual exercise.

[3]The transfers from the State come, for example, through non-State taxed earnings derived from the public management of integrated multifunctional (commercial and service) areas such as "Les Halles D'Aoste," also called "carport," that is visited by more than two million people of visitors every year or from the management of the Saint Vincent Casino (Noto and Meneghelli 2008).

[4]When we mention the rest of the world, we generally refer to the Italian State, for example, in relation to the shock on the transfers from the central government. On the other hand, when we examine, for example, the impact on capital we refer to any economic agent not resident in the Val d'Aosta region.

5 Counterfactual Analysis

We now discuss the results of the simulation producing a historical counterfactual growth path generated by a restrictive behaviour of the Italian government regarding the flow of transfers from the State to the region. We use our dynamic CGE model to evaluate what would have happened if, from 1992 to 2002, the Italian government had reduced by 20% the transfers both to families and the regional government as described in Table 5. This shock would have represented a feasible government intervention because this intervention would have not touched the highly politically sensitive vested right of the regional autonomy. The shock implies that in 1992 the level of transfers to the representative consumer would have decreased from 13.3 to 11% of the regional GDP while transfers to the regional government would have decreased from 10.4 to 8.5% of the regional GDP.

The direct effect of such a shock concerns the level of savings of all three economic actors: the representative consumer, the regional government and the rest of the world. As shown in Table 6, the shock produces a simultaneous increase in net capital inflows and a reduction in revenues perceived by the representative household and the regional government. Given that private consumption is equal to a fraction of the disposable income, the reduction of private and public savings is lower than the increase in the savings with respect to the rest of the world. In the presence of a neoclassical macro closure, the shock would have produced a strongly positive (crowding-in) effect on investments and, thus, on capital accumulation and the economic growth. In contrast, given that our model considers an intermediate macro closure, the effect on investments, even if it is still positive, is lower. Consequently, the macroeconomic equilibrium between aggregate savings and investments is achieved through an increase in the unemployment rate. In particular, as shown in Table 8, investments would have increased (with respect to the situation without the shock) by 9% in 1992 and by 19% in 2002.[5] The reduction in national transfers would have produced a negative impact on employment. In particular, in 1992, the unemployment rate would have increased by 6.7 p.p. for high skilled individuals, by 2.9 p.p. for medium skilled individuals and by 21.8 p.p. for low skilled individuals. This is crucial policy information that would have not been available if we had adopted a neoclassical closure that would have redistributed, if feasible, all excess supply of labour across sectors through a reduction in equilibrium wages.

The effect on savings produces an effect on the accumulation of capital. As shown in Table 7, the capital stock owned by the representative agent slightly decreases, while the regional government accumulates an important public debt and the foreign capital increases considerably. The impact on the regional GDP is negative. In 1992, given that the capital stock is a predetermined variable, the increase in the unemployment rate would produce a reduction in real GDP (−2.7%). The effect remains negative even if both investments and the capital stock available in the economy increase (Tables 8, 9 and 10).

[5]In contrast, using the standard neoclassical macro closure they would have increased by 12% in 1992 and by 158% in 2002.

Table 5 Shock on transfers from the rest of the world to the representative household and to the regional government

	Tax^{RoW}			Tax^{RoW}/GDP			$Transf^{RoW}$			$Transf^{RoW}/GDP$		
	Before	After	Var (%)	Before (%)	After (%)	Var (%)	Before	After	Var (%)	Before (%)	After (%)	Var (%)
1992	297.8	238.2	−20.0	13.3	10.9	−2.5	231.2	185.0	−20.0	10.4	8.4	−1.9
1993	302.7	242.1	−20.0	13.2	10.7	−2.4	237.8	190.3	−20.0	10.4	8.4	−1.9
1994	307.6	246.0	−20.0	13.0	10.6	−2.4	244.4	195.5	−20.0	10.3	8.4	−1.9
1995	312.4	249.9	−20.0	12.8	10.4	−2.4	251.0	200.8	−20.0	10.3	8.3	−1.9
1996	317.3	253.8	−20.0	12.5	10.1	−2.4	257.6	206.1	−20.0	10.1	8.2	−1.9
1997	322.2	257.7	−20.0	12.2	9.8	−2.4	264.2	211.4	−20.0	10.0	8.1	−1.9
1998	327.1	261.6	−20.0	11.8	9.5	−2.3	270.8	216.6	−20.0	9.8	7.8	−1.9
1999	331.9	265.5	−20.0	11.4	9.1	−2.3	277.4	221.9	−20.0	9.6	7.6	−2.0
2000	336.8	269.4	−20.0	11.0	8.7	−2.4	284.0	227.2	−20.0	9.3	7.3	−2.0
2001	341.7	273.3	−20.0	10.6	8.2	−2.4	290.6	232.5	−20.0	9.0	7.0	−2.0
2002	346.5	277.2	−20.0	10.1	7.8	−2.4	297.2	237.8	−20.0	8.7	6.7	−2.0

Table 6 Effect on savings

	S^H			S^G			S^{RoW}		
	Before	After	Var (%)	Before	After	Var (%)	Before	After	Var (%)
1992	369.3	353.0	−4.4	10.9	−65.4		−37.5	86.6	
1993	373.7	354.8	−5.0	16.2	−84.2		−20.8	140.0	
1994	377.9	356.5	−5.6	22.6	−106.8		−0.5	202.8	
1995	381.9	358.1	−6.2	30.3	−133.8		23.7	275.9	
1996	385.7	359.5	−6.8	39.3	−165.7		52.0	360.2	
1997	389.3	360.8	−7.3	49.7	−203.2		84.8	456.7	
1998	392.6	362.0	−7.8	61.6	−246.5		122.2	565.9	
1999	395.8	363.3	−8.2	75.2	−295.9		164.6	688.0	
2000	398.8	364.7	−8.5	90.6	−351.2		212.4	823.0	
2001	401.6	366.4	−8.8	107.8	−412.3		265.7	970.6	
2002	404.3	368.4	−8.9	127.1	−478.5		325.0	1130.5	

Table 7 Effect on capital

	K^H			K^G			K^{RoW}			*Total capital*		
	Before	After	Var (%)	Before	After	Var (%)	Before	After	Var (%)	Before	After	Var (%)
1992	672.4	672.4	0.0	11.3	11.3		10.6	10.6		694.3	694.3	0.0
1993	688.3	685.9	−0.3	12.2	2.0		5.1	21.7		705.5	709.6	0.6
1994	703.9	699.0	−0.7	13.7	−9.3		2.0	39.3		719.7	728.9	1.3
1995	719.4	711.5	−1.1	16.1	−23.1		1.9	64.3		737.4	752.7	2.1
1996	734.7	723.6	−1.5	19.3	−39.8		5.0	97.8		759.0	781.7	3.0
1997	749.8	735.3	−1.9	23.7	−59.8		11.7	140.9		785.1	816.3	4.0
1998	764.6	746.4	−2.4	29.1	−83.8		22.5	194.5		816.3	857.1	5.0
1999	779.2	757.1	−2.8	36.0	−112.3		37.8	259.8		852.9	904.6	6.1
2000	793.4	767.4	−3.3	44.3	−145.9		58.1	337.9		895.8	959.4	7.1
2001	807.4	777.1	−3.8	54.3	−184.9		83.7	429.6		945.4	1021.8	8.1
2002	821.1	786.5	−4.2	66.1	−230.0		115.3	536.0		1002.4	1092.5	9.0

Table 8 Effect on some macroeconomic variables

	GDP			*Disposable income*			*Consumption*			*Investments*		
	Before	After	Var (%)	Before	After	Var (%)	Before	After	Var (%)	Before	After	Var (%)
1992	2233.3	2172.0	−2.7	2078.5	1 986.8	−4.4	1709.1	1633.8	−4.4	342.6	374.2	9.2
1993	2295.8	2225.3	−3.1	2138.0	2 030.3	−5.0	1764.4	1675.4	−5.0	369.1	410.7	11.3
1994	2367.3	2289.1	−3.3	2198.6	2 074.4	−5.6	1820.7	1717.9	−5.6	400.0	452.5	13.1
1995	2448.9	2364.3	−3.5	2260.1	2 119.2	−6.2	1878.2	1761.1	−6.2	435.9	500.2	14.8
1996	2541.8	2452.0	−3.5	2322.6	2 164.8	−6.8	1936.9	1805.3	−6.8	477.0	554.0	16.1
1997	2647.5	2553.2	−3.6	2385.9	2 211.4	−7.3	1996.6	1850.6	−7.3	523.7	614.3	17.3
1998	2767.3	2669.2	−3.5	2450.2	2 259.2	−7.8	2057.6	1897.2	−7.8	576.5	681.4	18.2
1999	2902.7	2801.6	−3.5	2515.6	2 309.0	−8.2	2119.8	1945.7	−8.2	635.7	755.4	18.8
2000	3055.3	2952.1	−3.4	2582.1	2 361.4	−8.5	2183.4	1996.7	−8.5	701.7	836.5	19.2
2001	3226.8	3123.0	−3.2	2650.2	2 417.6	−8.8	2248.6	2051.3	−8.8	775.1	924.7	19.3
2002	3419.0	3316.3	−3.0	2720.0	2 478.4	−8.9	2315.7	2110.0	−8.9	856.3	1020.4	19.2

Table 9 Effect on the unemployment rate by skill level

	Total			*High skilled*			*Medium skilled*			*Low skilled*		
	Before (%)	After (%)	Var p.p. (%)	Before (%)	After (%)	Var p.p. (%)	Before (%)	After (%)	Var p.p. (%)	Before (%)	After (%)	Var p.p. (%)
1992	3.1	8.6	5.5	1.6	8.3	6.7	2.5	5.4	2.9	12.7	34.5	21.8
1993	3.1	9.7	6.6	1.6	9.4	7.8	2.5	5.9	3.4	13.0	40.4	27.5
1994	3.1	10.5	7.4	1.6	10.3	8.7	2.5	6.4	3.9	13.2	46.3	33.1
1995	3.0	11.2	8.1	1.6	11.0	9.4	2.5	6.8	4.2	13.5	52.1	38.6
1996	3.0	11.7	8.7	1.6	11.5	9.9	2.6	7.1	4.5	13.7	57.7	44.0
1997	3.0	12.1	9.1	1.6	11.8	10.2	2.6	7.3	4.8	13.9	63.2	49.3
1998	3.0	12.4	9.4	1.6	12.0	10.3	2.6	7.5	4.9	14.2	68.4	54.2
1999	3.0	12.5	9.6	1.6	12.0	10.4	2.6	7.6	5.0	14.4	73.3	58.8
2000	2.9	12.6	9.6	1.7	11.9	10.2	2.6	7.7	5.0	14.7	77.7	63.0
2001	2.9	12.4	9.5	1.7	11.6	10.0	2.7	7.7	5.0	14.9	81.5	66.6
2002	2.9	12.2	9.3	1.7	11.3	9.6	2.7	7.6	4.9	15.2	84.8	69.6

Table 10 Percentage variation in the value added of the economic sectors

	Agriculture (%)	Extractive industry (%)	Metallurgic industry (%)	Mechanic industry (%)	Chemical industry (%)	Agri-food and textile (%)	Other manufacturing industries (%)	Construction (%)
1992	−3.6	1.8	−7.3	−7.3	−3.4	−4.0	−0.9	6.6
1993	−3.9	2.1	−10.0	−8.9	−3.9	−4.6	−0.7	8.6
1994	−4.1	2.4	−12.5	−10.2	−4.3	−5.1	−0.4	10.8
1995	−4.2	2.5	−14.9	−11.2	−4.6	−5.5	−0.1	13.1
1996	−4.2	2.3	−17.4	−11.9	−4.9	−5.9	0.4	15.6
1997	−4.1	1.8	−20.0	−12.3	−5.0	−6.2	1.0	18.2
1998	−3.8	0.8	−22.9	−12.2	−5.1	−6.4	1.6	21.1
1999	−3.4	−1.0	−26.4	−11.5	−5.0	−6.4	2.4	24.3
2000	−2.7	−3.7	−30.8	−9.7	−4.8	−6.0	3.4	27.7
2001	−1.7	−7.8	−36.5	−5.8	−4.4	−5.2	4.6	31.6
2002	−0.4	−13.7	−44.7	2.6	−3.7	−3.7	6.2	36.2

(continued)

Table 10 (continued)

	Electricity (%)	Wholesale trade (%)	Hotels (%)	Transports and communication	Credit and insurance (%)	Public administration (%)	Total (%)
1992	−2.5	−3.7	−3.4	−3.6	−3.1	−3.0	−2.7
1993	−2.9	−4.1	−3.8	−4.1	−3.5	−3.3	−3.1
1994	−3.3	−4.6	−4.2	−4.6	−3.8	−3.5	−3.3
1995	−3.6	−4.9	−4.6	−4.9	−4.0	−3.6	−3.5
1996	−4.0	−5.2	−4.9	−5.2	−4.2	−3.6	−3.6
1997	−4.3	−5.5	−5.2	−5.5	−4.3	−3.6	−3.6
1998	−4.6	−5.6	−5.4	−5.7	−4.4	−3.5	−3.6
1999	−5.0	−5.7	−5.7	−5.8	−4.5	−3.5	−3.6
2000	−5.5	−5.8	−5.9	−5.8	−4.5	−3.4	−3.5
2001	−6.1	−5.7	−6.1	−5.8	−4.4	−3.3	−3.5
2002	−6.8	−5.5	−6.2	−5.7	−4.2	−3.3	−3.4

Concerning the representative consumer, the effect of the shock would be negative in terms of disposable income and consumption (−4.4% in 1992 and −8.9% in 2002). At the sectoral level, the effect is negative except for those sectors where investments represent an important component of the total demand, i.e. the mechanic industry, other manufacturing industries and, more importantly, the construction sector.

6 Conclusions

In this chapter, we developed a regional dynamic general equilibrium model for the Italian region Valle D'Aosta. The model is based on two SAMs (for 1963 and 2002) and is calibrated using a novel technique which consists in determining the dynamic path of most of the parameters of the model in order to perfectly reproduce both the initial and the final SAM. Thus, the dynamic model covers a 40 years period capturing the historical development path followed by regional sectors and institutions which is of central importance in understanding the nature of the current economic conditions (Engerman and Sokoloff 2000; Acemoglu and Robinson 2000, 2012; Acemoglu et al. 2001, 2005).

The historical calibration gives us the opportunity to implement a counterfactual analysis (by comparing the path actually followed by the regional economy with alternative policy scenarios) which cannot be applied using standard dynamic general equilibrium techniques.

In particular, we analyse the effects of a reduction, from 1992 to 2002, of the transfers from the Italian government both to families and the regional government. We find that this kind of shock would have had a negative effect on regional GDP and unemployment (especially for low skilled workers) even though investments increase thanks to a positive crowding-in effect.

The results of our counterfactual simulations clearly show that the parachute provided by the Italian government's generous transfer policy, especially in the '70s and '80s, was critical in helping the Val d'Aosta transition from an industrial to a mainly public service economy without incurring in major social costs. The open question is asking whether this transition, that led to an obese public administration also because not sufficiently accompanied by active policies especially in tourism, land saving, human capital and technology intensive sectors, was the most efficient path to development.

Acknowledgements The authors would like to thank Mirko Meneghelli, Jean Mercenier, Marco Missaglia, Lucio Scandizzo, and Sherman Robinson for their contributions to this work. Any error or omissions are sole responsibilities of the authors. Paper presented at the workshop "Social Accounting Matrices and Computable General Equilibrium Models: State of the Art and Prospects for the Future," Rome, January 25, 2017.

References

Acemoglu, D., & Robinson, J. (2000). Political losers as a barrier to economic development. *American Economic Review, 90,* 126–144.
Acemoglu, D., & Robinson, J. (2012). *Why nations fail. The origins of power, prosperity and poverty*. Crown Business.
Acemoglu, D., Johnson, S., & Robinson, J. A. (2001). The colonial origins of comparative development: An empirical investigation. *American Economic Review, 91,* 1369–1401.
Acemoglu, D., Johnson, S., & Robinson, J. (2005). The rise of Europe: Atlantic trade, institutional change, and economic growth. *American Economic Review, 95*(3), 546–579.
Arndt, C., Robinson, S., & Tarp, F. (2002). Parameter estimation for a computable general equilibrium model: A maximum entropy approach. *Economic Modelling, 19,* 375–398.
Dixon, P. B., & Rimmer, M. T. (1999). *Dynamic general equilibrium modelling for forecasting and policy: A practical guide and documentation of MONASH*. Amsterdam: North Holland.
Engerman, S. L., & Sokoloff, K. L. (2000). History lessons. Institutions, factor endowments, and paths of development in the new world. *Journal of Economic Perspectives, 3*(14), 217–232.
Finizia, A., Magnani, R., Perali, F., Polinori, P., & Salvioni, C. (2005). Il Modello MEG-ISMEA per l'analisi delle politiche agricole. *La Questione Agraria, 4,* 75–101.
Gass, S. I. (1983). Decision-aiding models: Validation, assessment, and related issues for policy analysis. *Operations Research, 31*(4), 603–631.
Istat. (2004a). *Conti economici delle imprese*. Roma.
Istat. (2004b). *Conti economici regionali*. Roma.
Istat. (2005). Il valore della moneta in Italia dal 1861 al 2004. In *Informazioni settori prezzi* (Vol. 24). Roma.
Kleijnen, J. P. C. (1999). Validation of models: Statistical techniques and data availability. In *Proceeding of the 1999 Winter Simulation Conference*.
Knutsen, C. H. (2013). Democracy, state capacity, and economic growth. *World Development, 43,* 1–18.
Leonida, L., Petraglia, C., & Murillo-Zamorano, L. R. (2004). Total factor productivity and the convergence hypothesis in the Italian regions. *Applied Economics, 36,* 2187–2193.
Löfgren, H., Harris R., & Robinson, S. (2001). *A standard computable general equilibrium (CGE) model in GAMS*. Trade and Macroeconomics Discussion Paper No. 75 (Washington, D.C.: International Food Policy Research Institute).
Lovo, S., Noto, S., & Perali, F. (2008). Un'analisi dello sviluppo della Valle d'Aosta in prospettiva storica con l'ausilio delle matrici di contabilità sociale. *Rivista di Politica Economica, 98*(1), 249–294.
Magnani, R. (2015). *The solow growth model revisited*. Introducing Keynesian Involuntary Unemployment, CEPN Working Papers, HAL.
Marrocu, M., Paci R., & Pala, R. (2000). *Estimation of total factor productivity for regions and sectors in Italy. A panel cointegration approach*. Contributi di Ricerca CRENoS, 00/16.
Noto, S., & Meneghelli, M. (2008). La crescita in Valle d'Aosta 1963-2002. In S. Noto (Ed.), *La Valle d'Aosta e l'Europa* (Vol. 1, pp. 133–175). Olschki: Firenze.
Thurlow, J. (2004). *A dynamic computable general equilibrium (CGE) model for South Africa: Extending the static IFPRI model*. TIPS Working paper 1-2004.

The manufacturer's authorised representative in the EU is Springer Nature Customer Service Centre GmbH, Europaplatz 3, 69115 Heidelberg, Germany. If you have any concerns regarding our products, please contact ProductSafety@springernature.com

Printed and bound by CPI Group (UK) Ltd, Croydon, CR0 4YY
30/09/2026
02225601-0017